Legal Notice

This book is copyright 2019 with all rights reserved. It is illegal to copy, distribute, or create derivative works from this book in whole or in part or to contribute to the copying, distribution, or creating of derivative works of this book.

For information on bulk purchases and licensing agreements, please email

support@SATPrepGet800.com

ISBN-13: 978-0-9998117-7-1

Also Available from Dr. Steve Warner

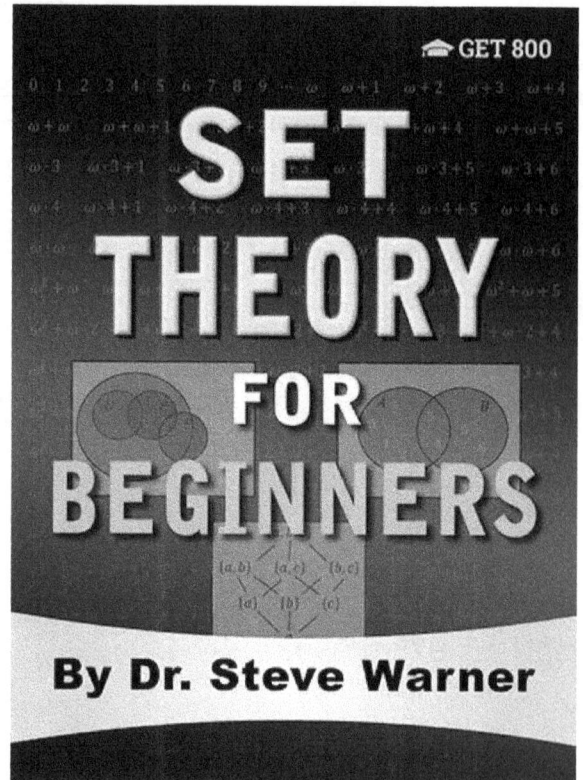

PURE MATHEMATICS FOR BEGINNERS

⬆ GET 800

Logic
Set Theory
Abstract Algebra
Number Theory
Real Analysis
Topology
Complex Analysis
Linear Algebra

By Dr. Steve Warner

SET THEORY FOR BEGINNERS

⬆ GET 800

By Dr. Steve Warner

CONNECT WITH DR. STEVE WARNER

www.facebook.com/SATPrepGet800

www.youtube.com/TheSATMathPrep

www.twitter.com/SATPrepGet800

www.linkedin.com/in/DrSteveWarner

www.pinterest.com/SATPrepGet800

plus.google.com/+SteveWarnerPhD

Topology
for Beginners

A Rigorous Introduction to Set Theory, Topological Spaces, Continuity, Separation, Countability, Metrizability, Compactness, Connectedness, Function Spaces, and Algebraic Topology

Dr. Steve Warner

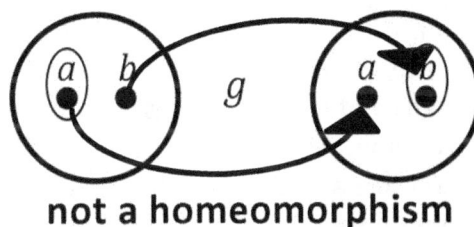

homeomorphism   not a homeomorphism

© 2019, All Rights Reserved

This book was written to provide a basic but rigorous introduction to topology. Additionally, a few more advanced topics are presented later in the book.

For students: There are no prerequisites for this book. The content is completely self-contained. Students with a bit of mathematical knowledge may have an easier time getting through some of the material, but no such knowledge is necessary to read this book.

More important than mathematical knowledge is "mathematical maturity." Although there is no single agreed upon definition of mathematical maturity, one reasonable way to define it is as "one's ability to analyze, understand, and communicate mathematics." A student with a higher level of mathematical maturity will be able to move through this book more quickly than a student with a lower level of mathematical maturity.

Whether your level of mathematical maturity is low or high, if you are just starting out in topology, then you're in the right place. If you read this book the "right way," then your level of mathematical maturity will continually be increasing. This increased level of mathematical maturity will not only help you to succeed in advanced math courses, but it will improve your general problem solving and reasoning skills. This will make it easier to improve your performance in college, in your professional life, and on standardized tests.

So, what is the "right way" to read this book? Simply reading each lesson from end to end without any further thought and analysis is not the best way to read the book. You will need to put in some effort to have the best chance of absorbing and retaining the material. When a new theorem is presented, don't just jump right to the proof and read it. Think about what the theorem is saying. Try to describe it in your own words. Do you believe that it is true? If you do believe it, can you give a convincing argument that it is true? If you do not believe that it is true, try to come up with an example that shows it is false, and then figure out why your example does not contradict the theorem. Pick up a pen or pencil. Draw some pictures, come up with your own examples, and try to write your own proof.

You may find that this book goes into more detail than other topology books when explaining examples, discussing concepts, and proving theorems. This was done so that any student can read this book, and not just students that are naturally gifted in mathematics. So, it is up to you as the student to try to answer questions before they are answered for you. When a new definition is given, try to think of your own examples before looking at those presented in the book. And when the book provides an example, do not just accept that it satisfies the given definition. Convince yourself. Prove it.

Each lesson is followed by a Problem Set. The problems in each Problem Set have been organized into five levels of difficulty, followed by one or more Challenge Problems. Level 1 problems are the easiest and Level 5 problems are the most difficult, except for the Challenge Problems. If you want to get just a small taste of topology, then you can work on the easier problems. If you want to achieve a deeper understanding of the material, take some time to struggle with the harder problems.

For instructors: This book can be used as an undergraduate text or an introductory graduate text in topology. The subject is developed slowly with an emphasis early on of developing skill with proof writing.

Lessons 1 through 4 provide a rigorous treatment of the basic set theory that anyone studying topology should know. All the basics of sets, subsets, set operations, relations, partitions, functions, and equinumerosity are included. If students have already seen this material in a prerequisite to your course, then this material can be reviewed quickly or skipped altogether.

Lesson 5 covers induction and a formal treatment of the natural numbers, integers, rational numbers, real numbers, and complex numbers. The formal treatments of these number systems can be included or omitted, depending on the taste of each individual instructor. On the other hand, I would spend at least a little time reviewing induction, as it can be a useful tool in many topological proofs.

Lesson 6 provides an introduction to groups, rings, and fields, with an emphasis on the specific structures that will serve as examples throughout this book. Ordered rings and fields and the completeness of the reals are covered as well.

Lessons 7 and 8 cover the basic topology of the real numbers and complex numbers, as well as Euclidean and unitary spaces. Lesson 7 provides basic concepts such as distance and openness in these spaces, whereas Lesson 8 discusses limits and continuity in these spaces.

Lessons 9 through 15 generalize the concepts discussed in Lessons 7 and 8 to arbitrary topological spaces. Lesson 9 covers all the basics including open and closed sets, bases and subbases, subspaces, and product spaces. The separation axioms in Lesson 10 allow us to rule out topological spaces that behave in an undesirable way. Lesson 11 introduces metrizable spaces, which generalize the Euclidean and unitary spaces from Lesson 7 in a way that many of the properties from those spaces are not lost. Lesson 12 covers compactness, Lesson 13 covers continuity and homeomorphisms, Lesson 14 covers connectedness, and Lesson 15 covers Banach spaces of functions.

Finally, in Lesson 16, we provide an introduction to the subject of algebraic topology. In particular, we introduce homotopy theory.

Students will have access to solutions to all problems in the Problem Sets at the end of each lesson, except for the Challenge Problems. These Challenge Problems can be used for graded assignments. I would recommend giving students at least several days to a week to work on each one. Most students will find them quite difficult.

The author welcomes all feedback from instructors. Any suggestions will be considered for future editions of the book. The author would also love to hear about the various courses that are created using these lessons. Feel free to email Dr. Steve Warner with any feedback at

steve@SATPrepGet800.com

LESSON 1
SETS AND SUBSETS

Describing Sets

A **set** is simply a collection of "objects." These objects can be numbers, letters, colors, animals, funny quotes, or just about anything else you can imagine. We will usually refer to the objects in a set as the **members** or **elements** of the set.

If a set consists of a small number of elements, we can describe the set simply by listing the elements in the set in curly braces, separating elements by commas.

Example 1.1:

1. {rabbit, crocodile} is the set consisting of two elements: *rabbit* and *crocodile.*

2. {tree, astronaut, yellow, green, motorcycle, trumpet} is the set consisting of six elements: *tree, astronaut, yellow, green, motorcycle,* and *trumpet.*

3. $\{0, 2, 4, 6, 8, 10\}$ is the set consisting of six elements: 0, 2, 4, 6, 8, and 10. The elements in this set happen to be *numbers*.

A set is determined by its elements and not the order in which the elements are presented. For example, the set $\{2, 6, 4, 0, 10, 8\}$ is the same as the set $\{0, 2, 4, 6, 8, 10\}$.

Also, the set $\{0, 0, 2, 4, 4, 6, 6, 6, 8, 8, 10\}$ is the same as the set $\{0, 2, 4, 6, 8, 10\}$. If we are describing a set by listing its elements, the most natural way to do this is to list each element just once.

We will usually name sets using capital letters such as A, B, C,..., and so on. For example, we might write $A = \{a, b, c\}$. So, A is the set consisting of the elements a, b, and c.

Example 1.2: Consider the sets $X = \{x, y, z\}$, $Y = \{y, z, x\}$, $Z = \{x, y, x, z, x\}$. Then X, Y, and Z all represent the same set. We can write $X = Y = Z$.

We use the symbol $\in$ for the membership relation (we will define the term "relation" more carefully in Lesson 3). So, $x \in A$ means "x is an element of A," whereas $x \notin A$ means "x is **not** an element of A." We will often simply say "x is in A," and "x is not in A," respectively.

Example 1.3: Let $B = \{c, z, 5, \Delta, \Gamma\}$. Then $c \in B$, $z \in B$, $5 \in B$, $\Delta \in B$, and $\Gamma \in B$.

If a set consists of many elements, we can use **ellipses** (...) to help describe the set. For example, the set consisting of the natural numbers between 12 and 1548, inclusive, can be written $\{12, 13, 14, \dots, 1547, 1548\}$ ("inclusive" means that we include 12 and 1548). The ellipses between 14 and 1547 are there to indicate that there are elements in the set that we are not explicitly mentioning.

Ellipses can also be used to help describe **infinite sets**. The set of **natural numbers** can be written $\mathbb{N} = \{0, 1, 2, 3, \dots\}$, and the set of **integers** can be written $\mathbb{Z} = \{\dots, -4, -3, -2, -1, 0, 1, 2, 3, 4, \dots\}$.

Note: Some mathematicians exclude 0 from the set of natural numbers. In this book, 0 will always be included. Symbolically, $0 \in \mathbb{N}$.

Example 1.4:

1. The even natural numbers can be written $\mathbb{E} = \{0, 2, 4, 6, \ldots\}$.

2. The odd natural numbers can be written $\mathbb{O} = \{1, 3, 5, \ldots\}$.

3. The even integers can be written $2\mathbb{Z} = \{\ldots, -6, -4, -2, 0, 2, 4, 6, \ldots\}$.

4. The positive integers can be written $\mathbb{Z}^+ = \{1, 2, 3, 4, \ldots\}$ or $\mathbb{N}^+ = \{1, 2, 3, 4, \ldots\}$ (the positive integers and the positive natural numbers describe the same set).

A set can also be described by a certain property P that all its elements have in common. In this case, we can use the **set-builder notation** $\{x | P(x)\}$ to describe the set. The expression $\{x | P(x)\}$ can be read "the set of all x such that the property $P(x)$ is true." Note that the symbol "|" is read as "such that."

As a simple example, let's consider the property of being a beverage. Formally, we can let $P(x)$ be the statement "x is a beverage." Symbolically, we can write $\{x | P(x)\} = \{x \mid x \text{ is a beverage}\}$. In words, this set can be described as "the set of all x such that x is a beverage." Water is an element of this set because water is a beverage. In other words, if we replace x by "water," then $P(x)$ is a true statement. Symbolically, we can write "water $\in \{x \mid x \text{ is a beverage}\}$." Tiger is not an element of this set because a tiger is not a beverage. Symbolically, we can write "tiger $\notin \{x \mid x \text{ is a beverage}\}$." However, tiger would be in the set $\{x \mid x \text{ is an animal}\}$. Tiger would also be in the set $\{x \mid x \text{ is a cat that roars}\}$.

Let's now turn to an example involving numbers.

Example 1.5: Let's look at a few different ways that we can describe the set $\{0, 2, 4, 6, 8, 10\}$. We have already seen that reordering and/or repeating elements does not change the set. For example, $\{2, 2, 0, 4, 10, 8, 8, 8, 6\}$ describes the same set. Here are a few more descriptions using set-builder notation:

- $\{n \mid n \text{ is an even natural number less than or equal to } 10\}$

- $\{n \in \mathbb{Z} \mid n \text{ is even}, 0 \le n < 12\}$

- $\{2k \mid k = 0, 1, 2, 3, 4, 5\}$

The first expression in the bulleted list can be read "the set of n such that n is an even natural number less than or equal to 10."

The second expression can be read "the set of integers n such that n is even and n is between 0 and 12, including 0, but excluding 12. Note that the abbreviation "$n \in \mathbb{Z}$" can be read "n is in the set of integers," or more succinctly, "n is an integer."

The third expression can be read "the set of $2k$ such that k is 0, 1, 2, 3, 4 or 5."

Example 1.6:

1. The set of **rational numbers** is $\mathbb{Q} = \left\{ \frac{a}{b} \mid a, b \in \mathbb{Z} \text{ and } b \neq 0 \right\}$. In words, $\mathbb{Q}$ is "the set of quotients a over b such that a and b are integers and b is not zero." Some examples of rational numbers are $\frac{0}{5}, \frac{1}{3}, \frac{2}{5}$, and $\frac{-6}{7}$. We identify rational numbers $\frac{a}{b}$ and $\frac{c}{d}$ whenever $ad = bc$. For example, $\frac{1}{2} = \frac{3}{6}$ because $1 \cdot 6 = 2 \cdot 3$. We also abbreviate the rational number $\frac{a}{1}$ as a. In this way, we can think of every integer as a rational number. For example, we have $\frac{0}{5} = \frac{0}{1}$ (because $0 \cdot 1 = 5 \cdot 0$), and therefore, we can abbreviate $\frac{0}{5}$ as 0. Similarly, we can abbreviate $\frac{15}{3}$ as 5 (because $\frac{15}{3} = \frac{5}{1}$).

2. In addition to the sets $\mathbb{N}$ (the natural numbers), $\mathbb{Z}$ (the integers), and $\mathbb{Q}$ (the rational numbers), in this book we will be very interested in the set of **real numbers**, $\mathbb{R}$. There are many equivalent ways of formally defining the real numbers and we will look at one such method in Lesson 5.

   For now, let's go with a naïve definition of the real numbers. We first define a **digit** to be one of the symbols $0, 1, 2, 3, 4, 5, 6, 7, 8$, or 9. We then define $\mathbb{R}$ to be the set of numbers of the form $x. y$, where $x \in \mathbb{Z}$ and y is an infinite "string" of digits without a **tail of 9's** (meaning there are infinitely many digits in the string that are **not** 9). Symbolically, we have

   $$\mathbb{R} = \{x. y \mid x \in \mathbb{Z} \text{ and } y \text{ is an infinite string of digits without a tail of 9's}\}.$$

   Some examples of real numbers are $0.000 \ldots, 0.333 \ldots, -16.000 \ldots$, and $1.010010001 \ldots$ We will generally delete tails of 0's. So, we would write $0.000 \ldots$ as 0 and $-16.000 \ldots$ as -16. We will not consider $53.023999999 \ldots$ to be a real number because of the tail of 9's (an alternative approach would be to identify $53.023999999 \ldots$ with 53.024). We can visualize the set of real numbers with the **real line**.

   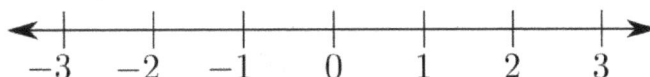

   There is a fairly simple algorithm that allows us to identify every rational number as a real number. For example, the rational number $\frac{3}{2}$ can be represented as the real number 1.5 and the rational number $\frac{2}{3}$ can be represented as the real number $0.66666 \ldots$ I leave the details of this algorithm for the interested reader to explore (see Problem 18 below). Any real number that does **not** correspond to a rational number in this way is called an **irrational number**.

3. The **complex numbers** are defined as $\mathbb{C} = \{a + bi \mid a, b \in \mathbb{R}\}$. In words, $\mathbb{C}$ is "the set of $a + bi$ such that a and b are real numbers." Some examples of complex numbers are $0 + 0i, -2 + 0i$, $2.3 - 5i = 2.3 + (-5)i$, and $4.235235235 \ldots + 51.2020020002 \ldots i$. We will abbreviate $0 + 0i$ as 0, $a + 0i$ as a, and $0 + bi$ as bi. For example, $-2 + 0i = -2$. By identifying $a + 0i$ as a, we can think of every real number as a complex number. Complex numbers of the form bi are called **pure imaginary numbers**.

   If we identify $1 = 1 + 0i$ with the ordered pair $(1, 0)$, and we identify $i = 0 + 1i$ with the ordered pair $(0, 1)$, then it is natural to write the complex number $a + bi$ as the point (a, b). Here is a reasonable justification for this: $a + bi = a(1, 0) + b(0, 1) = (a, 0) + (0, b) = (a, b)$

In this way, we can visualize a complex number as a point in **The Complex Plane**. A portion of the Complex Plane is shown to the right with several complex numbers displayed as points of the form (x, y).

The Complex Plane is formed by taking two copies of the real line and placing one horizontally and the other vertically. The horizontal copy of the real line is called the x-axis or the **real axis** (labeled x in the figure) and the vertical copy of the real line is called the y-axis or **imaginary axis** (labeled y in the figure). The two axes intersect at the point $(0, 0)$. This point is called the **origin**.

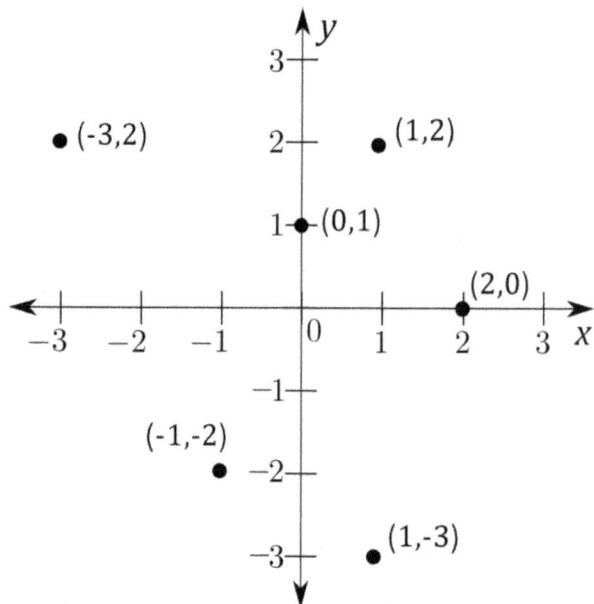

The **empty set** is the unique set with no elements. We use the symbol $\emptyset$ to denote the empty set (some authors use the symbol { } instead).

If A is a finite set, we define the **cardinality** of A, written $|A|$, to be the number of elements of A. For example, $|\{a, b\}| = 2$.

Example 1.7: Let $A = \{\text{canvas}, \text{flower}, \text{amoeba}\}$, $B = \{c, d, d\}$, $C = \{25, 26, 27, \ldots, 3167, 3168\}$, $D = \{\{x\}, \{x, x\}, \{x, x, x\}\}$, and $E = \emptyset$. Then $|A| = 3$, $|B| = 2$, $|C| = 3144$, $|D| = 1$, and $|E| = 0$.

Notes: (1) The set A consists of the three elements "canvas," "flower," and "amoeba."

(2) The set B consists of just two elements: c and d. Remember that $\{c, d, d\} = \{c, d\}$.

(3) The number of consecutive integers from m to n, inclusive, is $\boldsymbol{n - m + 1}$. For set C, we have $m = 25$ and $n = 3168$. Therefore, $|C| = 3168 - 25 + 1 = 3144$.

(4) I call the formula "$n - m + 1$" the **fence-post formula**. If you construct a 3-foot fence by placing a fence-post every foot, then the fence will consist of 4 fence-posts ($3 - 0 + 1 = 4$).

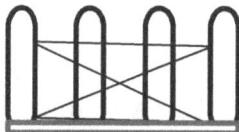

(5) Since $\{x, x\} = \{x\}$ and $\{x, x, x\} = \{x\}$, it follows that $D = \{\{x\}, \{x\}, \{x\}\} = \{\{x\}\}$. So, D consists of the single element $\{x\}$.

(6) Remember that $\emptyset$ (pronounced "the empty set") is the unique set with no elements.

Subsets and Proper Subsets

We say that a set A is a **subset** of a set B, written $A \subseteq B$, if every element of A is an element of B.

Example 1.8:

1. Let $A = \{x, y\}$ and $B = \{x, y, z\}$. The only elements of A are x and y. Since x and y are also elements of B, we see that $A \subseteq B$.

   Notice that $B \not\subseteq A$ (B is **not** a subset of A) because $z \in B$, but $z \notin A$.

2. Let $\mathbb{N} = \{0, 1, 2, 3, \ldots\}$ be the set of natural numbers and let $\mathbb{Z} = \{\ldots, -3, -2, -1, 0, 1, 2, 3, 4, \ldots\}$ be the set of integers. Since every natural number is an integer, $\mathbb{N} \subseteq \mathbb{Z}$.

3. As we saw in Example 1.6, by making appropriate identifications, we have the following sequence of inclusions:

$$\mathbb{N} \subseteq \mathbb{Z} \subseteq \mathbb{Q} \subseteq \mathbb{R} \subseteq \mathbb{C}.$$

   We will see in Theorem 1.14 below that if $A \subseteq B$ and $B \subseteq C$, then $A \subseteq C$ (we say that $\subseteq$ is a **transitive** relation). In this way we see that we have many other inclusions such as $\mathbb{N} \subseteq \mathbb{Q}$, $\mathbb{N} \subseteq \mathbb{R}, \ldots$, and so on.

To the right we see a physical representation of $A \subseteq B$. This figure is called a **Venn diagram**. These types of diagrams are very useful to help visualize relationships among sets. Notice that set A lies completely inside set B. We assume that all the elements of A and B lie in some **universal set** U.

As an example, let's let U be the set of all species of animals. If we let A be the set of species of cats and we let B be the set of species of mammals, then we have $A \subseteq B \subseteq U$, and we see that the Venn diagram to the right gives a visual representation of this situation. (Note that every cat is a mammal and every mammal is an animal.)

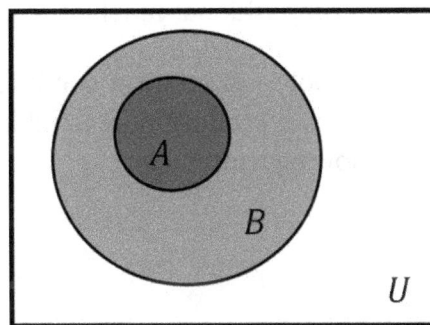

$A \subseteq B$

We say that A is a **proper subset** of B, written $A \subset B$ (or sometimes $A \subsetneq B$), if $A \subseteq B$, but $A \neq B$. For example, $\mathbb{N} \subset \mathbb{Z}$, whereas $\mathbb{N} \not\subset \mathbb{N}$ (although $\mathbb{N} \subseteq \mathbb{N}$).

Note: The definition of proper subset is not very important. It just gives us a convenient way to discuss all the subsets of a specific set except for the set itself. For example, it is quite cumbersome to say "Find all subsets of A, but exclude the set A." It's nice to be able to rephrase this as "Find all proper subsets of A."

Let's look at the definition of $\subseteq$ (subset) in a bit more detail.

Once again, we write $A \subseteq B$ if every element of A is an element of B. That is, $A \subseteq B$ if, for every x, $x \in A$ implies $x \in B$. Symbolically, we can write the following:

$$\forall x (x \in A \rightarrow x \in B)$$

Notes: (1) The symbol $\forall$ is called a **universal quantifier**, and it is pronounced "For all."

(2) The logical expression $\forall x (x \in A \rightarrow x \in B)$ can be translated into English as "For all x, if x is an element of A, then x is an element of B."

(3) To show that a set A is a subset of a set B, we need to show that the expression $\forall x(x \in A \rightarrow x \in B)$ is true. If the set A is finite and the elements are listed, we can just check that each element of A is also an element of B. However, if the set A is described by a property, say $A = \{x|P(x)\}$, we may need to craft an argument more carefully. We can begin by taking an **arbitrary but specific element** a from A and then arguing that this element a is in B.

What could we possibly mean by an arbitrary but specific element? Aren't the words "arbitrary" and "specific" antonyms? Well, by arbitrary, we mean that we don't know which element we are choosing – it's just some element a that satisfies the property P. So, we are just assuming that $P(a)$ is true. However, once we choose this element a, we use this same a for the rest of the argument, and that is what we mean by it being specific.

(4) "$p \rightarrow q$" is an example of a **statement** in **propositional logic**. It is usually read as "if p, then q" or "p implies q." The letters p and q are called **propositional variables**, and we generally assign a truth value of T (for true) or F (for false) to each propositional variable. Formally, we define a **truth assignment** of a list of propositional variables to be a choice of T or F for each propositional variable in the list.

The symbol $\rightarrow$ is called a **conditional** or **implication**. It is one example of a **logical connective** (it *connects* two propositional variables). The rules for determining the truth value for $p \rightarrow q$ are given by the following truth table:

p	q	$p \rightarrow q$
T	T	T
T	F	F
F	T	T
F	F	T

For example, if p and q are both assigned the truth value T, then the truth value of $p \rightarrow q$ is also T, as can be seen by the first row of the above truth table. We can write $T \rightarrow T \equiv T$. The symbol "$\equiv$" can be read "**is logically equivalent to**." Similarly, we have

$$T \rightarrow F \equiv F \qquad F \rightarrow T \equiv T \qquad F \rightarrow F \equiv T$$

Observe that the only time $p \rightarrow q$ can be false is if p is true and q is false. So, one way to prove that $p \rightarrow q$ is true is to assume that p is true and then provide a logically correct argument that q must also be true.

If we let p represent the statement "$x \in A$" and we let q represent the statement "$x \in B$," then $p \rightarrow q$ represents the statement "if $x \in A$, then $x \in B$." As stated in the last paragraph, one way to prove that this statement is true is to assume that $x \in A$ is true and then provide a logically correct argument that $x \in B$ must also be true.

Basic Theorems Involving Subsets

Let's try to prove our first theorem using the definition of a subset together with Note 3 above about arbitrary but specific elements.

Theorem 1.9: Every set A is a subset of itself.

14

Analysis: Before writing the proof, let's think about our strategy. We want to prove $A \subseteq A$. In other words, we want to show $\forall x (x \in A \to x \in A)$. So, we will take an arbitrary but specific $a \in A$ and then argue that $a \in A$. But that's pretty obvious, isn't it? In this case, the property we're describing is precisely the conclusion we are looking for. Here are the details.

Proof of Theorem 1.9: Let A be a set and let $a \in A$. Then $a \in A$. So, $a \in A \to a \in A$ is true. Since a was an arbitrary element of A, $\forall x (x \in A \to x \in A)$ is true. Therefore, $A \subseteq A$. □

Notes: (1) The proof begins with the **opening statement** "Let A be a set and let $a \in A$." In general, the opening statement states what is given in the problem and/or fixes any arbitrary but specific objects that we will need.

(2) The proof ends with the **closing statement** "Therefore, $A \subseteq A$." In general, the closing statement states the result.

(3) Everything between the opening statement and the closing statement is known as the **argument**.

(4) We place the symbol □ at the end of the proof to indicate that the proof is complete.

(5) Consider the logical statement $p \to p$. This statement is always true ($T \to T \equiv T$ and $F \to F \equiv T$). $p \to p$ is an example of a tautology. A **tautology** is a statement that is true for every possible truth assignment of the propositional variables.

(6) If we let p represent the statement $a \in A$, by Note 5, we see that $a \in A \to a \in A$ is always true.

Alternate proof of Theorem 1.9: Let A be a set and let $a \in A$. Since $p \to p$ is a tautology, we have that $a \in A \to a \in A$ is true. Since a was arbitrary, $\forall x (x \in A \to x \in A)$ is true. Therefore, $A \subseteq A$. □

Let's prove another basic but important theorem.

Theorem 1.10: The empty set is a subset of every set.

Analysis: This time we want to prove $\emptyset \subseteq A$. In other words, we want to show $\forall x (x \in \emptyset \to x \in A)$. Since $x \in \emptyset$ is always false (the empty set has no elements), $x \in \emptyset \to x \in A$ is always true.

In general, if p is a false statement, then we say that $p \to q$ is **vacuously true**.

Proof of Theorem 1.10: Let A be a set. The statement $x \in \emptyset \to x \in A$ is vacuously true for any x, and so, $\forall x (x \in \emptyset \to x \in A)$ is true. Therefore, $\emptyset \subseteq A$. □

Note: The opening statement is "Let A be a set," the closing statement is "Therefore, $\emptyset \subseteq A$," and the argument is everything in between.

Example 1.11: Let $C = \{a, b, c\}$, $D = \{a, c\}$, $E = \{b, c\}$, $F = \{b, d\}$, and $G = \emptyset$. Then $D \subseteq C$ and $E \subseteq C$. Also, since **the empty set is a subset of every set**, we have $G \subseteq C$, $G \subseteq D$, $G \subseteq E$, $G \subseteq F$, and $G \subseteq G$. **Every set is a subset of itself**, and so, $C \subseteq C$, $D \subseteq D$, $E \subseteq E$, and $F \subseteq F$.

Note: Below are possible Venn diagrams for this problem. The diagram on the left shows the relationship between the sets C, D, E, and F. Notice how D and E are both subsets of C, whereas F is not a subset of C. Also, notice how D and E overlap, E and F overlap, but there is no overlap between D and F (they have no elements in common). The diagram on the right shows the proper placement of the elements. Here, I chose the universal set to be $U = \{a, b, c, d, e, f, g\}$. This choice for the universal set is somewhat arbitrary. Any set containing $\{a, b, c, d\}$ would do.

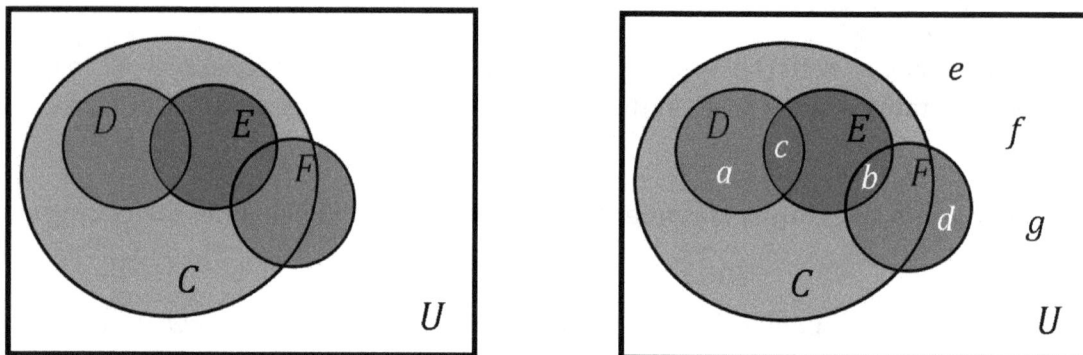

Power Sets

If A is a set, then the **power set** of A, written $\mathcal{P}(A)$, is the set of all subsets of A. In set-builder notation, we write $\mathcal{P}(A) = \{B \mid B \subseteq A\}$.

Example 1.12: The set $A = \{a, b\}$ has 2 elements and 4 subsets. The subsets of A are $\emptyset$, $\{a\}$, $\{b\}$, and $\{a, b\}$. It follows that $\mathcal{P}(A) = \{\emptyset, \{a\}, \{b\}, \{a, b\}\}$.

The set $B = \{a, b, c\}$ has 3 elements and 8 subsets. The subsets of B are $\emptyset$, $\{a\}$, $\{b\}$, $\{c\}$, $\{a, b\}$, $\{a, c\}$, $\{b, c\}$, and $\{a, b, c\}$. It follows that $\mathcal{P}(B) = \{\emptyset, \{a\}, \{b\}, \{c\}, \{a, b\}, \{a, c\}, \{b, c\}, \{a, b, c\}\}$.

Let's draw a **tree diagram** for the subsets of each of the sets A and B.

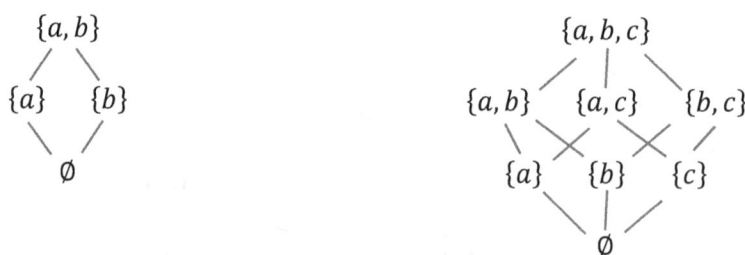

The tree diagram on the left is for the subsets of the set $A = \{a, b\}$. We start by writing the set $A = \{a, b\}$ at the top. On the next line we write the subsets of cardinality 1 ($\{a\}$ and $\{b\}$). On the line below that we write the subsets of cardinality 0 (just $\emptyset$). We draw a line segment between any two sets when the smaller (lower) set is a subset of the larger (higher) set. So, we see that $\emptyset \subseteq \{a\}$, $\emptyset \subseteq \{b\}$, $\{a\} \subseteq \{a, b\}$, and $\{b\} \subseteq \{a, b\}$. There is actually one more subset relationship, namely $\emptyset \subseteq \{a, b\}$ (and of course each set displayed is a subset of itself). We didn't draw a line segment from $\emptyset$ to $\{a, b\}$ to avoid unnecessary clutter. Instead, we can simply trace the path from $\emptyset$ to $\{a\}$ to $\{a, b\}$ (or from $\emptyset$ to $\{b\}$ to $\{a, b\}$). We are using a property called **transitivity** here (see Theorem 1.14 below).

The tree diagram on the right is for the subsets of $B = \{a, b, c\}$. Observe that from top to bottom we write the subsets of B of cardinality 3, then 2, then 1, and then 0. We then draw the appropriate line segments, just as we did for $A = \{a, b\}$.

How many subsets does a set of cardinality n have? Let's start by looking at some examples.

Example 1.13: A set with 0 elements must be $\emptyset$, and this set has exactly 1 subset (the only subset of the empty set is the empty set itself).

A set with 1 element has 2 subsets, namely $\emptyset$ and the set itself.

In the last example, we saw that a set with 2 elements has 4 subsets, and we also saw that a set with 3 elements has 8 subsets.

Do you see the pattern yet? $1 = 2^0, 2 = 2^1, 4 = 2^2, 8 = 2^3$. So, we see that a set with 0 elements has 2^0 subsets, a set with 1 element has 2^1 subsets, a set with 2 elements has 2^2 subsets, and a set with 3 elements has 2^3 subsets.

A reasonable guess would be that a set with n elements has $\mathbf{2^n}$ subsets. You will be asked to prove this result later (Problem 24 in Problem Set 5). We can also say that if $|A| = n$, then $|\mathcal{P}(A)| = 2^n$.

Transitivity of the Subset Relation

Let's get back to the transitivity mentioned above in Example 1.8 and in our discussion of tree diagrams.

Theorem 1.14: Let A, B, and C be sets such that $A \subseteq B$ and $B \subseteq C$. Then $A \subseteq C$.

Proof: Suppose that A, B, and C are sets with $A \subseteq B$ and $B \subseteq C$, and let $a \in A$. Since $A \subseteq B$ and $a \in A$, it follows that $a \in B$. Since $B \subseteq C$ and $a \in B$, it follows that $a \in C$. Since a was an arbitrary element of A, we have shown that every element of A is an element of C. That is, $\forall x(x \in A \rightarrow x \in C)$ is true. Therefore, $A \subseteq C$. □

Note: To the right we have a Venn diagram illustrating Theorem 1.14.

Theorem 1.14 tells us that the relation $\subseteq$ is **transitive**. Since $\subseteq$ is transitive, we can write things like $A \subseteq B \subseteq C \subseteq D$, and without explicitly saying it, we know that $A \subseteq C, A \subseteq D$, and $B \subseteq D$.

Example 1.15: The membership relation $\in$ is an example of a relation that is **not** transitive. For example, let $A = \{0\}$, $B = \{0, 1, \{0\}\}$, and $C = \{x, y, \{0, 1, \{0\}\}\}$. Observe that $A \in B$ and $B \in C$, but $A \notin C$.

$$A \subseteq B \subseteq C$$

$$\{0\} \in \{0, 1, \{0\}\} \in \{x, y, \{0, 1, \{0\}\}\}$$

Notes: (1) The set A has just 1 element, namely 0.

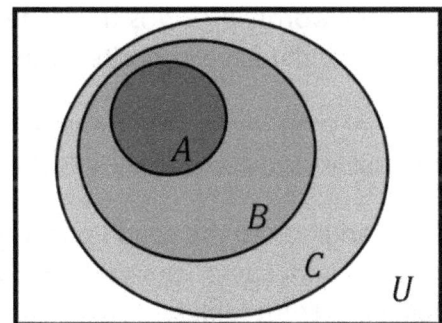

(2) The set B has 3 elements, namely 0, 1, and $\{0\}$. But wait! $A = \{0\}$. So, $A \in B$. The set A is circled twice in the above image.

(3) The set C also has 3 elements, namely, x, y, and $\{0,1,\{0\}\}$. But wait! $B = \{0, 1, \{0\}\}$. So, $B \in C$. The set B has a rectangle around it twice in the above image.

(4) Since $A \neq x$, $A \neq y$, and $A \neq \{0, 1, \{0\}\}$, we see that $A \notin C$.

(5) Is it clear that $\{0\} \notin C$? $\{0\}$ is in a set that's in C (namely, B), but $\{0\}$ is not itself in C.

(6) Here is a more basic example showing that $\in$ is not transitive: $\emptyset \in \{\emptyset\} \in \{\{\emptyset\}\}$, but $\emptyset \notin \{\{\emptyset\}\}$

The only element of $\{\{\emptyset\}\}$ is $\{\emptyset\}$.

Equality of Sets

Two sets A and B are **equal**, written $A = B$, if they have the same elements. Symbolically, we can write the following:

$$\forall x(x \in A \leftrightarrow x \in B)$$

Notes: (1) "$p \leftrightarrow q$" is another example of a **statement** in **propositional logic**. It is usually read as "p if and only if q."

The logical connective $\leftrightarrow$ is called a **biconditional**. The rules for determining the truth value for $p \leftrightarrow q$ are given by the following truth table:

p	q	$p \leftrightarrow q$
T	T	T
T	F	F
F	T	F
F	F	T

In other words, $p \leftrightarrow q$ is true when p and q have the same truth value (both T or both F) and false when p and q have opposite truth values (one T and the other F).

If we let p represent the statement "$x \in A$" and we let q represent the statement "$x \in B$," then $p \leftrightarrow q$ represents the statement "$x \in A$ if and only if $x \in B$."

(2) In addition to the conditional ($\rightarrow$) and biconditional ($\leftrightarrow$), there are three additional commonly used logical connectives: the **conjunction** ($\wedge$), the **disjunction** ($\vee$), and the **negation** ($\neg$). They have the following truth tables:

p	q	$p \wedge q$
T	T	T
T	F	F
F	T	F
F	F	F

p	q	$p \vee q$
T	T	T
T	F	T
F	T	T
F	F	F

p	$\neg p$
T	F
F	T

(3) We say that two statements are **logically equivalent** if every truth assignment of the propositional variables appearing in either statement (or both statements) leads to the same truth value for both statements.

It is easy to verify that $p \leftrightarrow q$ is logically equivalent to $(p \rightarrow q) \wedge (q \rightarrow p)$. To see this, we check that all possible truth assignments for p and q lead to the same truth value for the two statements. For example, if p and q are both true, then

$$p \leftrightarrow q \equiv \text{T} \leftrightarrow \text{T} \equiv \text{T} \quad \text{and} \quad (p \rightarrow q) \wedge (q \rightarrow p) \equiv (\text{T} \rightarrow \text{T}) \wedge (\text{T} \rightarrow \text{T}) \equiv \text{T} \wedge \text{T} \equiv \text{T}.$$

As another example, if p is true and q is false, then

$$p \leftrightarrow q \equiv \text{T} \leftrightarrow \text{F} \equiv \text{F} \quad \text{and} \quad (p \rightarrow q) \wedge (q \rightarrow p) \equiv (\text{T} \rightarrow \text{F}) \wedge (\text{F} \rightarrow \text{T}) \equiv \text{F} \wedge \text{T} \equiv \text{F}.$$

The reader should check the other two truth assignments for p and q.

(4) Letting p be the statement $x \in A$, letting q be the statement $x \in B$, and replacing $p \leftrightarrow q$ by the logically equivalent statement $(p \rightarrow q) \wedge (q \rightarrow p)$ gives us

$$\forall x(x \in A \leftrightarrow x \in B) \text{ if and only if } \forall x((x \in A \rightarrow x \in B) \wedge (x \in B \rightarrow x \in A)).$$

(5) It is also true that $\forall x(p(x) \wedge q(x))$ is logically equivalent to $\forall x(p(x)) \wedge \forall x(q(x))$. And so, we have

$$\forall x(x \in A \leftrightarrow x \in B) \text{ if and only if } \forall x(x \in A \rightarrow x \in B) \text{ and } \forall x(x \in B \rightarrow x \in A).$$

In other words, to show that $A = B$, we can instead show that $A \subseteq B$ and $B \subseteq A$.

The statement "$A = B$ if and only if $A \subseteq B$ and $B \subseteq A$" is usually called the **Axiom of Extensionality**. It is often easiest to prove that two sets are equal by showing that each one is a subset of the other.

Example 1.16: Let $A = \{n \in \mathbb{N} \mid n < 100\}$ and let $B = \{n \in \mathbb{Z} \mid 0 \leq n \leq 99\}$. Let's use the Axiom of Extensionality to prove that $A = B$.

Note: We are assuming that $<$ and $\leq$ are defined in the usual way on $\mathbb{N}$ and $\mathbb{Z}$. In the proof, we will freely use the properties of these relations that we know to be true. For example, 0 is the least element of $\mathbb{N}$ with respect to $<$ and $\leq$. As another example, in both $\mathbb{N}$ and $\mathbb{Z}$, we have $n < k + 1$ if and only if $n \leq k$.

Proof: We first prove that $A \subseteq B$. Let $n \in A$. Then $n \in \mathbb{N}$ and $n < 100$. Since $n \in \mathbb{N}$ and $\mathbb{N} \subseteq \mathbb{Z}$, it follows that $n \in \mathbb{Z}$. Since $n \in \mathbb{N}$ and the least element of $\mathbb{N}$ is 0, $0 \leq n$. Since $n < 100$ and $n \in \mathbb{N}$, we have $n \leq 99$. Therefore, $n \in B$. Since $n \in A$ was arbitrary, $\forall n(n \in A \rightarrow n \in B)$. Therefore, we have shown $A \subseteq B$.

We now prove that $B \subseteq A$. Let $n \in B$. Then $n \in \mathbb{Z}$ and $0 \leq n \leq 99$. Since $n \in \mathbb{Z}$ and $0 \leq n$, it follows that $n \in \mathbb{N}$. Since $n \leq 99$ and $99 < 100$, it follows that $n < 100$. Therefore, $n \in A$. Since $n \in B$ was arbitrary, $\forall n(n \in B \rightarrow n \in A)$. Therefore, we have shown $B \subseteq A$.

Since $A \subseteq B$ and $B \subseteq A$, we have $A = B$. $\qquad \square$

Problem Set 1

Full solutions to these problems are available for free download here:

www.SATPrepGet800.com/TFBZLF

LEVEL 1

1. Determine whether each of the following statements is true or false:

    (i) $7 \in \{7\}$

    (ii) $\delta \in \{\gamma, \delta, \epsilon\}$

    (iii) $-3 \in \{3\}$

    (iv) $3 \in \mathbb{Z}$

    (v) $-42 \in \mathbb{N}$

    (vi) $\frac{15}{17} \in \mathbb{Q}$

    (vii) $\emptyset \subseteq \{0, 1, 2\}$

    (viii) $\{\delta\} \subseteq \{\delta, \Delta\}$

    (ix) $\{x, y, z\} \subseteq \{x, y, z\}$

    (x) $\{a, b, \{c, d\}\} \subseteq \{a, b, c, d\}$

2. Determine the cardinality of each of the following sets:

    (i) $\{\square, \Delta, \Gamma\}$

    (ii) $\{0, 1, 2, 3, 4, 5\}$

    (iii) $\{1, 2, \dots, 37\}$

    (iv) $\left\{\frac{1}{2}, \frac{1}{3}, \dots, \frac{1}{11}\right\}$

3. Provide a single example of a set A with the following properties: (i) $A \subset \mathbb{C}$ (A is a *proper* subset of $\mathbb{C}$); (ii) A is infinite; (iii) A contains real numbers; and (iv) A contains complex numbers that are not real.

LEVEL 2

4. Compute the power set of each of the following sets:

    (i) $\emptyset$

    (ii) $\{5\}$

    (iii) $\{m, n\}$

    (iv) $\{\emptyset, \{\emptyset\}\}$

    (v) $\{\{\emptyset\}\}$

20

5. Determine whether each of the following statements is true or false:

    (i) $3 \in \emptyset$

    (ii) $\emptyset \in \{a, b\}$

    (iii) $\emptyset \in \emptyset$

    (iv) $\emptyset \in \{\emptyset, \{\emptyset\}\}$

    (v) $\{\emptyset\} \in \emptyset$

    (vi) $\{\emptyset\} \in \{\emptyset\}$

    (vii) $\emptyset \subseteq \emptyset$

    (viii) $\emptyset \subseteq \{\emptyset\}$

    (ix) $\{\emptyset\} \subseteq \emptyset$

    (x) $\{\emptyset\} \subseteq \{\emptyset\}$

    (xi) $\mathbb{Q} \subseteq \mathbb{C}$

    (xii) $3 \in \{2k \mid k = 1, 2, 3, 4, 5, 6\}$

6. Determine the cardinality of each of the following sets:

    (i) $\{a, a, b, c, d, d, d\}$

    (ii) $\{\{1, 2\}, \{3,4,5\}\}$

    (iii) $\{5, 6, 7, \dots, 2122, 2123\}$

LEVEL 3

7. Determine the cardinality of each of the following sets:

    (i) $\{\{\{a, b\}\}\}$

    (ii) $\{\{0, 1\}, 0, \{0\}, \{0, \{0, 1, 2\}\}\}$

    (iii) $\{a, \{a\}, \{a, a\}, \{a, a, a, a\}, \{a, a, \{a\}\}, \{a, \{a\}, \{a\}\}\}$

8. How many subsets does $\{a, b, c, d\}$ have? Draw a tree diagram for the subsets of $\{a, b, c, d\}$.

9. Let A, B, C, D, and E be sets such that $A \subseteq B$, $B \subseteq C$, $C \subseteq D$, and $D \subseteq E$. Prove that $A \subseteq E$.

LEVEL 4

10. A relation R is **reflexive** if $\forall x (xRx)$ and **symmetric** if $\forall x \forall y (xRy \rightarrow yRx)$. Show that $\subseteq$ is reflexive, but $\in$ is not. Then decide if each of $\subseteq$ and $\in$ is symmetric.

11. Determine whether each of the following statements is true or false:

    (i) $c \in \{a, \{c\}\}$

    (ii) $\{\Delta\} \in \{\delta, \Delta\}$

    (iii) $\{1\} \in \{1, a, 2, b\}$

    (iv) $\emptyset \in \{\{\emptyset\}\}$

    (v) $\{\{\emptyset\}\} \in \emptyset$

12. We say that a set A is **transitive** if $\forall x (x \in A \rightarrow x \subseteq A)$. Determine if each of the following sets is transitive:

    (i) $\emptyset$

    (ii) $\{\emptyset\}$

    (iii) $\{\{\emptyset\}\}$

    (iv) $\{\emptyset, \{\emptyset\}\}$

    (v) $\{\emptyset, \{\emptyset\}, \{\{\emptyset\}\}\}$

    (vi) $\{\{\emptyset\}, \{\emptyset, \{\emptyset\}\}\}$

LEVEL 5

13. Let A and B be sets with $A \subseteq B$. Prove that $\mathcal{P}(A) \subseteq \mathcal{P}(B)$.

14. Prove that if A is a transitive set, then $\mathcal{P}(A)$ is also a transitive set (see Problem 12 above for the definition of a transitive set).

15. Let $P(x)$ be the property $x \notin x$. Prove that $\{x | P(x)\}$ cannot be a set.

16. Let $A = \{a, b, c, d\}$, $B = \{X \mid X \subseteq A \wedge d \notin X\}$, and $C = \{X \mid X \subseteq A \wedge d \in X\}$. Show that there is a natural one-to-one correspondence between the elements of B and the elements of C. Then generalize this result to a set with $n + 1$ elements for $n > 0$.

CHALLENGE PROBLEMS

17. Let A and B be sets with $A \subseteq B$ and B transitive. Prove that $\mathcal{P}(A) \subseteq \mathcal{P}\big(\mathcal{P}(B)\big)$. (See Problem 12 above for the definition of a transitive set.)

18. Prove that there is a natural one-to-one correspondence between the elements of $\mathbb{Q}$ and the elements of a proper subset of $\mathbb{R}$.

LESSON 2
OPERATIONS ON SETS

Basic Set Operations

The **union** of the sets A and B, written $A \cup B$, is the set of elements that are in A or B (or both).

$$A \cup B = \{x \mid x \in A \text{ or } x \in B\}$$

The **intersection** of A and B, written $A \cap B$, is the set of elements that are simultaneously in A and B.

$$A \cap B = \{x \mid x \in A \text{ and } x \in B\}$$

The following Venn diagrams for the union and intersection of two sets can be useful for visualizing these operations.

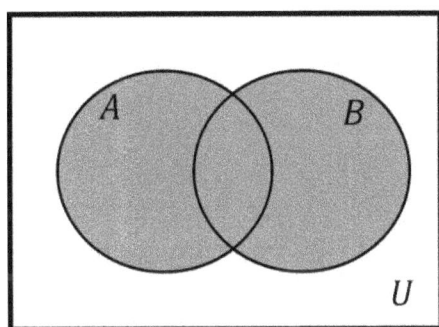

$A \cup B$

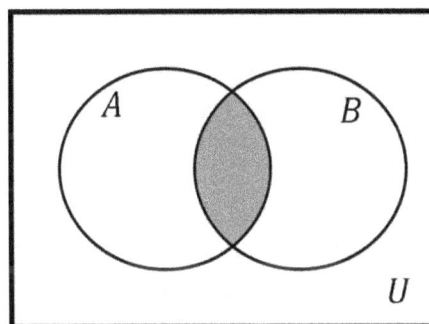

$A \cap B$

The **difference** $A \setminus B$ is the set of elements that are in A and not in B.

$$A \setminus B = \{x \mid x \in A \text{ and } x \notin B\}$$

The **symmetric difference** between A and B, written $A \, \Delta \, B$, is the set of elements that are in A or B, but not both.

$$A \, \Delta \, B = (A \setminus B) \cup (B \setminus A)$$

Let's also look at Venn diagrams for the difference and symmetric difference of two sets.

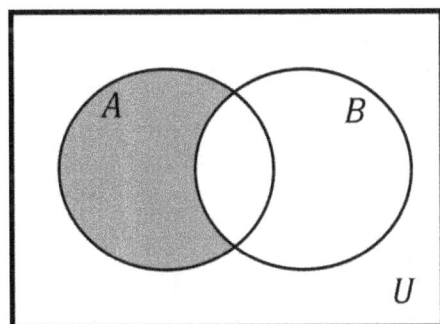

$A \setminus B$

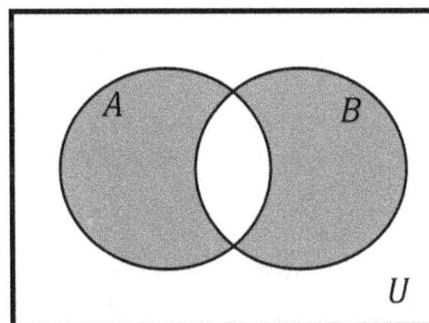

$A \, \Delta \, B$

23

Example 2.1: Let $A = \{0, 1, 2, 3, 4\}$ and $B = \{3, 4, 5, 6\}$. We have

1. $A \cup B = \{0, 1, 2, 3, 4, 5, 6\}$.
2. $A \cap B = \{3, 4\}$.
3. $A \setminus B = \{0, 1, 2\}$.
4. $B \setminus A = \{5, 6\}$.
5. $A \, \Delta \, B = \{0, 1, 2\} \cup \{5, 6\} = \{0, 1, 2, 5, 6\}$.

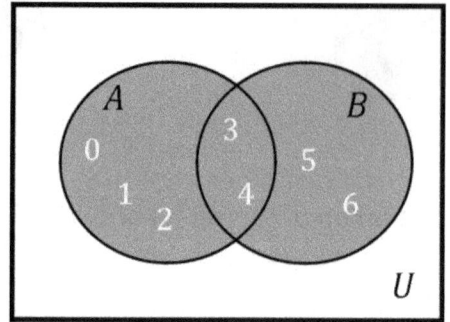

Example 2.2: Recall that the set of natural numbers is $\mathbb{N} = \{0, 1, 2, 3, \ldots\}$ and the set of integers is $\mathbb{Z} = \{\ldots, -4, -3, -2, -1, 0, 1, 2, 3, 4, \ldots\}$. Observe that in this case, $\mathbb{N} \subseteq \mathbb{Z}$. We have

1. $\mathbb{N} \cup \mathbb{Z} = \mathbb{Z}$.
2. $\mathbb{N} \cap \mathbb{Z} = \mathbb{N}$.
3. $\mathbb{N} \setminus \mathbb{Z} = \emptyset$.
4. $\mathbb{Z} \setminus \mathbb{N} = \{\ldots, -4, -3, -2, -1\} = \mathbb{Z}^-$. ($\mathbb{Z}^-$ is "the set of negative integers.")
5. $\mathbb{N} \, \Delta \, \mathbb{Z} = \emptyset \cup \mathbb{Z}^- = \mathbb{Z}^-$.

Note: Whenever A and B are sets and $B \subseteq A$, then $A \cup B = A$, $A \cap B = B$, and $B \setminus A = \emptyset$. We will prove the first and third of these three facts in Theorems 2.6 and 2.7 below, respectively. You will be asked to prove the second in Problem 7 below.

Example 2.3: Let $\mathbb{E} = \{0, 2, 4, 6, \ldots\}$ be the set of even natural numbers and let $\mathbb{O} = \{1, 3, 5, 7, \ldots\}$ be the set of odd natural numbers. We have

1. $\mathbb{E} \cup \mathbb{O} = \{0, 1, 2, 3, 4, 5, 6, 7, \ldots\} = \mathbb{N}$.
2. $\mathbb{E} \cap \mathbb{O} = \emptyset$.
3. $\mathbb{E} \setminus \mathbb{O} = \emptyset$.
4. $\mathbb{O} \setminus \mathbb{E} = \emptyset$.
5. $\mathbb{E} \, \Delta \, \mathbb{O} = \emptyset \cup \emptyset = \emptyset$.

In general, we say that sets A and B are **disjoint** or **mutually exclusive** if $A \cap B = \emptyset$. To the right is a Venn diagram for disjoint sets.

In Example 2.3 above, we saw that the sets $\mathbb{E}$ and $\mathbb{O}$ are disjoint.

Example 2.4: Consider the sets $A = \{a + bi \in \mathbb{C} \mid a, b \in \mathbb{Z}\}$ and $B = \{a + bi \in \mathbb{C} \mid a \notin \mathbb{Q}\}$. Then A and B are disjoint. To see this, suppose that $a + bi \in A \cap B$. Then $a + bi \in A$, and so, $a \in \mathbb{Z}$. Since $\mathbb{Z} \subseteq \mathbb{Q}$, $a \in \mathbb{Q}$. Also, $a + bi \in B$. So, $a \notin \mathbb{Q}$. Since we cannot have both $a \in \mathbb{Q}$ and $a \notin \mathbb{Q}$, we must have $A \cap B = \emptyset$.

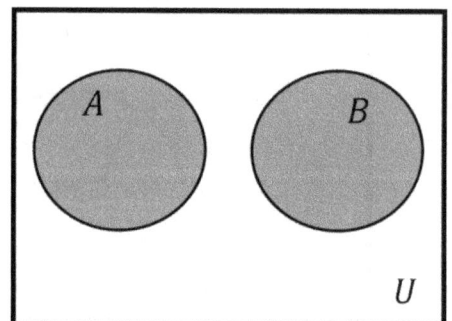

$A \cap B = \emptyset$

Let's prove some theorems involving unions of sets. You will be asked to prove the analogous results for intersections of sets in Problems 6 and 7 below.

Theorem 2.5: If A and B are sets, then $A \subseteq A \cup B$.

Before going through the proof, look once more at the Venn diagram above for $A \cup B$ and convince yourself that this theorem should be true.

Proof of Theorem 2.5: Suppose that A and B are sets and let $x \in A$. Then $x \in A$ or $x \in B$. Therefore, $x \in A \cup B$. Since x was an arbitrary element of A, we have shown that every element of A is an element of $A \cup B$. That is, $\forall x(x \in A \rightarrow x \in A \cup B)$ is true. Therefore, $A \subseteq A \cup B$. □

Note: If p is a true statement, then $p \vee q$ (p or q) is true no matter what the truth value of q is. In the second sentence of the proof above, we are using this fact with p being the statement $x \in A$ and q being the statement $x \in B$.

We will use this same reasoning in the second paragraph of the next proof as well.

Theorem 2.6: $B \subseteq A$ if and only if $A \cup B = A$.

Before going through the proof, it's a good idea to draw a Venn diagram for $B \subseteq A$ and convince yourself that this theorem should be true.

Proof of Theorem 2.6: Suppose that $B \subseteq A$ and let $x \in A \cup B$. Then $x \in A$ or $x \in B$. If $x \in A$, then $x \in A$ (trivially). If $x \in B$, then since $B \subseteq A$, it follows that $x \in A$. Since x was an arbitrary element of $A \cup B$, we have shown that every element of $A \cup B$ is an element of A. That is, $\forall x(x \in A \cup B \rightarrow x \in A)$ is true. Therefore, $A \cup B \subseteq A$. By Theorem 2.5, $A \subseteq A \cup B$. Since $A \cup B \subseteq A$ and $A \subseteq A \cup B$, it follows that $A \cup B = A$.

Now, suppose that $A \cup B = A$ and let $x \in B$. Since $x \in B$, it follows that $x \in A$ or $x \in B$. Therefore, $x \in A \cup B$. Since $A \cup B = A$, we have $x \in A$. Since x was an arbitrary element of B, we have shown that every element of B is an element of A. That is, $\forall x(x \in B \rightarrow x \in A)$. Therefore, $B \subseteq A$. □

Theorem 2.7: Let A and B be sets. If $B \subseteq A$, then $B \setminus A = \emptyset$.

We will use an **indirect proof** to prove Theorem 2.7. Specifically, we will use a **proof by contrapositive**.

The contrapositive of the conditional statement $p \rightarrow q$ is the statement $\neg q \rightarrow \neg p$. These two statements are logically equivalent. To see this, we check that all possible truth assignments for p and q lead to the same truth value for the two statements. For example, if p and q are both true, then $p \rightarrow q \equiv T \rightarrow T \equiv T$ and $\neg q \rightarrow \neg p \equiv F \rightarrow F \equiv T$. The reader should check the other three truth assignments for p and q.

The contrapositive of the statement "If $B \subseteq A$, then $B \setminus A = \emptyset$" is "If $B \setminus A \neq \emptyset$, then $B \nsubseteq A$." So, we will prove Theorem 2.7 by assuming that $B \setminus A \neq \emptyset$ and using this to show that $B \nsubseteq A$.

Proof of Theorem 2.7: Let A and B be sets such that $B \setminus A \neq \emptyset$. Since $B \setminus A \neq \emptyset$, there is $a \in B \setminus A$. Then $a \in B$ and $a \notin A$. So, $\forall x(x \in B \rightarrow x \in A)$ is false. Therefore, $B \nsubseteq A$. □

Properties of Unions and Intersections

Unions, intersections, and set differences have many nice algebraic properties such as

1. **Commutativity:** $A \cup B = B \cup A$ and $A \cap B = B \cap A$.

2. **Associativity:** $(A \cup B) \cup C = A \cup (B \cup C)$ and $(A \cap B) \cap C = A \cap (B \cap C)$.

3. **Distributivity:** $A \cap (B \cup C) = (A \cap B) \cup (A \cap C)$ and $A \cup (B \cap C) = (A \cup B) \cap (A \cup C)$.

4. **De Morgan's Laws:** $C \setminus (A \cup B) = (C \setminus A) \cap (C \setminus B)$ and $C \setminus (A \cap B) = (C \setminus A) \cup (C \setminus B)$.

5. **Idempotent Laws:** $A \cup A = A$ and $A \cap A = A$.

As an example, let's prove that the operation of forming unions is associative. You will be asked to prove that the other properties hold in the problems below.

Theorem 2.8: The operation of forming unions is associative.

Note: Before beginning the proof, let's draw Venn diagrams of the situation to convince ourselves that the theorem is true.

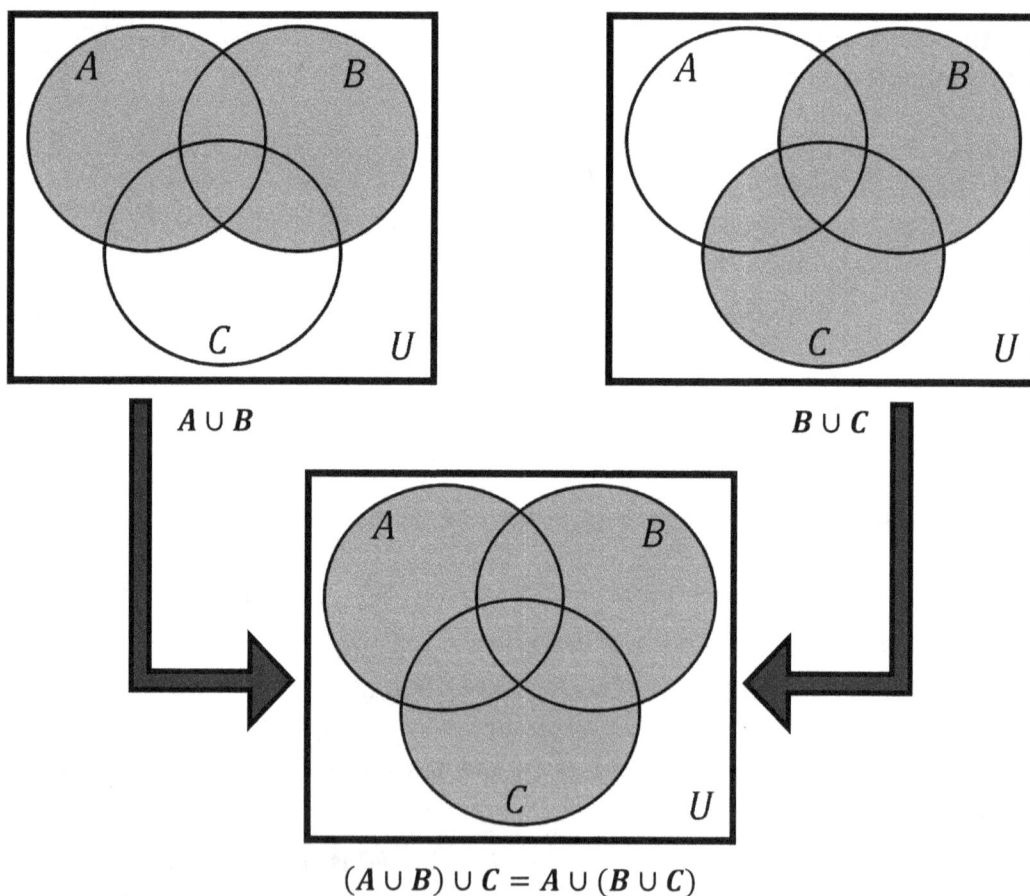

$A \cup B$

$B \cup C$

$(A \cup B) \cup C = A \cup (B \cup C)$

Proof of Theorem 2.8: Let A, B, and C be sets, and let $x \in (A \cup B) \cup C$. Then $x \in A \cup B$ or $x \in C$. If $x \in C$, then $x \in B$ or $x \in C$. So, $x \in B \cup C$. Then $x \in A$ or $x \in B \cup C$. So, $x \in A \cup (B \cup C)$. If, on the other hand, $x \in A \cup B$, then $x \in A$ or $x \in B$. If $x \in A$, then $x \in A$ or $x \in B \cup C$. So, $x \in A \cup (B \cup C)$. If $x \in B$, then $x \in B$ or $x \in C$. So, $x \in B \cup C$. Then $x \in A$ or $x \in B \cup C$. So, $x \in A \cup (B \cup C)$. Since x was arbitrary, we have shown $\forall x\big(x \in (A \cup B) \cup C \to x \in A \cup (B \cup C)\big)$. Therefore, we have shown that $(A \cup B) \cup C \subseteq A \cup (B \cup C)$.

A similar argument can be used to show $A \cup (B \cup C) \subseteq (A \cup B) \cup C$ (the reader should write out the details).

Since $(A \cup B) \cup C \subseteq A \cup (B \cup C)$ and $A \cup (B \cup C) \subseteq (A \cup B) \cup C$, $(A \cup B) \cup C = A \cup (B \cup C)$, and therefore, the operation of forming unions is associative. □

Remember that associativity allows us to drop parentheses. So, we can now simply write $A \cup B \cup C$ when taking the union of the three sets A, B, and C.

Arbitrary Unions and Intersections

We will often be interested in taking unions and intersections of more than two sets Therefore, we make the following more general definitions.

Let X be a nonempty set of sets.

$$\cup X = \{y \mid \text{there is } Y \in X \text{ with } y \in Y\} \qquad \text{and} \qquad \cap X = \{y \mid \text{for all } Y \in X, y \in Y\}.$$

If you're having trouble understanding what these definitions are saying, you're not alone. The notation probably looks confusing, but the ideas behind these definitions are very simple. You have a whole bunch of sets (possibly infinitely many). To take the union of all these sets, you simply throw all the elements together into one big set. To take the intersection of all these sets, you take only the elements that are in every single one of those sets.

Example 2.9:

1. Let A and B be sets and let $X = \{A, B\}$. Then

$$\cup X = \{y \mid \text{there is } Y \in X \text{ with } y \in Y\} = \{y \mid y \in A \text{ or } y \in B\} = A \cup B.$$
$$\cap X = \{y \mid \text{for all } Y \in X, y \in Y\} = \{y \mid y \in A \text{ and } y \in B\} = A \cap B.$$

2. Let A, B, and C be sets, and let $X = \{A, B, C\}$. Then

$$\cup X = \{y \mid \text{there is } Y \in X \text{ with } y \in Y\} = \{y \mid y \in A, y \in B, \text{or } y \in C\} = A \cup B \cup C.$$
$$\cap X = \{y \mid \text{for all } Y \in X, y \in Y\} = \{y \mid y \in A, y \in B, \text{and } y \in C\} = A \cap B \cap C.$$

3. Let $X = \big\{\{-n, \dots, -3, -2, -1, 0, 1, 2, 3, 4, \dots, n\} \mid n \in \mathbb{N}\big\}$. Then

$$\cup X = \{y \mid \text{there is } Y \in X \text{ with } y \in Y\}$$
$$= \{y \mid \text{there is } n \in \mathbb{N} \text{ with } y \in \{-n, \dots, -3, -2, -1, 0, 1, 2, 3, 4, \dots, n\}\} = \mathbb{Z}.$$
$$\cap X = \{y \mid \text{for all } Y \in X, y \in Y\}$$
$$= \{y \mid \text{for all } n \in \mathbb{N}, y \in \{-n, \dots, -3, -2, -1, 0, 1, 2, 3, 4, \dots, n\}\} = \{0\}.$$

Notes: (1) Examples 1 and 2 give a good idea of what $\bigcup X$ and $\bigcap X$ look like when X is finite. More generally, if $X = \{A_1, A_2, \ldots, A_n\}$, then $\bigcup X = A_1 \cup A_2 \cup \cdots \cup A_n$ and $\bigcap X = A_1 \cap A_2 \cap \cdots \cap A_n$.

(2) As a specific example of Note 1, let $A_1 = \{\ldots, -3, -2, -1, 0, 1, 2, 3, 4, 5\}$, $A_2 = \{0, 1, 2, 3, 4, 5\}$, $A_3 = \{2, 3, 4, 5\}$, and $A_4 = \{4, 5, 6, 7, \ldots, 98, 99\}$. Let $X = \{A_1, A_2, A_3, A_4\}$. Then

$$\bigcup X = A_1 \cup A_2 \cup A_3 \cup A_4 = \{\ldots, -3, -2, -1, 0, 1, 2, 3, \ldots, 98, 99\}.$$

$$\bigcap X = A_1 \cap A_2 \cap A_3 \cap A_4 = \{4, 5\}.$$

If you have trouble seeing how to compute the intersection, it may help to take the intersections two at a time:

$A_1 \cap A_2 = A_2 = \{0, 1, 2, 3, 4, 5\}$ because $A_2 \subseteq A_1$.

$\{0, 1, 2, 3, 4, 5\} \cap A_3 = A_3 = \{2, 3, 4, 5\}$ because $A_3 \subseteq \{0, 1, 2, 3, 4, 5\}$.

$\{2, 3, 4, 5\} \cap A_4 = \{2, 3, 4, 5\} \cap \{4, 5, 6, 7, \ldots, 98, 99\} = \{4, 5\}$.

(3) Let's prove carefully that $\{y \mid \text{there is } n \in \mathbb{N} \text{ with } y \in \{-n, \ldots, -3, -2, -1, 0, 1, 2, 3, 4, \ldots, n\}\} = \mathbb{Z}$.

For convenience, let's let $A = \{y \mid \text{there is } n \in \mathbb{N} \text{ with } y \in \{-n, \ldots, -3, -2, -1, 0, 1, 2, 3, 4, \ldots, n\}\}$.

If $y \in A$, then there is $n \in \mathbb{N}$ with $y \in \{-n, \ldots, -3, -2, -1, 0, 1, 2, 3, 4, \ldots, n\}$. In particular, $y \in \mathbb{Z}$. Since $y \in A$ was arbitrary, we have shown that $A \subseteq \mathbb{Z}$.

Let $y \in \mathbb{Z}$. Then $y \in \{-n, \ldots, -3, -2, -1, 0, 1, 2, 3, 4, \ldots, n\}$, where $n = y$ if $y \geq 0$ and $n = -y$ if $y < 0$ (in other words, n is the **absolute value** of y, written $n = |y|$). So, $y \in A$. Since $y \in \mathbb{Z}$ was arbitrary, we have shown that $\mathbb{Z} \subseteq A$.

Since $A \subseteq \mathbb{Z}$ and $\mathbb{Z} \subseteq A$, it follows that $A = \mathbb{Z}$.

(4) Let's also prove carefully that $\{y \mid \text{for all } n \in \mathbb{N}, y \in \{-n, \ldots, -3, -2, -1, 0, 1, 2, 3, 4, \ldots, n\}\} = \{0\}$.

For convenience, let's let $B = \{y \mid \text{for all } n \in \mathbb{N}, y \in \{-n, \ldots, -3, -2, -1, 0, 1, 2, 3, 4, \ldots, n\}\}$.

If $y \in B$, then for all $n \in \mathbb{N}$, $y \in \{-n, \ldots, -3, -2, -1, 0, 1, 2, 3, 4, \ldots, n\}$. In particular, $y \in \{0\}$. Since $y \in B$ was arbitrary, we have shown that $B \subseteq \{0\}$.

Now, let $y \in \{0\}$. Then $y = 0$. For all $n \in \mathbb{N}$, $0 \in \{-n, \ldots, -3, -2, -1, 0, 1, 2, 3, 4, \ldots, n\}$. So, $y \in B$. It follows that $\{0\} \subseteq B$.

Since $B \subseteq \{0\}$ and $\{0\} \subseteq B$, it follows that $B = \{0\}$.

(5) Note that the empty union is empty. Indeed, we have $\bigcup \emptyset = \{y \mid \text{there is } Y \in \emptyset \text{ with } y \in Y\} = \emptyset$.

If X is a nonempty set of sets, we say that X is **disjoint** if $\bigcap X = \emptyset$. We say that X is **pairwise disjoint** if for all $A, B \in X$ with $A \neq B$, A and B are disjoint. For example, if we let $X = \{(n, n + 1) \mid n \in \mathbb{Z}\}$, then X is both disjoint and pairwise disjoint.

Are the definitions of disjoint and pairwise disjoint equivalent? You will be asked to answer this question in Problem 5 below.

Problem Set 2

Full solutions to these problems are available for free download here:
www.SATPrepGet800.com/TFBZLF

LEVEL 1

1. Let $A = \{a, b, \Delta, \delta\}$ and $B = \{b, c, \delta, \gamma\}$. Determine each of the following:

    (i) $A \cup B$

    (ii) $A \cap B$

    (iii) $A \setminus B$

    (iv) $B \setminus A$

    (v) $A \Delta B$

2. Draw Venn diagrams for $(A \setminus B) \setminus C$ and $A \setminus (B \setminus C)$. Are these two sets equal for all sets A, B, and C? If so, prove it. If not, provide a counterexample.

LEVEL 2

3. Let $A = \left\{\emptyset, \{\emptyset, \{\emptyset\}\}\right\}$ and $B = \{\emptyset, \{\emptyset\}\}$. Compute each of the following:

    (i) $A \cup B$

    (ii) $A \cap B$

    (iii) $A \setminus B$

    (iv) $B \setminus A$

    (v) $A \Delta B$

4. Prove the following:

    (i)   The operation of forming unions is commutative.

    (ii) The operation of forming intersections is commutative.

    (iii) The operation of forming intersections is associative.

LEVEL 3

5. Prove or provide a counterexample:

    (i)   Every pairwise disjoint set of sets is disjoint.

    (ii) Every disjoint set of sets is pairwise disjoint.

6. Let A and B be sets. Prove that $A \cap B \subseteq A$.

LEVEL 4

7. Prove that $B \subseteq A$ if and only if $A \cap B = B$.

8. Let A, B, and C be sets. Prove each of the following:

   (i) $A \cap (B \cup C) = (A \cap B) \cup (A \cap C)$.

   (ii) $A \cup (B \cap C) = (A \cup B) \cap (A \cup C)$.

   (iii) $C \setminus (A \cup B) = (C \setminus A) \cap (C \setminus B)$.

   (iv) $C \setminus (A \cap B) = (C \setminus A) \cup (C \setminus B)$.

LEVEL 5

9. Let X be a nonempty set of sets. Prove the following:

   (i)   For all $A \in X$, $A \subseteq \bigcup X$.

   (ii)   For all $A \in X$, $\bigcap X \subseteq A$.

10. Let A be a set and let X be a nonempty set of sets. Prove each of the following:

   (i) $A \cap \bigcup X = \bigcup \{A \cap B \mid B \in X\}$

   (ii) $A \cup \bigcap X = \bigcap \{A \cup B \mid B \in X\}$

   (iii) $A \setminus \bigcup X = \bigcap \{A \setminus B \mid B \in X\}$

   (iv) $A \setminus \bigcap X = \bigcup \{A \setminus B \mid B \in X\}$.

CHALLENGE PROBLEMS

11. Let X be a nonempty set of sets. Prove that $\mathcal{P}(\bigcap X) = \bigcap \{\mathcal{P}(A) \mid A \in X\}$.

12. Let X be a nonempty set of sets. Prove that $\mathcal{P}(\bigcup X) = \bigcup \{\mathcal{P}(A) \mid A \in X\}$ if and only if $\bigcup X \in X$.

LESSON 3
RELATIONS

Cartesian Products

An **unordered pair** is a set with 2 elements. Recall, that a set doesn't change if we write the elements in a different order or if we write the same element multiple times. For example, $\{x, y\} = \{y, x\}$ and $\{x, x\} = \{x\}$.

We now define the **ordered pair** (x, y) in such a way that (y, x) will **not** be the same as (x, y). The simplest way to define a set with this property is as follows:

$$(x, y) = \{\{x\}, \{x, y\}\}$$

Let's show that with this definition, the ordered pair behaves as we would expect.

Theorem 3.1: $(x, y) = (z, w)$ if and only if $x = z$ and $y = w$.

Part of the proof of this theorem is a little trickier than expected. Assuming that $(x, y) = (z, w)$, there are actually two cases to consider: $x = y$ and $x \neq y$. If $x = y$, then (x, y) is a set with just one element. Indeed, $(x, x) = \{\{x\}, \{x, x\}\} = \{\{x\}, \{x\}\} = \{\{x\}\}$. So, the only element of (x, x) is $\{x\}$. Watch carefully how this plays out in the proof.

Proof of Theorem 3.1: First suppose that $x = z$ and $y = w$. Then by direct substitution, $\{x\} = \{z\}$ and $\{x, y\} = \{z, w\}$. So, $(x, y) = \{\{x\}, \{x, y\}\} = \{\{z\}, \{z, w\}\} = (z, w)$.

Conversely, suppose that $(x, y) = (z, w)$. Then $\{\{x\}, \{x, y\}\} = \{\{z\}, \{z, w\}\}$. There are two cases to consider.

Case 1: If $x = y$, then $\{\{x\}, \{x, y\}\} = \{\{x\}\}$. So, $\{\{x\}\} = \{\{z\}, \{z, w\}\}$. It follows that $\{z\} = \{x\}$ and $\{z, w\} = \{x\}$. Since $\{z, w\} = \{x\}$, we must have $z = x$ and $w = x$. Therefore, $x, y, z,$ and w are all equal. In particular, $x = z$ and $y = w$.

Case 2: If $x \neq y$, then $\{x, y\}$ is a set with two elements. So, $\{x, y\}$ cannot be equal to $\{z\}$ (because $\{z\}$ has just one element). Therefore, we must have $\{x, y\} = \{z, w\}$. It then follows that $\{x\} = \{z\}$. So, we have $x = z$. Since $x = z$ and $\{x, y\} = \{z, w\}$, we must have $y = w$. $\square$

Note: (x, y) is an abbreviation for the set $\{\{x\}, \{x, y\}\}$. In the study of set theory, every object can be written as a set like this. It's often convenient to use abbreviations, but we should always be aware that if necessary, we can write any object in its unabbreviated form.

We can extend the idea of an ordered pair to an **ordered k-tuple**. An ordered 3-tuple (also called an **ordered triple)** is defined by $(x, y, z) = ((x, y), z)$, an ordered 4-tuple is $(x, y, z, w) = ((x, y, z), w)$, and so on. For a general k-tuple, we will use a single letter with subscripts for the variable names. For example, using the letter x, we can write a k-tuple as $(x_1, x_2, \ldots, x_k)$.

Example 3.2: Let's write the ordered triple (x, y, z) in its unabbreviated form (take a deep breath!).

$$(x, y, z) = ((x, y), z) = \{\{(x, y)\}, \{(x, y), z\}\} = \Big\{\big\{\{\{x\}, \{x, y\}\}\big\}, \big\{\{\{x\}, \{x, y\}\}, z\big\}\Big\}$$

The **Cartesian product** of the sets A and B, written $A \times B$ is the set of ordered pairs (a, b) with $a \in A$ and $b \in B$. Symbolically, we have

$$A \times B = \{(a, b) \mid a \in A \wedge b \in B\}.$$

Observe that if A and B are finite sets with $|A| = m$ and $|B| = n$, then $|A \times B| = mn$.

Example 3.3:

1. Let $A = \{a, b, c\}$ and $B = \{0, 1\}$. Then $A \times B = \{(a, 0), (a, 1), (b, 0), (b, 1), (c, 0), (c, 1)\}$. Note that $|A| = 3$, $|B| = 2$, and $|A \times B| = 3 \cdot 2 = 6$.

2. Let $C = \emptyset$ and $D = \{x, y, z, w, u, v\}$. Then $C \times D = \emptyset$ (since there are no elements in C, there can be no elements in $C \times D$). Note that $|C| = 0$, $|D| = 6$, and $|C \times D| = 0 \cdot 6 = 0$.

3. $\mathbb{N} \times \mathbb{Z} = \{(m, n) \mid m \in \mathbb{N} \wedge n \in \mathbb{Z}\}$. For example, $(5, -3) \in \mathbb{N} \times \mathbb{Z}$, whereas $(-3, 5) \notin \mathbb{N} \times \mathbb{Z}$ (although it is in $\mathbb{Z} \times \mathbb{N}$). We can visualize $\mathbb{N} \times \mathbb{Z}$ as follows:

$$\ldots, (0, -3), (0, -2), (0, -1), (0, 0), (0, 1), (0, 2), (0, 3), \ldots$$
$$\ldots, (1, -3), (1, -2), (1, -1), (1, 0), (1, 1), (1, 2), (1, 3), \ldots$$
$$\ldots, (2, -3), (2, -2), (2, -1), (2, 0), (2, 1), (2, 2), (2, 3), \ldots$$
$$\vdots \qquad \vdots \qquad \vdots$$

4. $\mathbb{R} \times \mathbb{R} = \{(x, y) \mid x, y \in \mathbb{R}\}$. We can visualize elements of $\mathbb{R} \times \mathbb{R}$ as points in the **Cartesian plane.** A portion of the Cartesian plane is shown to the right. The elements $(3, 2)$ and $(-1, -2)$ of $\mathbb{R} \times \mathbb{R}$ are displayed as points.

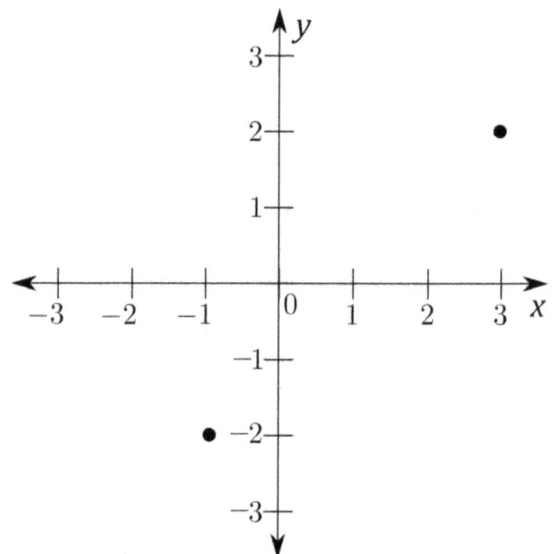

   We form the Cartesian plane by taking two copies of the real line and placing one horizontally and the other vertically, exactly as we did for the Complex Plane in part 3 of Example 1.6. The horizontal copy of the real line is called the x-axis (labeled x in the figure) and the vertical copy of the real line is called the y-axis (labeled y in the figure). The two axes intersect at the point $(0, 0)$. This point is called the **origin.**

   Notice that visually the Cartesian plane $\mathbb{R} \times \mathbb{R}$ is indistinguishable from the Complex Plane.

We can extend the definition of the Cartesian product to more than two sets in the obvious way:

$$A \times B \times C = \{(a, b, c) \mid a \in A \wedge b \in B \wedge c \in C\}$$
$$A \times B \times C \times D = \{(a, b, c, d) \mid a \in A \wedge b \in B \wedge c \in C \wedge d \in D\}$$

32

Observe that if A, B, and C are finite sets with $|A| = m$, $|B| = n$, and $|C| = k$, then we have $|A \times B \times C| = mnk$.

In general, the **Cartesian product** of the sets $A_1, A_2, \ldots, A_n$ is

$$A_1 \times A_2 \times \cdots \times A_n = \{(a_1, a_2, \ldots, a_n) \mid a_1 \in A_1 \wedge a_2 \in A_2 \wedge \cdots \wedge a_n \in A_n\}.$$

If $A_1, A_2, \ldots, A_n$ are finite, then the cardinality of $A_1 \times A_2 \times \cdots \times A_n$ is the product of the cardinalities of $A_1, A_2, \ldots, A_n$. Symbolically, we have $|A_1 \times A_2 \times \cdots \times A_n| = |A_1| \cdot |A_2| \cdots |A_n|$.

Remark: In Lesson 9 we will define the Cartesian product of infinitely many sets.

Example 3.4:

1. $\{\alpha\} \times \{\beta\} \times \{\gamma\} \times \{\delta\} = \{(\alpha, \beta, \gamma, \delta)\}$.

   Note that $|\{\alpha\} \times \{\beta\} \times \{\gamma\} \times \{\delta\}| = |\{\alpha\}| \cdot |\{\beta\}| \cdot |\{\gamma\}| \cdot |\{\delta\}| = 1 \cdot 1 \cdot 1 \cdot 1 = 1$.

2. $\{0\} \times \{0, 1\} \times \{1\} \times \{0, 1\} \times \{0\} = \{(0, 0, 1, 0, 0), (0, 0, 1, 1, 0), (0, 1, 1, 0, 0), (0, 1, 1, 1, 0)\}$.

   Note that $|\{0\} \times \{0, 1\} \times \{1\} \times \{0, 1\} \times \{0\}| = 1 \cdot 2 \cdot 1 \cdot 2 \cdot 1 = 4$.

3. $\{0\} \times \mathbb{Z} \times \mathbb{N} \times \mathbb{R} = \{(0, m, n, x) \mid m \in \mathbb{Z} \wedge n \in \mathbb{N} \wedge x \in \mathbb{R}\}$.

We abbreviate Cartesian products of sets with themselves using exponents.

$$A^2 = A \times A \qquad A^3 = A \times A \times A \qquad A^4 = A \times A \times A \times A \qquad A^n = \underbrace{A \times A \times \cdots \times A}_{n \text{ times}}$$

Example 3.5:

1. $\mathbb{Z}^2 = \mathbb{Z} \times \mathbb{Z} = \{(x, y) \mid x, y \in \mathbb{Z}\}$ is the set of ordered pairs of integers. A few sample elements in $\mathbb{Z}^2$ are $(0, 0)$, $(-1, 2)$, $(15, -106)$ and $(-53, -53)$.

2. $\mathbb{N}^5 = \mathbb{N} \times \mathbb{N} \times \mathbb{N} \times \mathbb{N} \times \mathbb{N} = \{(a, b, c, d, e) \mid a, b, c, d, e \in \mathbb{N}\}$ is the set of ordered 5-tuples of natural numbers. A few sample elements in $\mathbb{N}^5$ are $(0, 0, 0, 0, 0)$, $(1, 1, 2, 3, 3)$, $(0, 1, 17, 86, 0)$ and $(1000, 2529, 8, 900, 106)$.

3. $\{0, 1\}^2 = \{0, 1\} \times \{0, 1\} = \{(0, 0), (0, 1), (1, 0), (1, 1)\}$.

4. $\{0, 1\}^3 = \{0, 1\} \times \{0, 1\} \times \{0, 1\}$

   $\qquad = \{(0, 0, 0), (0, 0, 1), (0, 1, 0), (0, 1, 1), (1, 0, 0), (1, 0, 1), (1, 1, 0), (1, 1, 1)\}$.

5. $\mathbb{R}^2 = \mathbb{R} \times \mathbb{R} = \{(x, y) \mid x, y \in \mathbb{R}\}$ was discussed in part 4 of Example 3.3 above. We will also be interested in larger Cartesian products of $\mathbb{R}$ such as $\mathbb{R}^3 = \{(x, y, z) \mid x, y, z \in \mathbb{R}\}$, and more generally, $\mathbb{R}^n = \{(x_1, x_2, \ldots, x_n) \mid x_1, x_2, \ldots, x_n \in \mathbb{R}\}$.

6. $\mathbb{C}^n = \{(z_1, z_2, \ldots, z_n) \mid z_1, z_2, \ldots, z_n \in \mathbb{C}\}$ is the set of ordered n-tuples of complex numbers. For example, $\left(3 + 2i, -1 + \frac{5}{7}i\right) \in \mathbb{C}^2$ and $\left(1 - i, 2 + \frac{1}{2}i, -5i, 0.253\right) \in \mathbb{C}^4$.

Binary Relations

A **binary relation** on a set A is a subset of $A^2 = A \times A$. Symbolically, we have

$$R \text{ is a binary relation on } A \text{ if and only if } R \subseteq A \times A.$$

We will usually abbreviate $(a, b) \in R$ as aRb.

Remark: The statement $R \subseteq A \times A$ is equivalent to the statement $R \in \mathcal{P}(A \times A)$. It follows that for a finite set A, the number of binary relations on A is $|\mathcal{P}(A \times A)|$.

Example 3.6:

1. Let $R = \{(a, b) \in \mathbb{N} \times \mathbb{N} \mid a < b\}$. For example, we have $(0, 1) \in R$ because $0 < 1$. However, $(1, 1) \notin R$ because $1 \not< 1$. We abbreviate $(0, 1) \in R$ by $0R1$.

   Observe that $R \subseteq \mathbb{N} \times \mathbb{N}$, and so, R is a binary relation on $\mathbb{N}$.

   We would normally use the name $<$ for this relation R. So, we have $(0, 1) \in <$, which we abbreviate as $0 < 1$, and we have $(1, 1) \notin <$, which we abbreviate as $1 \not< 1$.

2. There are binary relations $<, \leq, >, \geq$ defined on $\mathbb{N}, \mathbb{Z}, \mathbb{Q}$, and $\mathbb{R}$. For example, if we consider $> \subseteq \mathbb{Z}^2$, we have $(13, -7) \in >$, or equivalently, $13 > -7$.

3. Let $A = \{a\}$. Since $|A| = 1$, we have $|A \times A| = 1 \cdot 1 = 1$. So, $|\mathcal{P}(A \times A)| = 2^1 = 2$. So, there are 2 binary relations on A. They are $R_1 = \emptyset$ and $R_2 = \{(a, a)\}$.

4. Let $B = \{0, 1\}$. Since $|B| = 2$, we have $|B \times B| = 2 \cdot 2 = 4$. So, $|\mathcal{P}(B \times B)| = 2^4 = 16$. So, there are 16 binary relations on B. A few examples are $R_1 = \emptyset$, $R_2 = \{(0, 0)\}$, $R_3 = \{(0, 1)\}$, and $R_4 = \{(0, 0), (0, 1)\}$. Can you list the rest of them?

5. Let A be a set and let R be the binary relation on A defined by $R = \{(a, b) \in A \times A \mid a \in b\}$. R is known as the **membership relation**, and it is usually denoted by $\in$. So, if $a, b \in A$ and a is a member of b, we can write $(a, b) \in \in$, which we will usually abbreviate as $a \in b$. As a specific example, let $A = \{\emptyset, \{\emptyset\}, \{\{\emptyset\}\}\}$. Then $(\emptyset, \{\emptyset\}) \in \in$, or equivalently, $\emptyset \in \{\emptyset\}$. Similarly, we have $(\{\emptyset\}, \{\{\emptyset\}\}) \in \in$, or equivalently, $\{\emptyset\} \in \{\{\emptyset\}\}$.

6. Let $R = \{((a, b), (c, d)) \in (\mathbb{N} \times \mathbb{N})^2 \mid a + d = b + c\}$. Then R is a binary relation on $\mathbb{N} \times \mathbb{N}$. For example, we have $(5, 0)R(6, 1)$ because $5 + 1 = 0 + 6$. However, we see that $(5, 0)\not R(6, 2)$ because $5 + 2 \neq 0 + 6$.

7. Let $R = \{((a, b), (c, d)) \in (\mathbb{Z} \times \mathbb{Z}^*)^2 \mid ad = bc\}$. (Recall that $\mathbb{Z}^*$ is the set of *nonzero* integers.) Then R is a binary relation on $\mathbb{Z} \times \mathbb{Z}^*$. For example, $(1, 2)R(2, 4)$ because $1 \cdot 4 = 2 \cdot 2$. However, $(1, 2)\not R(2, 5)$ because $1 \cdot 5 \neq 2 \cdot 2$. Compare this to the rational number system (see part 1 of Example 1.6), where we have $\frac{1}{2} = \frac{2}{4}$ because $1 \cdot 4 = 2 \cdot 2$, but $\frac{1}{2} \neq \frac{2}{5}$ because $1 \cdot 5 \neq 2 \cdot 2$.

The **domain** of a binary relation R, written dom R, is $\{x \mid \exists y (xRy)\}$. The **range** of a binary relation, written ran R, is $\{y \mid \exists x (xRy)\}$. The **field** of a binary relation R is dom $R \cup$ ran R.

Notes: (1) The symbol $\exists$ is called an **existential quantifier**, and it is pronounced "There exists" or "There is."

(2) The expression $\exists y(xRy)$ can be translated into English as "There exists a y such that xRy." Similarly, the expression $\exists x(xRy)$ can be translated into English as "There exists an x such that xRy." In general, if $P(x)$ is some property, then the expression $\exists x\big(P(x)\big)$ can be translated into English as "There exists an x such that $P(x)$."

Example 3.7:

1. Let $R = \{(a,b) \in \mathbb{N} \times \mathbb{N} \mid a < b\}$. Then dom $R = \mathbb{N}$, ran $R = \mathbb{N}$, and field $R = \mathbb{N} \cup \mathbb{N} = \mathbb{N}$.

2. Let $B = \{0, 1, 2, 3\}$ and $R = \{(0,2), (0,3), (1,3)\}$. Then dom $R = \{0, 1\}$, ran $R = \{2, 3\}$, and field $R = \{0, 1\} \cup \{2, 3\} = \{0, 1, 2, 3\} = B$.

3. Let $R = \{\big((a,b),(c,d)\big) \in (\mathbb{N} \times \mathbb{N})^2 \mid a + d = b + c\}$. Then dom $R = \mathbb{N} \times \mathbb{N}$, ran $R = \mathbb{N} \times \mathbb{N}$, and field $R = (\mathbb{N} \times \mathbb{N}) \cup (\mathbb{N} \times \mathbb{N}) = \mathbb{N} \times \mathbb{N}$.

We say that a binary relation R on a set A is

- **reflexive** if for all $a \in A$, aRa.

- **symmetric** if for all $a, b \in A$, aRb implies bRa.

- **transitive** if for all $a, b, c \in A$, aRb and bRc imply aRc.

- **antireflexive** if for all $a \in A$, $a\not Ra$.

- **antisymmetric** if for all $a, b \in A$, aRb and bRa imply $a = b$.

Example 3.8:

1. Let A be any set and let $R = \{(a,b) \in A^2 \mid a = b\}$. Then R is reflexive ($a = a$), symmetric (if $a = b$, then $b = a$), transitive (if $a = b$ and $b = c$, then $a = c$), and antisymmetric (trivially). If $A \neq \emptyset$, then this relation is not antireflexive because $a \neq a$ is false for any $a \in A$.

2. The binary relations $\leq$ and $\geq$ defined in the usual way on $\mathbb{Z}$ are transitive (if $a \leq b$ and $b \leq c$, then $a \leq c$, and similarly for $\geq$), reflexive ($a \leq a$ and $a \geq a$), and antisymmetric (if $a \leq b$ and $b \leq a$, then $a = b$, and similarly for $\geq$). These relations are not symmetric. For example, $1 \leq 2$, but $2 \not\leq 1$). These relations are not antireflexive. For example, $1 \leq 1$ is true.

   Any relation that is transitive, reflexive, and antisymmetric is called a **partial ordering**.

3. The binary relations $<$ and $>$ defined on $\mathbb{Z}$ are transitive (if $a < b$ and $b < c$, then $a < c$, and similarly for $>$), antireflexive ($a \not< a$ and $a \not> a$), and antisymmetric (this is vacuously true because $a < b$ and $b < a$ can never occur). These relations are not symmetric (for example, $1 < 2$, but $2 \not< 1$). These relations are not reflexive (for example, $1 < 1$ is false).

   Any relation that is transitive, antireflexive, and antisymmetric is called a **strict partial ordering**.

4. Let $R = \{(0,0), (0,2), (2,0), (2,2), (2,3), (3,2), (3,3)\}$ be a binary relation on $\mathbb{N}$. Then it is easy to see that R is symmetric. R is not reflexive because $1 \in \mathbb{N}$, but $(1,1) \notin R$ (however, if we were to consider R as a relation on $\{0, 2, 3\}$ instead of on $\mathbb{N}$, then R **would** be reflexive). R is not transitive because we have $(0,2), (2,3) \in R$, but $(0,3) \notin R$. R is not antisymmetric because we have $(2,3), (3,2) \in R$ and $2 \neq 3$. R is not antireflexive because $(0,0) \in R$ (and also, $(2,2) \in R$ an $(3,3) \in R$).

n-ary Relations

We can extend the idea of a binary relation on a set A to an **n-ary relation** on A. For example, a 3-ary relation (or **ternary relation**) on A is a subset of $A^3 = A \times A \times A$. More generally, we have that R is an n-ary relation on A if and only if $R \subseteq A^n$. A **1-ary relation** (or **unary relation**) on A is just a subset of A.

Example 3.9:

1. $\mathbb{R}$ is a unary relation on $\mathbb{C}$ because $\mathbb{R} \subseteq \mathbb{C}$.

2. Let $R = \{(x, y, z) \in \mathbb{Z}^3 \mid x + y = z\}$. Then R is a ternary (or 3-ary) relation on $\mathbb{Z}$. We have, for example, $(1, 2, 3) \in R$ (because $1 + 2 = 3$) and $(1, 2, 4) \notin R$ (because $1 + 2 \neq 4$).

3. Let C be the set of all colors. For example, blue $\in C$, pink $\in C$, and violet $\in C$. Let $S = \{(a, b, c) \in C^3 \mid$ when a and b are combined in equal quantities, the result is $c\}$. Then S is a ternary relation on C. We have, for example, $(\text{red}, \text{yellow}, \text{orange}) \in S$.

4. Let $T = \{(a, b, c, d, e) \in \mathbb{N}^5 \mid ab + c = de\}$. Then T is a 5-ary relation on $\mathbb{N}$. We have, for example, $(1, 2, 8, 5, 2) \in T$ $(1 \cdot 2 + 8 = 5 \cdot 2)$ and $(1, 1, 1, 1, 1) \notin T$ $(1 \cdot 1 + 1 \neq 1 \cdot 1)$.

Orderings

A binary relation $\leq$ on a set A is a **partial ordering** on A if $\leq$ is reflexive, antisymmetric, and transitive on A. If we replace "reflexive" by "antireflexive," then we call the relation a **strict partial ordering** on A (we would normally use the symbol $<$ instead of $\leq$ for a strict partial ordering).

A **partially ordered set** (or **poset**) is a pair $(A, \leq)$, where A is a set and $\leq$ is a partial ordering on A. Similarly, a **strict poset** is a pair $(A, <)$, where A is a set and $<$ is a strict partial ordering on A.

Note: If $<$ is a strict partial ordering on a set A, then $\forall a, b \in A(a < b \rightarrow b \not< a)$. To see this, let $a < b$. Since $<$ is antireflexive, $a \neq b$. Since $<$ is antisymmetric, $b < a$ would lead to $a = b$. Since we cannot have both $a \neq b$ and $a = b$, we must have $b \not< a$.

Example 3.10:

1. The usual ordering $\leq$ on $\mathbb{Z} = \{\ldots, -3, -2, -1, 0, 1, 2, 3, \ldots\}$ is a partial ordering, and the ordering $<$ on $\mathbb{Z}$ is a strict partial ordering. See Example 3.8 (parts 2 and 3).

2. If A is a set, then $(\mathcal{P}(A), \subseteq)$ is a poset. Since every set is a subset of itself, $\subseteq$ is reflexive (see Theorem 1.9). If $X, Y \in \mathcal{P}(A)$ with $X \subseteq Y$ and $Y \subseteq X$, then $X = Y$ (see the end of Lesson 1). So, $\subseteq$ is antisymmetric. By Theorem 1.14, $\subseteq$ is transitive.

   See the tree diagrams at the end of Example 1.12 for visual representations of this poset when $A = \{a, b\}$ and $A = \{a, b, c\}$.

   Similarly, $(\mathcal{P}(A), \subset)$ is a strict poset (the relation here is the **proper subset** relation). The relation $\subset$ is antireflexive because no set is a proper subset of itself. If $X \subset Y$ and $Y \subset X$, then $X \subseteq Y$ and $Y \subseteq X$, and so, $X = Y$ (again, see the end of Lesson 1). Finally, suppose that $X \subset Y$ and $Y \subset Z$. Then $X \subseteq Y$ and $Y \subseteq Z$, and so, by Theorem 1.14, $X \subseteq Z$. Suppose toward contradiction that $X = Z$. Then $Z \subseteq X$ and again, by Theorem 1.14, $Z \subseteq Y$. Since $Y \subseteq Z$ and $Z \subseteq Y$, it follows that $Y = Z$, contradicting our assumption that Y is a *proper* subset of Z. So, $X \neq Z$, and therefore, $X \subset Z$.

Note that in the argument above, to prove that $X \neq Z$, we assumed that $X = Z$, and then used a logically valid argument to derive the statement $Y = Z$, which we already knew was false. This is known as a proof by contradiction. We will use this kind of argument in Theorems 3.12 and 3.13 below as well. See the notes after the proof of Theorem 4.12 in Lesson 4 for a more detailed explanation as to how this type of argument works. Theorem 5.6 in Lesson 5 provides another good example.

3. Let $(A, \leq_A)$ be a partially ordered set, $B \subseteq A$, and $\leq_B = \{(x, y) \mid x, y \in B \wedge x \leq_A y\}$. Then $(B, \leq_B)$ is also a partially ordered set. Let's check this carefully.

   To see that $\leq_B$ is reflexive on B, let $x \in B$. Since $B \subseteq A$, $x \in A$. Since $\leq_A$ is reflexive on A, $x \leq_A x$. Since $x \in B$, $x \leq_B x$.

   To see that $\leq_B$ is antisymmetric on B, let $x, y \in B$ with $x \leq_B y$ and $y \leq_B x$. Then $x \leq_A y$ and $y \leq_A x$. Since $\leq_A$ is antisymmetric on A, $x = y$.

   Finally, to see that $\leq_B$ is transitive on B, let $x, y, z \in B$ with $x \leq_B y$ and $y \leq_B z$. Then $x \leq_A y$ and $y \leq_A z$. Since $\leq_A$ is transitive on A, $x \leq_A z$. Since $x, z \in B$, $x \leq_B z$.

4. Let $X = \{x \mid x \text{ is a word in the English language}\}$ and define the **dictionary order** on X as follows: $x <_D y$ if x appears before y alphabetically. Then $(X, <_D)$ is a strict poset. In this poset, we have aardvark $<_D$ antelope (because a = a and a appears before n alphabetically), we have stranger $<_D$ violin (because s appears before v alphabetically), and we have dragon $<_D$ drainage (because d = d, r = r, a = a and g appears before i alphabetically).

   We can use a similar idea to define the **dictionary order** on products of posets (and strict posets).

   For example, the dictionary order $<_D$ can be defined on $\mathbb{Z} \times \mathbb{Q}$ by $(a, b) <_D (c, d)$ if and only if either $a <_\mathbb{Z} c$ or both $a = c$ and $b <_\mathbb{Q} d$, where $<_\mathbb{Z}$ and $<_\mathbb{Q}$ are the usual strict partial orderings on $\mathbb{Z}$ and $\mathbb{Q}$, respectively. It is easy (but a bit tedious) to verify that $(\mathbb{Z} \times \mathbb{Q}, <_D)$ is a strict poset. I leave the verification to the reader.

   In the strict poset $(\mathbb{Z} \times \mathbb{Q}, <_D)$, we have $(4, 7) <_D (9, -2)$ because $4 <_\mathbb{Z} 9$ (notice that the second coordinates are irrelevant because the first coordinates are not equal). We also have $\left(3, \frac{1}{2}\right) <_D \left(3, \frac{3}{4}\right)$ because $3 = 3$ and $\frac{1}{2} <_\mathbb{Q} \frac{3}{4}$.

   We can visualize this particular dictionary order in a Cartesian plane as solid vertical lines passing through each integer value along the x-axis. Each point is less than any point higher than it on the same vertical line and each point is also less than any point on a vertical line to the right of that point (regardless of the height).

   In the figure below, we see that $(1, -2) <_D \left(1, \frac{3}{2}\right)$ because $\left(1, \frac{3}{2}\right)$ is above $(1, -2)$ on the same vertical line. We also have $(1, -2) <_D (3, -3)$ because $(3, -3)$ is to the right of $(1, -2)$ (note that it does **not** matter that $(3, -3)$ is below $(1, -2)$).

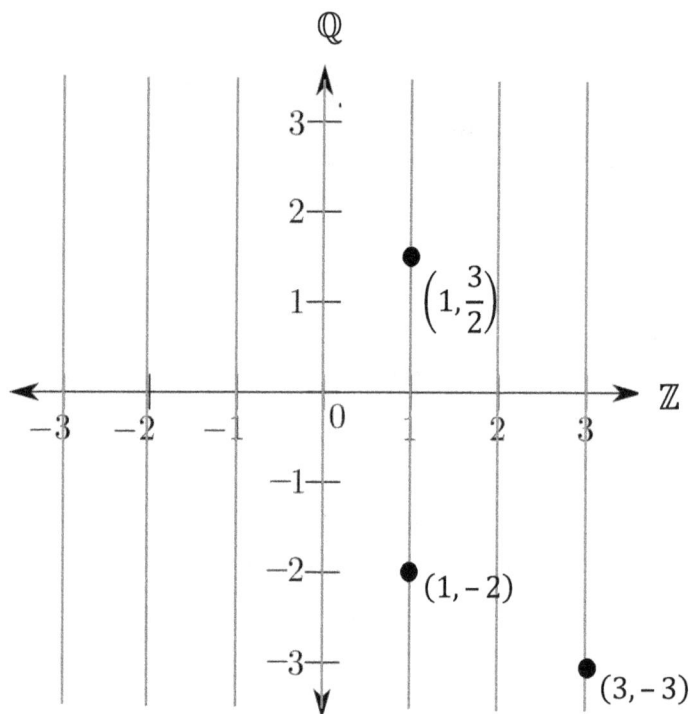

As another example, in the dictionary order of $\mathbb{R}^5$, we have $(1,3,5,2,11) <_D (1,3,5,4,10)$ because $1 = 1$, $3 = 3$, $5 = 5$, and $2 <_\mathbb{R} 4$.

As one more example, let $A = \{a, b\}$, consider the strict poset $(\mathcal{P}(A), \subset)$ (the relation here is the **proper subset** relation), and let $<_D$ be the corresponding dictionary order on $\left(\mathcal{P}(A)\right)^2 = \mathcal{P}(A) \times \mathcal{P}(A)$. We have $(\emptyset, \{a\}) <_D (\{b\}, \{b\})$ because $\emptyset \subset \{b\}$. Also, we have $(\{a\}, \{b\}) <_D (\{a\}, \{a, b\})$ because $\{a\} = \{a\}$ and $\{b\} \subset \{a, b\}$.

Let $(A, \leq)$ be a poset. We say that $a, b \in A$ are **comparable** if $a \leq b$ or $b \leq a$. The poset satisfies the **comparability condition** if every pair of elements in A are comparable. A poset that satisfies the comparability condition is called a **linearly ordered set** (or **totally ordered set**). Similarly, a **strict linearly ordered set** $(A, <)$ satisfies **trichotomy**: If $a, b \in A$, then $a < b$, $a = b$, or $b < a$.

Example 3.11:

1. $(\mathbb{N}, \leq)$, $(\mathbb{Z}, \leq)$, $(\mathbb{Q}, \leq)$, and $(\mathbb{R}, \leq)$ are linearly ordered sets. Similarly, $(\mathbb{N}, <)$, $(\mathbb{Z}, <)$, $(\mathbb{Q}, <)$, and $(\mathbb{R}, <)$ are strict linearly ordered sets.

2. If A has at least two elements, then $(\mathcal{P}(A), \subset)$ is **not** a strict linearly ordered set. Indeed, if $a, b \in A$ with $a \neq b$, then $\{a\} \not\subset \{b\}$ and $\{b\} \not\subset \{a\}$. See either of the tree diagrams at the end of Example 1.12.

   Similarly, $\left(\left(\mathcal{P}(A)\right)^2, <_D\right)$, where $<_D$ is the dictionary order is **not** a strict linearly ordered set. Indeed, $(\{a\}, \emptyset)$ and $(\{b\}, \emptyset)$ are not comparable with respect to $<_D$ because $\{a\} \not\subset \{b\}$ and $\{b\} \not\subset \{a\}$, just like we observed above. Similarly, $(\{a\}, \{a\})$ and $(\{a\}, \{b\})$ are not comparable.

3. If $(A, <_A)$ and $(B, <_B)$ are strict linearly ordered sets, then the dictionary order $<_D$ on $A \times B$ is also a strict linearly ordered set. To see this, let $(a, b), (c, d) \in A \times B$. If $a <_A c$, then we have $(a, b) <_D (c, d)$. If $c <_A a$, then $(c, d) <_D (a, b)$. If $a = c$, then we look at the second coordinates. In this case, if $b <_B d$, then $(a, b) <_D (c, d)$, and if $d <_B b$, then $(c, d) <_D (a, b)$. Otherwise, $a = c$ and $b = d$, and therefore, $(a, b) = (c, d)$.

We can modify a partial ordering slightly to get a strict partial ordering. We do this as follows:

Theorem 3.12: Let A be a set and let $\leq_A$ be a partial ordering on A. Define the binary relation $<_A$ on A by $a <_A b$ if and only if $a \leq_A b$ and $a \neq b$. Then $<_A$ is a strict partial ordering on A.

Proof: Suppose that $a <_A b$ and $b <_A c$. Then $a \leq_A b$ and $b \leq_A c$. Since $\leq_A$ is transitive, $a \leq_A c$. Also, $a \neq b$ and $b \neq c$. Suppose toward contradiction that $a = c$. Since $b \leq_A c$, we have $b \leq_A a$. Since $\leq_A$ is antisymmetric, $a = b$, contrary to our assumption. Therefore, $a \neq c$. Since $a \leq_A c$ and $a \neq c$, we have $a <_A c$. This shows that $<_A$ is transitive.

Now, suppose that $a <_A b$ and $b <_A a$. Then $a \leq_A b$ and $b \leq_A a$. Since $\leq_A$ is antisymmetric, $a = b$. Therefore, $<_A$ is antisymmetric.

Finally, by definition, $a <_A a$ is false. Therefore, $<_A$ is antireflexive. Since $<_A$ is transitive, antisymmetric, and antireflexive, $<_A$ is a strict partial ordering on A. □

Similarly, we can modify a strict partial ordering to get a partial ordering.

Theorem 3.13: Let A be a set and let $<_A$ be a strict partial ordering on A. Define the binary relation $\leq_A$ on A by $a \leq_A b$ if and only if $a <_A b$ or $a = b$. Then $\leq_A$ is a partial ordering on A.

Proof: Suppose that $a \leq_A b$ and $b \leq_A c$. If $a = b$, then we have $a \leq_A c$ by direct substitution. Similarly, if $b = c$, we have $a \leq_A c$ by direct substitution. If $a <_A b$ and $b <_A c$, then $a <_A c$ because $<_A$ is transitive. It follows that $a \leq_A c$. This shows that $\leq_A$ is transitive.

Now, suppose that $a \leq_A b$ and $b \leq_A a$. Assume toward contradiction that $a \neq b$. Then $a <_A b$ and $b <_A a$. Since $<_A$ is antisymmetric, $a = b$, contrary to our assumption. Therefore, $\leq_A$ is antisymmetric.

Finally, by definition, $a \leq_A a$ is true. Therefore, $\leq_A$ is reflexive. Since $\leq_A$ is transitive, antisymmetric, and reflexive, $\leq_A$ is a partial ordering on A. □

Intervals

A set I of real numbers is called an **interval** if any real number that lies between two numbers in I is also in I. Symbolically, we can write

$$\forall x, y \in I \; \forall z \in \mathbb{R} \; (x < z < y \rightarrow z \in I).$$

The expression above can be read "For all x, y in I and all $z \in \mathbb{R}$, if x is less than z and z is less than y, then z is in I."

A simple way to think of an interval is as a set of real numbers with no "gaps" or "holes." If x and y are in the interval, then everything between x and y is in the interval as well.

Example 3.14:

1. The set $A = \{0, 1\}$ is **not** an interval. A consists of just the two real numbers 0 and 1. There are infinitely many real numbers between 0 and 1. For example, the real number $\frac{1}{2}$ satisfies $0 < \frac{1}{2} < 1$, but $\frac{1}{2} \notin A$.

39

2. The set $(0, 1) = \{x \in \mathbb{R} \mid 0 < x < 1\}$ is an example of an **open interval**.

3. The set $[0, 1] = \{x \in \mathbb{R} \mid 0 \leq x \leq 1\}$ is an example of a **closed interval**.

4. $\mathbb{R}$ is an interval. This follows trivially from the definition. If we replace I by $\mathbb{R}$, we get $\forall x, y \in \mathbb{R} \; \forall z \in \mathbb{R} \; (x < z < y \rightarrow z \in \mathbb{R})$. In other words, if we start with two real numbers, and take a real number between them, then that number is a real number (which we already said).

When we are thinking of $\mathbb{R}$ as an interval, we sometimes use the notation $(-\infty, \infty)$ and refer to this as **the real line**. The following picture gives the standard geometric interpretation of the real line.

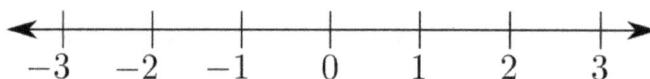

In addition to the real line, there are 8 other types of intervals.

Open Interval: $\qquad\qquad (a, b) = \{x \in \mathbb{R} \mid a < x < b\}$

Closed Interval: $\qquad\quad\; [a, b] = \{x \in \mathbb{R} \mid a \leq x \leq b\}$

Half-open Intervals: $\qquad (a, b] = \{x \in \mathbb{R} \mid a < x \leq b\} \qquad [a, b) = \{x \in \mathbb{R} \mid a \leq x < b\}$

Infinite Open Intervals: $\;\; (a, \infty) = \{x \in \mathbb{R} \mid x > a\} \qquad\quad (-\infty, b) = \{x \in \mathbb{R} \mid x < b\}$

Infinite Closed Intervals: $[a, \infty) = \{x \in \mathbb{R} \mid x \geq a\} \qquad\quad (-\infty, b] = \{x \in \mathbb{R} \mid x \leq b\}$

Warning: It is unfortunate that the notation "(a, b)" is used for both an open interval and an ordered pair of real numbers. Most of the time it will be clear which one of these we mean. However, occasionally we may be discussing open intervals and ordered pairs of real numbers at the same time. In these instances, it is important to pay attention to the details of the discussion.

Example 3.15:

1. The half-open interval $(-2, 1] = \{x \in \mathbb{R} \mid -2 < x \leq 1\}$ has the following graph:

2. The infinite open interval $(0, \infty) = \{x \in \mathbb{R} \mid x > 0\}$ has the following graph:

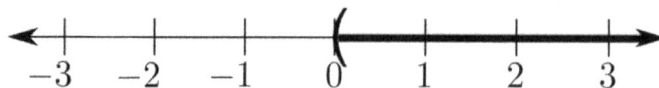

Note: If I is an interval of reals and $A \subseteq \mathbb{R}$, then we will use the notation $I \cap A$ for the corresponding interval as a subset of A. For example, $(-2, 1] \cap \mathbb{Q} = \{x \in \mathbb{Q} \mid -2 < x \leq 1\}$ is a half-open interval of rational numbers. It consists of only the rational numbers between -2 and 1, including 1 and excluding -2. We can visualize this interval of rational numbers in the same we visualize the corresponding interval of real numbers. If we wish, we may label the graph with a $\mathbb{Q}$ for extra clarification as follows:

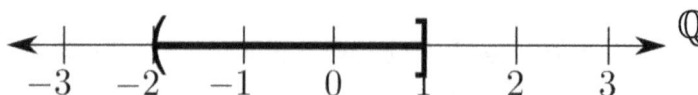

40

Example 3.16: Let $A = (-2, 1]$ and $B = (0, \infty)$. We have

1. $A \cup B = (-2, \infty)$
2. $A \cap B = (0, 1]$
3. $A \setminus B = (-2, 0]$
4. $B \setminus A = (1, \infty)$
5. $A \triangle B = (-2, 0] \cup (1, \infty)$

Note: If you have trouble seeing how to compute these, it may be helpful to draw the graphs of A and B lined up vertically, and then draw vertical lines through the endpoints of each interval.

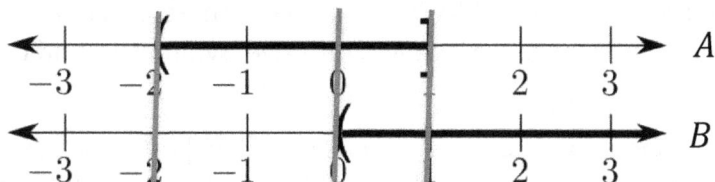

The results follow easily by combining these graphs into a single graph using the vertical lines as guides. For example, let's look at $A \cap B$ in detail. We're looking for all numbers that are in both A and B. The two rightmost vertical lines drawn passing through the two graphs above isolate all those numbers nicely. We see that all numbers between 0 and 1 are in the intersection. We should then think about the two endpoints 0 and 1 separately. $0 \notin B$ and therefore, 0 cannot be in the intersection of A and B. On the other hand, $1 \in A$ and $1 \in B$. Therefore, $1 \in A \cap B$. So, we see that $A \cap B = (0, 1]$.

Equivalence Relations

A binary relation R on a set A is an **equivalence relation** if R is reflexive, symmetric, and transitive.

Example 3.17:

1. The most basic equivalence relation on a set A is the relation $R = \{(a, b) \in A^2 \mid a = b\}$ (the **equality relation**). We already saw in part 1 of Example 3.8 that this relation is reflexive, symmetric and transitive.

2. Another trivial equivalence relation on a set A is the set A^2. Since every ordered pair (a, b) is in A^2, reflexivity, symmetry, and transitivity can never fail. We will refer to A^2 as the **trivial equivalence relation** on A.

3. We say that integers a and b have the same **parity** if they are both even or both odd. Define $\equiv_2$ on $\mathbb{Z}$ by $\equiv_2 = \{(a, b) \in \mathbb{Z}^2 \mid a$ and b have the same parity$\}$. It is easy to see that $\equiv_2$ is reflexive ($a \equiv_2 a$ because every integer has the same parity as itself), $\equiv_2$ is symmetric (if $a \equiv_2 b$, then a has the same parity as b, so b has the same parity as a, and therefore, $b \equiv_2 a$), and $\equiv_2$ is transitive (if $a \equiv_2 b$ and $b \equiv_2 c$, then a, b, and c all have the same parity, and so, $a \equiv_2 c$). Therefore, $\equiv_2$ is an equivalence relation.

4. An integer n is **divisible** by an integer m, written $m|n$, if there is another integer k such that $n = mk$. Another way to say that a and b have the same parity is to say that $b - a$ is divisible by 2, or equivalently, $2|b - a$. This observation allows us to generalize the notion of having the same parity. For example, $\equiv_3 = \{(a, b) \in \mathbb{Z}^2 \mid 3|b - a\}$ is an equivalence relation, and more generally, for each $n \in \mathbb{Z}^+$, $\equiv_n = \{(a, b) \in \mathbb{Z}^2 \mid n|b - a\}$ is an equivalence relation. I leave the proof that $\equiv_n$ is reflexive, symmetric, and transitive on $\mathbb{Z}$ as an exercise (see Problem 8 below).

5. Consider the relation $R = \{((a, b), (c, d)) \in (\mathbb{N} \times \mathbb{N})^2 \mid a + d = b + c\}$ defined in part 6 of Example 3.6. Since $a + b = b + a$, we see that $(a, b)R(a, b)$, and therefore, R is reflexive. If $(a, b)R(c, d)$, then $a + d = b + c$. Therefore, $c + b = d + a$, and so, $(c, d)R(a, b)$. Thus, R is symmetric. Finally, suppose that $(a, b)R(c, d)$ and $(c, d)R(e, f)$. Then $a + d = b + c$ and $c + f = d + e$. So, $a + d + c + f = b + c + d + e$. Therefore, $a + f = b + e$, and so, we have $(a, b)R(e, f)$. So, R is transitive. Since R is reflexive, symmetric, and transitive, it follows that R is an equivalence relation.

6. Consider the relation $R = \{((a, b), (c, d)) \in (\mathbb{Z} \times \mathbb{Z}^*)^2 \mid ad = bc\}$ defined in part 7 of Example 3.6. Since $ab = ba$, we see that $(a, b)R(a, b)$, and therefore, R is reflexive. If $(a, b)R(c, d)$, then $ad = bc$. Therefore, $cb = da$, and so, $(c, d)R(a, b)$. Thus, R is symmetric. Finally, suppose that $(a, b)R(c, d)$ and $(c, d)R(e, f)$. Then $ad = bc$ and $cf = de$. So, $adcf = bcde$. Therefore, $cd(af - be) = adcf - bcde = 0$. If $a = 0$, then $bc = 0$, and so, $c = 0$ (because $b \neq 0$). So, $de = 0$, and therefore, $e = 0$ (because $d \neq 0$). So, $af = be$ (because they're both 0). If $a \neq 0$, then $c \neq 0$. Therefore, $af - be = 0$, and so, $af = be$. Since $a = 0$ and $a \neq 0$ both lead to $af = be$, we have $(a, b)R(e, f)$. So, R is transitive. Since R is reflexive, symmetric, and transitive, it follows that R is an equivalence relation.

Let $\sim$ be an equivalence relation on a set S. If $x \in S$, the **equivalence class** of x, written $[x]$, is the set

$$[x] = \{y \in S \mid x \sim y\}.$$

Example 3.18:

1. Let A be a set and let $R = \{(a, b) \in A^2 \mid a = b\}$ be the equality relation. Then for each $a \in A$, $[a] = \{a\}$.

2. Let A be a set and let $R = A^2$ be the trivial equivalence relation. Then for each $a \in A$, $[a] = A$.

3. Consider the equivalence relation $\equiv_2 = \{(a, b) \in \mathbb{Z}^2 \mid a \text{ and } b \text{ have the same parity}\}$ on $\mathbb{Z}$. Let $\mathbb{E} = \{2k \mid k \in \mathbb{Z}\}$ be the set of even integers and let $\mathbb{O} = \{2k + 1 \mid k \in \mathbb{Z}\}$ be the set of odd integers. We have $[0] = \{y \in \mathbb{Z} \mid 0 \equiv_2 y\} = \mathbb{E}$. Observe that $[2] = [0]$, and in fact, if n is any even integer, then $[n] = [0] = \mathbb{E}$. Similarly, if n is any odd integer, then $[n] = [1] = \mathbb{O}$.

Partitions

Recall: (1) If X is a nonempty set of sets, we say that X is **pairwise disjoint** if for all $A, B \in X$ with $A \neq B$, A and B are disjoint ($A \cap B = \emptyset$).

(2) If X is a nonempty set of sets, then **union** X is defined by $\bigcup X = \{y \mid \text{there is } Y \in X \text{ with } y \in Y\}$.

A **partition** of a set S is a set of pairwise disjoint nonempty subsets of S whose union is S. Symbolically, $\boldsymbol{X}$ is a partition of S if and only if

$$\forall A \in \boldsymbol{X}(A \neq \emptyset \wedge A \subseteq S) \wedge \forall A, B \in \boldsymbol{X}(A \neq B \rightarrow A \cap B = \emptyset) \wedge \cup \boldsymbol{X} = S.$$

Example 3.19:

1. Let $\mathbb{E} = \{2k \mid k \in \mathbb{Z}\}$ be the set of even integers and let $\mathbb{O} = \{2k + 1 \mid k \in \mathbb{Z}\}$ be the set of odd integers. Then $\boldsymbol{X} = \{\mathbb{E}, \mathbb{O}\}$ is a partition of $\mathbb{Z}$. We can visualize this partition as follows:

$$\mathbb{Z} = \{\dots, -4, -2, 0, 2, 4, \dots\} \cup \{\dots, -3, -1, 1, 3, 5, \dots\}$$

2. Let $A = \{3k \mid k \in \mathbb{Z}\}$, $B = \{3k + 1 \mid k \in \mathbb{Z}\}$, and $C = \{3k + 2 \mid k \in \mathbb{Z}\}$. Then $\boldsymbol{X} = \{A, B, C\}$ is a partition of $\mathbb{Z}$. We can visualize this partition as follows:

$$\mathbb{Z} = \{\dots, -6, -3, 0, 3, 6, \dots\} \cup \{\dots, -5, -2, 1, 4, 7, \dots\} \cup \{\dots, -4, -1, 2, 5, 8, \dots\}$$

3. For each $n \in \mathbb{N}$, let $A_n = \{2n, 2n + 1\}$. Then $\boldsymbol{X} = \{A_n \mid n \in \mathbb{N}\}$ is a partition of $\mathbb{N}$. We can visualize this partition as follows:

$$\mathbb{N} = \{0, 1\} \cup \{2, 3\} \cup \{4, 5\} \cup \{6, 7\} \cup \{8, 9\} \cup \cdots$$

4. For each $n \in \mathbb{Z}$, let $A_n = \{(n, m) \mid m \in \mathbb{Z}\}$. Then $\boldsymbol{X} = \{A_n \mid n \in \mathbb{Z}\}$ is a partition of $\mathbb{Z} \times \mathbb{Z}$. We can visualize this partition as follows:

$$\vdots \qquad \qquad \vdots \qquad \qquad \vdots$$

$$\begin{aligned}
A_{-2} &= \{\dots, (-2, -3), (-2, -2), (-2, -1), (-2, 0), (-2, 1), (-2, 2), (-2, 3), \dots\} \\
A_{-1} &= \{\dots, (-1, -3), (-1, -2), (-1, -1), (-1, 0), (-1, 1), (-1, 2), (-1, 3), \dots\} \\
A_0 &= \{\dots, (\ 0, -3), (\ 0, -2), (\ 0, -1), (\ 0, 0), (\ 0, 1), (\ 0, 2), (\ 0, 3), \dots\} \\
A_1 &= \{\dots, (\ 1, -3), (\ 1, -2), (\ 1, -1), (\ 1, 0), (\ 1, 1), (\ 1, 2), (\ 1, 3), \dots\} \\
A_2 &= \{\dots, (\ 2, -3), (\ 2, -2), (\ 2, -1), (\ 2, 0), (\ 2, 1), (\ 2, 2), (\ 2, 3), \dots\}
\end{aligned}$$

$$\vdots \qquad \qquad \vdots \qquad \qquad \vdots$$

5. For each $a \in \mathbb{R}$, let $X_a = \{a + bi \mid b \in \mathbb{R}\}$. Then $\boldsymbol{X} = \{X_a \mid a \in \mathbb{R}\}$ is a partition of $\mathbb{C}$. This partition cannot be visualized as easily as the partition given in part 4 above. We will see in Lesson 4 that it is impossible to form a list of the real numbers and therefore, we cannot form a list of the elements of $\boldsymbol{X}$ or lists of the elements of each X_a.

6. The only partition of the one element set $\{a\}$ is $\{\{a\}\}$. The partitions of the two element set $\{a, b\}$ with $a \neq b$ are $\{\{a\}, \{b\}\}$ and $\{\{a, b\}\}$.

We will now explore the relationship between equivalence relations and partitions. Let's begin with an example.

Example 3.20: Consider the equivalence relation $\equiv_2$ from part 3 of Example 3.17, defined by $a \equiv_2 b$ if and only if a and b have the same parity, and the partition $\{\mathbb{E}, \mathbb{O}\}$ of $\mathbb{Z}$ from part 1 of Example 3.19. For this partition, we are thinking of $\mathbb{Z}$ as the union of the even and odd integers:

$$\mathbb{Z} = \{\dots, -4, -2, 0, 2, 4, \dots\} \cup \{\dots, -3, -1, 1, 3, 5, \dots\}$$

Observe that a and b are in the same member of the partition if and only if $a \equiv_2 b$ if and only if $[a] = [b]$. If n is any even integer, then $[n] = [0] = \mathbb{E}$ and if n is any odd integer, then $[n] = [1] = \mathbb{O}$.

43

Example 3.21: Recall that the **power set** of A, written $\mathcal{P}(A)$, is the set consisting of all subsets of A.

$$\mathcal{P}(A) = \{X \mid X \subseteq A\}$$

For example, if $A = \{a, b, c\}$, then $\mathcal{P}(A) = \{\emptyset, \{a\}, \{b\}, \{c\}, \{a, b\}, \{a, c\}, \{b, c\}, \{a, b, c\}\}$. We can define a binary relation $\sim$ on $\mathcal{P}(A)$ by $X \sim Y$ if and only if $|X| = |Y|$ (X and Y have the same number of elements). It is easy to see that $\sim$ is an equivalence relation on $\mathcal{P}(A)$. There are four equivalence classes.

$$[\emptyset] = \{\emptyset\} \qquad\qquad [\{a\}] = \{\{a\}, \{b\}, \{c\}\}$$

$$[\{a, b\}] = \{\{a, b\}, \{a, c\}, \{b, c\}\} \qquad\qquad [\{a, b, c\}] = \{\{a, b, c\}\}$$

Notes: (1) $\{a\} \sim \{b\} \sim \{c\}$ because each of these sets has one element. It follows that $\{a\}$, $\{b\}$, and $\{c\}$ are all in the same equivalence class. Above, we chose to use $\{a\}$ as the **representative** for this equivalence class. This is an arbitrary choice. In fact, $[\{a\}] = [\{b\}] = [\{c\}]$.

Similarly, $[\{a, b\}] = [\{a, c\}] = [\{b, c\}]$.

(2) The empty set is the only subset of A with 0 elements. Therefore, the equivalence class of $\emptyset$ contains only itself. Similarly, the equivalence class of $A = \{a, b, c\}$ contains only itself.

(3) Notice that the four equivalence classes are pairwise disjoint, nonempty, and their union is $\mathcal{P}(A)$. In other words, the equivalence classes form a partition of $\mathcal{P}(A)$.

Theorem 3.22: Let P be a partition of a set S. Then there is an equivalence relation $\sim$ on S for which the elements of P are the equivalence classes of $\sim$. Conversely, if $\sim$ is an equivalence relation on a set S, then the equivalence classes of $\sim$ form a partition of S.

You will be asked to prove Theorem 3.22 in Problem 16 below.

Important note: We will sometimes want to define relations or operations on equivalence classes. When we do this, we must be careful that what we are defining is **well-defined**. For example, consider the equivalence relation $\equiv_2$ on $\mathbb{Z}$, and let $X = \{[0], [1]\}$ be the set of equivalence classes.

Let's attempt to define a relation on X by $[x]R[y]$ if and only if $x < y$. Is $[0]R[1]$ true? It looks like it is because $0 < 1$. But this isn't the end of the story. Since $[0] = [2]$, if $[0]R[1]$, then we must also have $[2]R[1]$ (by a direct substitution). But $2 \not< 1$! So, $[2]R[1]$ is false. To summarize, $[0]R[1]$ should be true and $[2]R[1]$ should be false, but $[0]R[1]$ and $[2]R[1]$ represent the same statement. So, R is **not** a well-defined relation on X.

As another example, let's attempt to define an operation $+: X \times X \to X$ by $[x] + [y] = [x + y]$. This **is** a well-defined operation. It can be proved that the sum of two even integers is even, the sum of two odd integers is even, and the sum of an odd integer and an even integer is odd. These results can be used to show that the operation $+$ is well-defined. For example, if $[x] = [0]$ and $[y] = [0]$, then x and y are even. It follows that $x + y$ is even. So, $x + y \sim 0$, independent of our choices for x and y (as long as they are both even). The reader should check the other three cases to finish verifying that $+$ is well-defined on X.

Problem Set 3

Full solutions to these problems are available for free download here:

www.SATPrepGet800.com/TFBZLF

LEVEL 1

1. List the elements of $\{\Gamma, T, \Delta\} \times \{\gamma, \delta\}$.

2. For each set A below, evaluate (i) A^2; (ii) A^3; (iii) $\mathcal{P}(A)$.

$$1.\ A = \emptyset \qquad 2.\ A = \{\emptyset\} \qquad 3.\ A = \mathcal{P}(\{\emptyset\})$$

3. Let $C = (-\infty, 2]$ and $D = (-1, 3]$. Compute each of the following:

   (i) $C \cup D$

   (ii) $C \cap D$

   (iii) $C \setminus D$

   (iv) $D \setminus C$

   (v) $C \,\Delta\, D$

4. Find all partitions of the three-element set $\{a, b, c\}$ and the four-element set $\{a, b, c, d\}$.

5. Let $A = \{1, 2, 3, 4\}$ and let $R = \{(1,1), (1,3), (2,2), (2,4), (3,1), (3,3), (4,2), (4,4)\}$. Note that R is an equivalence relation on A. Find the equivalence classes of R.

LEVEL 2

6. Compute each of the following:

   (i) $\{0, 1, 2\}^3$

   (ii) $\{a, b\}^4$

7. Find the domain, range, and field of each of the following relations:

   (i) $R = \{(a, b), (c, d), (e, f), (f, a)\}$

   (ii) $S = \{(2k, 2t + 1) \mid k, t \in \mathbb{Z}\}$

8. Prove that for each $n \in \mathbb{Z}^+$, $\equiv_n$ (see part 4 of Example 3.17) is an equivalence relation on $\mathbb{Z}$.

LEVEL 3

9. Let A, B, C, and D be sets with $A \subseteq B$ and $C \subseteq D$. Prove that $A \times C \subseteq B \times D$.

10. Prove that there do not exist sets A and B such that the relation $<$ on $\mathbb{R}$ is equal to $A \times B$.

11. Let X be a set of equivalence relations on a nonempty set A. Prove that $\cap X$ is an equivalence relation on A.

LEVEL 4

12. Let $R = \{(x, y) \in \mathbb{R} \times \mathbb{R} \mid x - y \in \mathbb{Z}\}$. Prove that R is an equivalence relation on $\mathbb{R}$ and describe the equivalence classes of R.

13. Let A, B, C, and D be sets. Determine if each of the following statements is true or false. If true, provide a proof. If false, provide a counterexample.

    (i) $(A \times B) \cap (C \times D) = (A \cap C) \times (B \cap D)$

    (ii) $(A \times B) \cup (C \times D) = (A \cup C) \times (B \cup D)$

14. Let R be a relation on a set A. Determine if each of the following statements is true or false. If true, provide a proof. If false, provide a counterexample.

    (i)   If R is symmetric and transitive on A, then R is reflexive on A.

    (ii)   If R is antisymmetric on A, then R is not symmetric on A.

LEVEL 5

15. For $a, b \in \mathbb{N}$, we will say that a divides b, written $a|b$, if there is a natural number k such that $b = ak$. Notice that $|$ is a binary relation on $\mathbb{N}$. Prove that $(\mathbb{N}, |)$ is a partially ordered set, but it is not a linearly ordered set.

16. Let P be a partition of a set S. Prove that there is an equivalence relation $\sim$ on S for which the elements of P are the equivalence classes of $\sim$. Conversely, if $\sim$ is an equivalence relation on a set S, prove that the equivalence classes of $\sim$ form a partition of S.

CHALLENGE PROBLEM

17. Let $a, b, c, d \in \mathbb{N}$, let A and B be finite sets with $|A| = a$ and $|B| = b$, let $C = \left(\mathcal{P}(A \times B)\right)^c$, let $D = \left(\mathcal{P}(C^2)\right)^d$, and let Z be the set of relations on $C \times D$. Evaluate $|Z|$.

LESSON 4
FUNCTIONS AND EQUINUMEROSITY

Functions

Let A and B be sets. f is a **function** from A to B, written $f: A \to B$, if the following two conditions hold.

1. $f \subseteq A \times B$.

2. For all $a \in A$, there is a unique $b \in B$ such that $(a, b) \in f$.

Notes: (1) A function $f: A \to B$ is a binary relation on $A \cup B$.

(2) Not every binary relation on $A \cup B$ is a function from A to B. See part 2 of Example 4.1 below.

(3) The uniqueness in the second clause in the definition of a function above is equivalent to the statement "if $(a, b), (a, c) \in f$, then $b = c$."

(4) When we know that f is a function, we will abbreviate $(a, b) \in f$ by $f(a) = b$.

If $f: A \to B$, the **domain** of f, written dom f, is the set A, and the **range** of f, written ran f, is the set $\{f(a) \mid a \in A\}$. Observe that ran $f \subseteq B$. The set B is sometimes called the **codomain** of f.

Example 4.1:

1. $f = \{(0, a), (1, a)\}$ is a function with dom $f = \{0, 1\}$ and ran $f = \{a\}$. Instead of $(0, a) \in f$, we will usually write $f(0) = a$. Similarly, instead of $(1, a) \in f$, we will write $f(1) = a$. Here is a visual representation of this function.

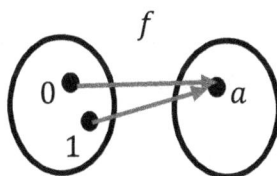

This function $f: \{0, 1\} \to \{a\}$ is called a **constant function** because the range of f consists of a single element.

Note also that f is a binary relation on the set $\{0, 1, a\}$.

2. If $a \neq b$, then $g = \{(0, a), (0, b)\}$ is **not** a function because it violates the second clause in the definition of being a function. It is, however, a binary relation on $\{0, a, b\}$.

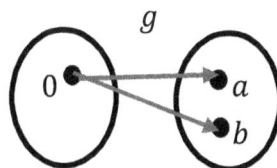

3. $h = \{(a, b) \mid a, b \in \mathbb{R} \wedge a > 0 \wedge a^2 + b^2 = 2\}$ is a relation on $\mathbb{R}$ that is **not** a function. $(1, 1)$ and $(1, -1)$ are both elements of h, violating the second clause in the definition of a function. See the figure below on the left. Notice how a vertical line hits the graph twice.

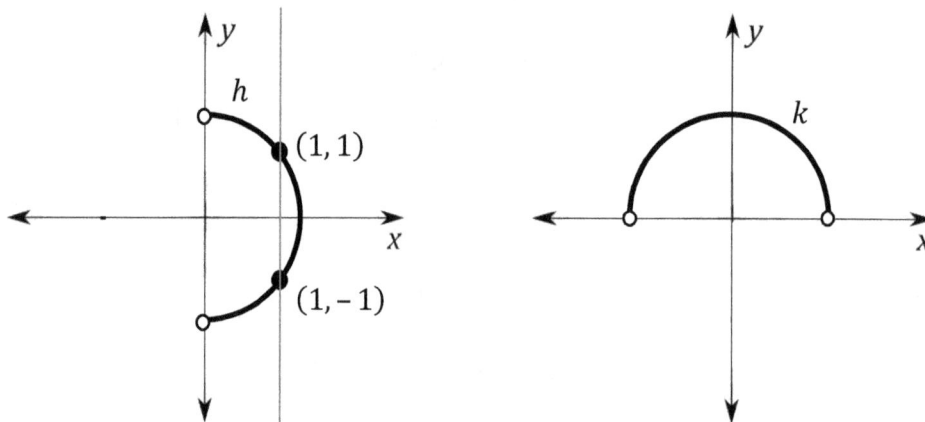

4. $k = \{(a, b) \mid a, b \in \mathbb{R} \wedge b > 0 \wedge a^2 + b^2 = 2\}$ **is** a function. See the figure above on the right. To see that the second clause in the definition of a function is satisfied, suppose that (a, b) and (a, c) are both in f. Then $a^2 + b^2 = 2$, $a^2 + c^2 = 2$, and b and c are both positive. It follows that $b^2 = c^2$, and since b and c are both positive, we have $b = c$.

We have dom $k = \left(-\sqrt{2}, \sqrt{2}\right)$ and ran $k = \left(0, \sqrt{2}\,\right]$. So, $k: \left(-\sqrt{2}, \sqrt{2}\right) \to \left(0, \sqrt{2}\,\right]$.

Note that if $a \in \mathbb{R}$, then $a^2 = a \cdot a$. Also, if $a \geq 0$ and $a^2 = b$, then $\sqrt{b} = a$.

5. $b = \{(z, w) \mid z, w \in \mathbb{C} \wedge w = z - 1\}$ is a function. By Note 4 above, we can describe b using the notation $b(z) = z - 1$.

If we write $z = x + yi$ and $b(z) = u + vi$, then $b(x + yi) = x + yi - 1 = (x - 1) + yi$ and we see that $u(x, y) = x - 1$ and $v(x, y) = y$.

b is an example of a complex-valued function. One way to visualize this function is to simply stay in the same plane and to analyze how a typical point moves or how a certain set is transformed. The function b takes the point (x, y) to the point $(x - 1, y)$. That is, each point is shifted one unit to the left. Similarly, if $S \subseteq \mathbb{C}$, then each point of the set S is shifted one unit to the left by the function b. Both these situations are demonstrated in the figure to the right. Observe how the point $(2, 2)$ is shifted to the point $(1, 2)$ and how each point of the

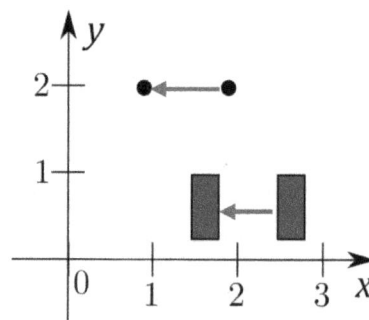

rightmost rectangle is shifted one point to the left to form the leftmost rectangle.

A second way to visualize this function is to draw two separate planes: an xy-plane and a uv-plane. We can then draw a point or a set in the xy-plane and its image under b in the uv-plane. In the figure below, we do this for the same point and the same rectangle as we did in the previous figure.

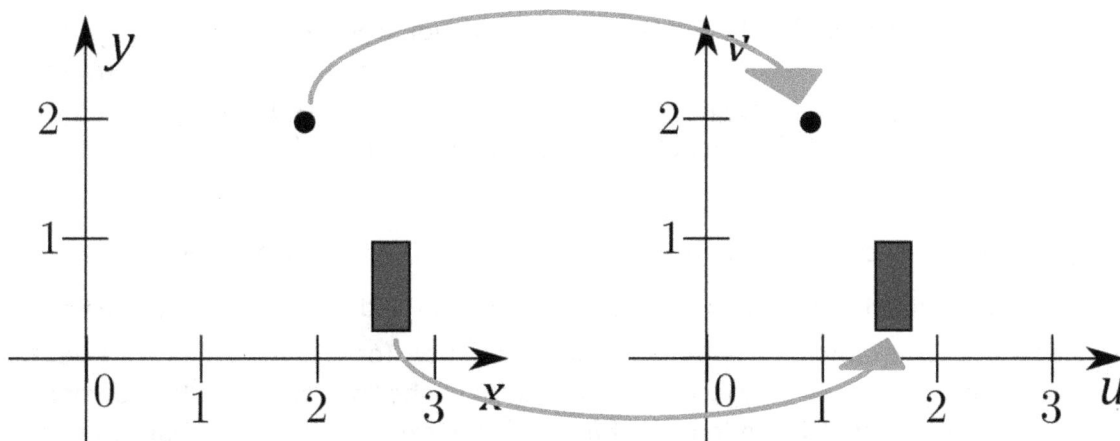

A function with domain $\mathbb{N}$ is called an **infinite sequence**. For example, let $f: \mathbb{N} \to \{0, 1\}$ be defined by $f(n) = \begin{cases} 0 & \text{if } n \text{ is even.} \\ 1 & \text{if } n \text{ is odd.} \end{cases}$ A nice way to visualize an infinite sequence is to list the "outputs" of the sequence in order in parentheses. So, we may write f as $(0, 1, 0, 1, 0, 1, \ldots)$. In general, if A is a nonempty set and $f: \mathbb{N} \to A$ is a sequence, then we can write f as $(f(0), f(1), f(2), \ldots)$.

Similarly, a **finite sequence** is a function with domain $\{0, 1, \ldots, n-1\}$ for some n. For example, the sequence $(0, 2, 4, 6, 8, 10)$ is the function $g: \{0, 1, 2, 3, 4, 5\} \to \mathbb{N}$ defined by $g(k) = 2k$. If the domain of a finite sequence is $\{0, 1, \ldots, n-1\}$, we say that n is the **length** of the sequence.

Observe how a finite sequence with domain $\{0, 1, \ldots, n-1\}$ and range A looks just like an n-tuple in A^n. In fact, it's completely natural to identify a finite sequence of length n with the corresponding n-tuple. So, $(0, 2, 4, 6, 8, 10)$ can be thought of as a 6-tuple from $\mathbb{N}^6$, or as the function $g: \{0, 1, 2, 3, 4, 5\} \to \mathbb{N}$ defined by $g(k) = 2k$.

Informally, we can think of an infinite sequence as an infinite length tuple. As one more example, $(1, 2, 4, 8, 16, 32, \ldots)$ represents the sequence $h: \mathbb{N} \to \mathbb{N}$ defined by $h(n) = 2^n$. If $f: \mathbb{N} \to X$ is defined by $f(n) = x_n$, then we may represent the sequence f using the notation $(x_n)_{n \in \mathbb{N}}$ or simply (x_n). For example, the sequence h can be represented as $(2^n)_{n \in \mathbb{N}}$ or (2^n). x_n is called the nth term of the sequence. The 0th term of the sequence h is 1, the first term of the sequence h is 2, and so on. In general, the nth term of the sequence h is 2^n.

Note: In the study of set theory, we define the natural numbers by letting $0 = \emptyset$, $1 = \{0\}$, $2 = \{0, 1\}$, $3 = \{0, 1, 2\}, \ldots$ and so on. In general, the natural number n is the set of all its predecessors. Specifically, $n = \{0, 1, 2, \ldots, n-1\}$. Using this notation, we can say that a finite sequence of length n is a function $f: n \to A$ for some set A. For example, the function g above has domain 6, so that $g: 6 \to \mathbb{N}$.

Example 4.2:

1. There is exactly one finite sequence of length 0, namely the empty sequence, $\emptyset$. The empty sequence can be described using function notation as $f: 0 \to X$, where X is any set.

2. The infinite sequence $(0, -1, 2, -3, 4, -5, \dots)$ is a function from $\mathbb{N}$ to $\mathbb{Z}$. If we name this function g, then we have that $g: \mathbb{N} \to \mathbb{Z}$ and g is defined by $g(n) = (-1)^n n$. So, the nth term of the sequence g is $(-1)^n n$ and we can represent the sequence as $((-1)^n n)$. This function g is an example of an *integer-valued function* (or $\mathbb{Z}$-*valued function*) because the codomain of g consists of only integers.

3. The infinite sequence $\left(\frac{1}{n+1}\right)$ is a function from $\mathbb{N}$ to $\mathbb{Q}$. The nth term of this sequence is $\frac{1}{n+1}$. If we name this function h, then we have that $h: \mathbb{N} \to \mathbb{Q}$ and h is defined by $h(n) = \frac{1}{n+1}$. This function h is an example of a *rational-valued function* (or $\mathbb{Q}$-*valued function*) because the codomain of h consists of only rational numbers. Since the "outputs" of h take on only positive values, we can "shrink" the codomain of h to $\mathbb{Q}^+$, the set of positive rational numbers. So, we can write $h: \mathbb{N} \to \mathbb{Q}^+$. We can visualize this sequence as follows:

$$\left(1, \frac{1}{2}, \frac{1}{3}, \frac{1}{4}, \frac{1}{5}, \dots\right)$$

4. The infinite sequence $\left(n^2 + \sqrt{n}i\right)$ is a function from $\mathbb{N}$ to $\mathbb{C}$. The nth term of this sequence is $n^2 + \sqrt{n}i$. If we name this function k, then we have that $k: \mathbb{N} \to \mathbb{C}$ and k is defined by $k(n) = n^2 + \sqrt{n}i$. This function k is an example of a *complex-valued function* (or $\mathbb{C}$-*valued function*) because the codomain of k consists of only complex numbers. We can visualize this sequence as follows: $\left(0, 1 + i, 4 + \sqrt{2}i, 9 + \sqrt{3}i, 16 + 2i, \dots\right)$

Injections, Surjections, and Bijections

A function $f: A \to B$ is **injective** (or **one-to-one**), written $f: A \hookrightarrow B$, if for all $a, b \in A$, if $a \neq b$, then $f(a) \neq f(b)$. In this case, we call f an **injection**.

Notes: (1) The contrapositive of the conditional statement $p \to q$ is the statement $\neg q \to \neg p$. These two statements are logically equivalent. See the analysis after the statement of Theorem 2.7.

(2) The contrapositive of the statement "If $a \neq b$, then $f(a) \neq f(b)$" is "If $f(a) = f(b)$, then $a = b$." So, we can say that a function $f: A \to B$ is injective if for all $a, b \in A$, if $f(a) = f(b)$, then $a = b$.

A function $f: A \to B$ is **surjective** (or **onto** B), written $f: A \twoheadrightarrow B$, if for all $b \in B$, there is an $a \in A$ such that $f(a) = b$. In this case, we call f a **surjection**.

A function $f: A \to B$ is **bijective**, written $f: A \cong B$ if f is both injective and surjective. In this case, we call f a **bijection**.

Example 4.3:

1. $f = \{(0, a), (1, a)\}$ from part 1 of Example 4.1 is **not** an injective function because $f(0) = a$, $f(1) = a$, and $0 \neq 1$. If we think of f as $f: \{0, 1\} \to \{a\}$, then f is surjective. However, if we think of f as $f: \{0, 1\} \to \{a, b\}$, then f is **not** surjective. So, surjectivity depends upon the codomain of the function.

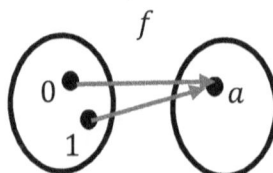

2. $k = \{(a, b) \mid a, b \in \mathbb{R} \land b > 0 \land a^2 + b^2 = 2\}$ from part 4 of Example 4.1 is **not** an injective function. For example, $(1, 1) \in k$ because $1^2 + 1^2 = 1 + 1 = 2$ and $(-1, 1) \in k$ because $(-1)^2 + 1^2 = 1 + 1 = 2$. Notice how a horizontal line hits the graph twice. If we think of k as a function from $\left(-\sqrt{2}, \sqrt{2}\right)$ to $\mathbb{R}^+$, then k is **not** surjective. For example, $2 \notin \operatorname{ran} k$ because for any $a \in \mathbb{R}$, $a^2 + 2^2 = a^2 + 4 \geq 4$, and so, $a^2 + 2^2$ cannot be equal to 2. However, if instead we consider k as a function with codomain $\left(0, \sqrt{2}\right]$, that is $k: \left(-\sqrt{2}, \sqrt{2}\right) \to \left(0, \sqrt{2}\right]$, then k **is** surjective. Indeed, if $0 < b \leq \sqrt{2}$, then $0 < b^2 \leq 2$, and so, $a^2 = 2 - b^2 \geq 0$. Therefore, $a = \sqrt{2 - b^2}$ is a real number such that $k(a) = b$.

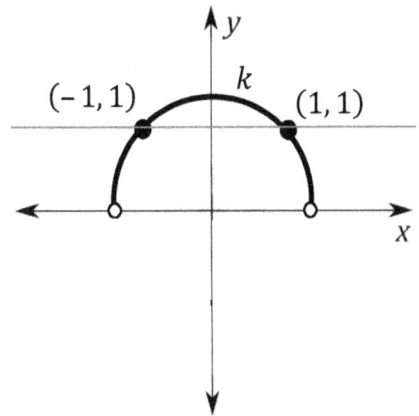

3. Define $g: \mathbb{R} \to \mathbb{R}$ by $g(x) = 7x - 3$. Then g is injective because if $g(a) = g(b)$, we then have $7a - 3 = 7b - 3$. Adding 3 to each side of this equation, we get $7a = 7b$, and then multiplying each side of this last equation by $\frac{1}{7}$, we get $a = b$ (see Lesson 5 for justification of these cancellation rules). Also, g is surjective because if $b \in \mathbb{R}$, then $\frac{b+3}{7} \in \mathbb{R}$ and

$$g\left(\frac{b+3}{7}\right) = 7\left(\frac{b+3}{7}\right) - 3 = (b + 3) - 3 = b + (3 - 3) = b + 0 = b$$

Therefore, g is bijective. See the image to the right for a visual representation of $\mathbb{R}^2$ and the graph of the function g.

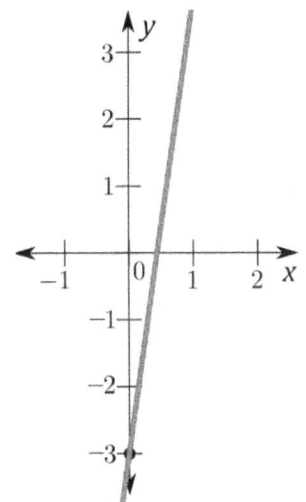

Notice that any vertical line will hit the graph of g exactly once because g is a function with domain $\mathbb{R}$. Also, any horizontal line will hit the graph exactly once because g is bijective. Injectivity ensures that each horizontal line hits the graph *at most* once and surjectivity ensures that each horizontal line hits the graph *at least* once.

This function g is an example of a *real-valued function* (or $\mathbb{R}$-*valued function*) because the codomain of g consists of only real numbers.

Inverse Functions

If $f: A \to B$ is bijective, we define $f^{-1}: B \to A$, the **inverse** of f, by $f^{-1} = \{(b, a) \mid (a, b) \in f\}$. In other words, for each $b \in B$, $f^{-1}(b) = $ "the unique $a \in A$ such that $f(a) = b$."

Notes: (1) Let $f: A \to B$ be bijective. Since f is surjective, for each $b \in B$, there is an $a \in A$ such that $f(a) = b$. Since f is injective, there is only one such value of a.

(2) The inverse of a bijective function is also bijective.

Example 4.4:

1. Define $f: \{0, 1\} \to \{a, b\}$ by $f = \{(0, a), (1, b)\}$. Then f is a bijection and $f^{-1}: \{a, b\} \to \{0, 1\}$ is defined by $f^{-1} = \{(a, 0), (b, 1)\}$. Observe that f^{-1} is also a bijection.

2. Let $\mathbb{E} = \{0, 2, 4, 6, 8, \dots\}$ be the set of even natural numbers and let $\mathbb{O} = \{1, 3, 5, 7, 9 \dots\}$ be the set of odd natural numbers. The function $f: \mathbb{E} \to \mathbb{O}$ defined by $f(n) = n + 1$ is a bijection with inverse $f^{-1}: \mathbb{O} \to \mathbb{E}$ defined by $f(n) = n - 1$.

3. If X and Y are sets, we define ^XY to be the set of functions from X to Y. Symbolically, we have

$$^XY = \{f \mid f: X \to Y\}$$

For example, if $A = \{a, b\}$ and $B = \{0, 1\}$, then ^AB has 4 elements (each element is a function from A to B). The elements are $f_1 = \{(a, 0), (b, 0)\}$, $f_2 = \{(a, 0), (b, 1)\}$, $f_3 = \{(a, 1), (b, 0)\}$, and $f_4 = \{(a, 1), (b, 1)\}$. Here is a visual representation of these four functions.

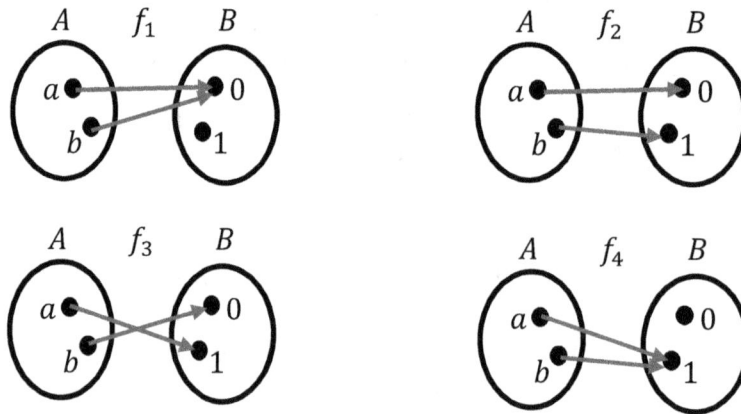

Define $F: {}^AB \to \mathcal{P}(A)$ by $F(f) = \{x \in A \mid f(x) = 1\}$.

So, $F(f_1) = \emptyset$, $F(f_2) = \{b\}$, $F(f_3) = \{a\}$, and $F(f_4) = \{a, b\}$.

Since $\mathcal{P}(A) = \{\emptyset, \{a\}, \{b\}, \{a, b\}\}$, we see that F is a bijection from ^AB to $\mathcal{P}(A)$.

The inverse of F is the function $F^{-1}: \mathcal{P}(A) \to {}^AB$ defined by $F^{-1}(C)(x) = \begin{cases} 0 & \text{if } x \notin C. \\ 1 & \text{if } x \in C. \end{cases}$

So, we see that $F^{-1}(\emptyset) = f_1$, $F^{-1}(\{b\}) = f_2$, $F^{-1}(\{a\}) = f_3$, and $F^{-1}(\{a, b\}) = f_4$.

4. For $A \neq \emptyset$ and $B = \{0, 1\}$, the function $F: {}^AB \to \mathcal{P}(A)$ defined by $F(f) = \{x \in A \mid f(x) = 1\}$ is always a bijection.

To see that F is injective, let $f, g \in {}^AB$ with $f \neq g$. Since f and g are different, there is some $a \in A$ such that either $f(a) = 0$, $g(a) = 1$ or $f(a) = 1$, $g(a) = 0$. **Without loss of generality,** (see Note 1 below) assume that $f(a) = 0$ and $g(a) = 1$. Since $f(a) = 0$, $a \notin F(f)$. Since $g(a) = 1$, $a \in F(g)$. So, $F(f) \neq F(g)$. Since $f \neq g$ implies $F(f) \neq F(g)$, F is injective.

To see that F is surjective, let $C \in \mathcal{P}(A)$, so that $C \subseteq A$. Define $f \in {}^AB$ by $f(x) = \begin{cases} 0 & \text{if } x \notin C. \\ 1 & \text{if } x \in C. \end{cases}$ Then $x \in F(f)$ if and only if $f(x) = 1$ if and only if $x \in C$. So, $F(f) = C$. Since $C \in \mathcal{P}(A)$ was arbitrary, F is surjective.

As in 3, the inverse of F is the function $F^{-1}: \mathcal{P}(A) \to {}^A B$ defined by $F^{-1}(C)(x) = \begin{cases} 0 & \text{if } x \notin C. \\ 1 & \text{if } x \in C. \end{cases}$

Notes: (1) In part 4 of Example 4.4, we used the expression "Without loss of generality." This expression can be used when an argument can be split up into 2 or more cases, and the proof of each of the cases is nearly identical.

In the example above, the two cases are (i) $f(a) = 0, g(a) = 1$ and (ii) $f(a) = 1, g(a) = 0$. The argument for case (ii) is the same as the argument for case (i), essentially word for word—only the roles of f and g are interchanged.

(2) Using the definition $n = \{0, 1, 2, \ldots, n - 1\}$, we have just shown that for any nonempty set A, there is a bijection $f: {}^A 2 \to \mathcal{P}(A)$.

Composite Functions

Given functions $f: A \to B$ and $g: B \to C$, the **composite** (or **composition**) of f and g, written $g \circ f: A \to C$, is defined by $(g \circ f)(a) = g(f(a))$ for all $a \in A$. Symbolically, we have

$$g \circ f = \{(a, c) \in A \times C \mid \text{There is a } b \in B \text{ such that } (a, b) \in f \text{ and } (b, c) \in g\}.$$

We can visualize the composition of two functions f and g as follows.

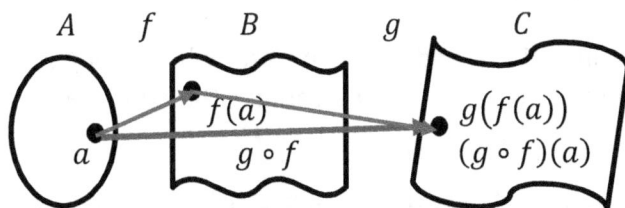

In the picture above, sets A, B, and C are drawn as different shapes simply to emphasize that they can all be different sets. Starting with an arbitrary element $a \in A$, we have an arrow showing a being mapped by f to $f(a) \in B$ and another arrow showing $f(a)$ being mapped by g to $g(f(a)) \in C$. There is also an arrow going directly from $a \in A$ to $(g \circ f)(a) = g(f(a))$ in C. However, note that the only way we know how to get from a to $(g \circ f)(a)$ is to first travel from a to $f(a)$, and then to travel from $f(a)$ to $g(f(a))$.

Example 4.5:

1. Define $f: \mathbb{R} \to \mathbb{R}$ by $f(x) = x + 3$ and $g: \mathbb{R} \to \mathbb{R}$ by $g(x) = (x - 1)^2$ Then $g \circ f: \mathbb{R} \to \mathbb{R}$ is defined by $(g \circ f)(x) = g(f(x)) = g(x + 3) = ((x + 3) - 1)^2 = (x + 2)^2$.

2. Define $f: \mathbb{Z} \to \mathbb{Q}$ by $f(n) = \frac{n}{2}$ and define $g: \mathbb{Q} \to \{0, 1\}$ by $g(n) = \begin{cases} 0 & \text{if } n \in \mathbb{Z}. \\ 1 & \text{if } n \in \mathbb{Q} \setminus \mathbb{Z}. \end{cases}$ We will show that $g \circ f: \mathbb{Z} \to \{0, 1\}$ is defined by $(g \circ f)(n) = \begin{cases} 0 & \text{if } n \text{ is even.} \\ 1 & \text{if } n \text{ is odd.} \end{cases}$

To see this, observe that if $n \in \mathbb{Z}$ is even, then there is an integer k such that $n = 2k$. It follows that $(g \circ f)(n) = g(f(n)) = g\left(\frac{n}{2}\right) = g\left(\frac{2k}{2}\right) = g(k) = 0$ because $k \in \mathbb{Z}$. If n is odd, then there is an integer k such that $n = 2k + 1$. So, $(g \circ f)(n) = g(f(n)) = g\left(\frac{n}{2}\right) = g\left(\frac{2k+1}{2}\right) = 1$ because $\frac{2k+1}{2} \in \mathbb{Q} \setminus \mathbb{Z}$. To see that $\frac{2k+1}{2} \in \mathbb{Q}$, simply observe that $2k + 1 \in \mathbb{Z}$, $2 \in \mathbb{Z}$, and $2 \neq 0$. To see that $\frac{2k+1}{2} \notin \mathbb{Z}$, first note that, $\frac{2k+1}{2} = \frac{2k}{2} + \frac{1}{2} = k + \frac{1}{2}$ (Check this!). Now, let $m = k + \frac{1}{2}$. If $m \in \mathbb{Z}$, then we would have $m - k \in \mathbb{Z}$ because when we subtract one integer from another, we always get an integer. It would then follow that $\frac{1}{2} \in \mathbb{Z}$, which we know it is not. Since assuming that $\frac{2k+1}{2} \in \mathbb{Z}$ would lead to the false statement $\frac{1}{2} \in \mathbb{Z}$, we know that the statement $\frac{2k+1}{2} \in \mathbb{Z}$ must be false.

Note: In part 2 of Example 4.5 above, to show that $m \notin \mathbb{Z}$, we began with the assumption that $m \in \mathbb{Z}$, and then used a logically valid argument to derive a false statement. This is known as a proof by contradiction. See the proof of Theorem 4.12 below for a formal proof by contradiction and a detailed explanation as to how it works. Theorem 5.6 provides another example.

It will be important to know that when we take the composition of bijective functions, we always get a bijective function. We will prove this in two steps. We will first show that the composition of injective functions is injective. We will then show that the composition of surjective functions is surjective.

Theorem 4.6: If $f: A \hookrightarrow B$ and $g: B \hookrightarrow C$, then $g \circ f: A \hookrightarrow C$.

Note: We are given that f and g are injections, and we want to show that $g \circ f$ is an injection. We can show this directly using the definition of injectivity, or we can use the contrapositive of the definition of injectivity. Let's do it both ways.

Direct proof of Theorem 4.6: Suppose that $f: A \hookrightarrow B$ and $g: B \hookrightarrow C$, and let $x, y \in A$ with $x \neq y$. Since f is injective, $f(x) \neq f(y)$. Since g is injective, $g(f(x)) \neq g(f(y))$. So, $(g \circ f)(x) \neq (g \circ f)(y)$. Since $x, y \in A$ were arbitrary, $g \circ f: A \hookrightarrow C$. □

Contrapositive proof of Theorem 4.6: Suppose that $f: A \hookrightarrow B$ and $g: B \hookrightarrow C$, let $x, y \in A$ and suppose that $(g \circ f)(x) = (g \circ f)(y)$. Then $g(f(x)) = g(f(y))$. Since g is injective, $f(x) = f(y)$. Since f is injective, $x = y$. Since $x, y \in A$ were arbitrary, $g \circ f: A \hookrightarrow C$. □

Theorem 4.7: If $f: A \mapsto B$ and $g: B \mapsto C$, then $g \circ f: A \mapsto C$.

Proof: Suppose that $f: A \mapsto B$ and $g: B \mapsto C$, and let $c \in C$. Since g surjective, there is $b \in B$ with $g(b) = c$. Since f is surjective, there is $a \in A$ with $f(a) = b$. So, $(g \circ f)(a) = g(f(a)) = g(b) = c$. Since $c \in C$ was arbitrary, $g \circ f$ is surjective. □

Corollary 4.8: If $f: A \cong B$ and $g: B \cong C$, then $g \circ f: A \cong C$.

Proof: Suppose that $f: A \cong B$ and $g: B \cong C$. Then f and g are injective. By Theorem 4.6, $g \circ f$ is injective. Also, f and g are surjective. By Theorem 4.7, $g \circ f$ is surjective. Since $g \circ f$ is both injective and surjective, $g \circ f$ is bijective. □

Note: A **corollary** is a theorem that follows easily from a theorem or theorems that have already been proved.

Identity Functions

If A is any set, then we define the **identity function** on A, written $i_A: A \to A$ by $i_A(a) = a$ for all $a \in A$. Note that the identity function on A is a bijection from A to itself.

Theorem 4.9: If $f: A \cong B$, then $f^{-1} \circ f = i_A$ and $f \circ f^{-1} = i_B$.

Proof: Let $a \in A$ with $f(a) = b$. Then $f^{-1}(b) = a$, and so, $(f^{-1} \circ f)(a) = f^{-1}(f(a)) = f^{-1}(b) = a$. Since $i_A(a) = a$, we see that $(f^{-1} \circ f)(a) = i_A(a)$. Since $a \in A$ was arbitrary, $f^{-1} \circ f = i_A$.

Now, let $b \in B$. Since $f: A \cong B$, there is a unique $a \in A$ with $f(a) = b$. Equivalently, $f^{-1}(b) = a$. We have $(f \circ f^{-1})(b) = f(f^{-1}(b)) = f(a) = b$ Since $i_B(b) = b$, we see that $(f \circ f^{-1})(b) = i_B(b)$. Since $b \in B$ was arbitrary, $f \circ f^{-1} = i_B$. $\qquad\square$

Images and and Inverse Images

If $f: X \to Y$ and $A \subseteq X$, then the **image of A under f** is the set $f[A] = \{f(x) \mid x \in A\}$. Similarly, if $B \subseteq Y$, then the **inverse image of B under f** is the set $f^{-1}[B] = \{x \in X \mid f(x) \in B\}$.

Example 4.10:

1. Let $f: \{a, b, c, d\} \to \{0, 1, 2\}$ be defined by $f = \{(a, 0), (b, 0), (c, 1), (d, 2)\}$. Let $A = \{a\}$, $B = \{a, b\}$, $C = \{a, c\}$, and $D = \{b, c, d\}$. Then $f[A] = \{0\}$, $f[B] = \{0\}$, $f[C] = \{0, 1\}$, and $f[D] = \{0, 1, 2\}$. Now, let $X = \{0\}$, $Y = \{0, 1\}$, $Z = \{0, 2\}$, and $W = \{0, 1, 2\}$. Then we have $f^{-1}[X] = \{a, b\}$, $f^{-1}[Y] = \{a, b, c\}$, $f^{-1}[Z] = \{a, b, d\}$, and $f^{-1}[W] = \{a, b, c, d\}$.

2. Define $f: \mathbb{R} \to \mathbb{R}$ by $f(x) = x^4$. Then we have $f[\mathbb{R}] = [0, \infty)$, $f(\{-2, 0, 3\}) = \{0, 16, 81\}$, $f[(-3, 2]] = [0, 81)$, $f^{-1}[\mathbb{R}] = \mathbb{R}$, $f^{-1}[\{16\}] = \{-2, 2\}$, $f^{-1}[[0, \infty)] = \mathbb{R}$, $f^{-1}[(-\infty, 0)] = \emptyset$, and $f^{-1}[(0, \infty)] = (-\infty, 0) \cup (0, \infty) = \mathbb{R} \setminus \{0\}$.

Equinumerosity

We say that two sets A and B are **equinumerous**, written $A \sim B$ if there is a bijection $f: A \cong B$.

It is easy to see that $\sim$ **is an equivalence relation**. For any set A, the identity function $i_A: A \to A$ is a bijection, showing that $\sim$ is reflexive. For sets A and B, if $f: A \cong B$, then $f^{-1}: B \cong A$, showing that $\sim$ is symmetric. For sets A, B, and C, if $f: A \cong B$ and $g: B \cong C$, then $g \circ f: A \cong C$ by Corollary 4.8, showing that $\sim$ is transitive.

Example 4.11:

1. Let $A = \{$anteater, elephant, giraffe$\}$ and $B = \{$apple, banana, orange$\}$. Then $A \sim B$. We can define a bijection $f: A \cong B$ by $f($anteater$) = $ apple, $f($elephant$) = $ banana, and $f($giraffe$) = $ orange. This is not the only bijection from A to B, but we need only find one (or prove one exists) to show that the sets are equinumerous.

2. At this point it should be easy to see that two finite sets are equinumerous if and only if they have the same number of elements. It should also be easy to see that a finite set can never be equinumerous with an infinite set.

3. Let $\mathbb{N} = \{0, 1, 2, 3, 4 \dots\}$ be the set of natural numbers and $\mathbb{E} = \{0, 2, 4, 6, 8 \dots\}$ the set of even natural numbers. Then $\mathbb{N} \sim \mathbb{E}$. We can actually see a bijection between these two sets just by looking at the sets themselves.

$$0 \quad 1 \quad 2 \quad 3 \quad 4 \quad 5 \quad 6 \dots$$
$$0 \quad 2 \quad 4 \quad 6 \quad 8 \quad 10 \quad 12 \dots$$

The function $f\colon \mathbb{N} \to \mathbb{E}$ defined by $f(n) = 2n$ is an explicit bijection. To see that f maps $\mathbb{N}$ into $\mathbb{E}$, just observe that if $n \in \mathbb{N}$, then $2n \in \mathbb{E}$ by the definition of an even integer ($a \in \mathbb{N}$ is even if there is $b \in \mathbb{N}$ with $a = 2b$). f is injective because if $f(n) = f(m)$, then $2n = 2m$, and so, $n = m$ (see part (viii) of Problem 4 from Problem Set 5 for details). Finally, f is surjective because if $n \in \mathbb{E}$, then there is $k \in \mathbb{N}$ such that $n = 2k$. So, $f(k) = 2k = n$.

4. $\mathbb{N} \sim \mathbb{Z}$ via the bijection $f\colon \mathbb{N} \cong \mathbb{Z}$ defined by $f(n) = \begin{cases} \dfrac{n}{2} & \text{if } n \text{ is even.} \\ -\dfrac{n+1}{2} & \text{if } n \text{ is odd.} \end{cases}$

You will be asked to show that f is a bijection in Problem 9 below. Let's look at this correspondence visually:

$$0 \quad 1 \quad 2 \quad 3 \quad 4 \quad 5 \quad 6 \dots$$
$$0 \quad -1 \quad 1 \quad -2 \quad 2 \quad -3 \quad 3 \dots$$

Many students get confused here because they are under the misconception that the integers should be written "in order." However, when checking to see if two sets are equinumerous, we **do not** include any other structure. In other words, we are just trying to "pair up" elements—it does not matter how we do so.

5. For A any nonempty set, $^A2 \sim \mathcal{P}(A)$. We showed this in part 4 of Example 4.4.

6. $[0, 1] \sim [0, 5]$ via the bijection $f\colon [0, 1] \to [0, 5]$ defined by $f(x) = 5x$. To see that f maps into $[0, 5]$, note that if $0 \le x \le 1$, then $0 \le 5x \le 5$. To see that f is injective, note that if $x \ne y$, then $5x \ne 5y$. Finally, to see that f is surjective, let $y \in [0, 5]$. Then $0 \le y \le 5$ and so, it follows that $0 \le \frac{y}{5} \le 1$ and $f\left(\frac{y}{5}\right) = 5 \cdot \frac{y}{5} = y$.

Countable and Uncountable Sets

We say that a set is **countable** if it is equinumerous with a subset of $\mathbb{N}$. It's easy to visualize a countable set because a bijection from a subset of $\mathbb{N}$ to a set A generates a list. For example, the set $\mathbb{E}$ can be listed as 0, 2, 4, 6, … and the set $\mathbb{Z}$ can be listed as 0, – 1, 1, – 2, 2, … (see Example 4.11 above).

There are two kinds of countable sets: finite sets and **denumerable** sets. We say that a set is denumerable if it is countably infinite.

At this point, you may be asking yourself if all infinite sets are denumerable. If this were the case, then we would simply have finite sets and infinite sets, and that would be the end of it. However, there are in fact infinite sets that are **not** denumerable. An infinite set that is not denumerable is **uncountable**.

Theorem 4.12 (Cantor's Theorem): If A is any set, then A is **not** equinumerous with $\mathcal{P}(A)$.

Analysis: How can we prove that A is not equinumerous with $\mathcal{P}(A)$? Well, we need to show that there **does not** exist a bijection from A to $\mathcal{P}(A)$. Recall that a bijection is a function which is both an injection and a surjection. So, we will attempt to show that there do not exist any surjections from A to $\mathcal{P}(A)$. To do this, we will take an arbitrary function $f: A \rightarrow \mathcal{P}(A)$, and then argue that f is not surjective. We will show that ran $f \neq \mathcal{P}(A)$ by finding a set $B \in \mathcal{P}(A) \setminus$ ran f. In words, we will find a subset of A that is **not** in the range of f.

Let's begin by looking at $\mathbb{N}$, the set of natural numbers. Given a specific function $f: \mathbb{N} \rightarrow \mathcal{P}(\mathbb{N})$, it's not too hard to come up with a set $B \in \mathcal{P}(\mathbb{N}) \setminus$ ran f. Let's choose a specific such f and use this example to try to come up with a procedure for describing the set B.

$$f(0) = \{\mathbf{0}, 1, 2, 3, 4, 5, 6, 7, 8, 9, 10, \dots\}$$
$$f(1) = \{0, \mathbf{1}, 3, 4, 5, 6, 7, 8, 9, 10, \dots\}$$
$$f(2) = \{0, 1, 4, 5, 6, 7, 8, 9, 10, \dots\}$$
$$f(3) = \{0, 1, 4, 6, 7, 8, 9, 10, \dots\}$$
$$f(4) = \{0, 1, \mathbf{4}, 6, 8, 9, 10, \dots\}$$

$$\dots$$

Technical note: Recall that a **prime number** is a natural number with **exactly** two factors, 1 and itself. The set of prime numbers looks like this: $\{2, 3, 5, 7, 11, 13, 17, \dots\}$. The function $f: \mathbb{N} \rightarrow \mathcal{P}(\mathbb{N})$ that we chose to use here is defined by $f(n) = \{k \in \mathbb{N} \mid k$ is not equal to one of the first n prime numbers$\}$. Notice how $f(0)$ is just the set $\mathbb{N}$ of all natural numbers, $f(1)$ is the set of all natural numbers except 2 (we left out the first prime), $f(2)$ is the set of all natural numbers except 2 and 3 (we left out the first two primes), and so on.

Observe that the "inputs" of our function are natural numbers, and the "outputs" are sets of natural numbers. So, it's perfectly natural to ask the question "Is n in $f(n)$?"

For example, we see that $0 \in f(0)$, $1 \in f(1)$, and $4 \in f(4)$ (indicated in bold in the definition of the function above). However, we also see that $2 \notin f(2)$ and $3 \notin f(3)$.

Let's let B be the set of natural numbers n that are **not** inside their images. Symbolically, we have

$$B = \{n \in \mathbb{N} \mid n \notin f(n)\}.$$

Which natural numbers are in the set B? Well, we already said that $0 \in f(0)$. It follows that $0 \notin B$. Similarly, $1 \notin B$ and $4 \notin B$, but $2 \in B$ and $3 \in B$.

Why did we choose to define B this way? The reason is because we are trying to make sure that B cannot be equal to $f(n)$ for every n. Since $0 \in f(0)$, but $0 \notin B$, it follows that $f(0)$ and B are different sets because they differ by at least one element, namely 0. Similarly, since $1 \in f(1)$, but $1 \notin B$, B cannot be equal to $f(1)$. What about 2? Well $2 \notin f(2)$, but $2 \in B$. Therefore, $B \neq f(2)$ as well... and so on down the line. We intentionally chose to make B disagree with $f(n)$ for every natural number n, ensuring that B will not be in the range of f.

I think we are now ready to prove the theorem.

Proof of Theorem 4.12: Let $f: A \to \mathcal{P}(A)$, and let $B = \{a \in A \mid a \notin f(a)\}$. Suppose toward contradiction that $B \in \operatorname{ran} f$. Then there is $a \in A$ with $f(a) = B$. But then we have $a \in B$ if and only if $a \notin f(A)$ if and only if $a \notin B$. This contradiction tells us that $B \notin \operatorname{ran} f$, and so, f is not surjective. Since $f: A \to \mathcal{P}(A)$ was arbitrary, there does not exist a surjection from A to $\mathcal{P}(A)$, and therefore, there is no bijection from A to $\mathcal{P}(A)$. So, A is not equinumerous with $\mathcal{P}(A)$. $\qquad\square$

Notes: (1) The proof given here is a **proof by contradiction**. A proof by contradiction works as follows:

1. We assume the negation of what we are trying to prove.

2. We use a logically valid argument to derive a statement which is false.

3. Since the argument was logically valid, the only possible error is our original assumption. Therefore, the negation of our original assumption must be true.

(2) In this problem we are trying to prove that A is **not** equinumerous with $\mathcal{P}(A)$. The negation of this statement is that A **is** equinumerous with $\mathcal{P}(A)$, and so that is what we assume. Since $A \sim \mathcal{P}(A)$, there is a bijection $f: A \to \mathcal{P}(A)$. So, f is a surjection, which means that every subset of $\mathcal{P}(A)$ is in the range of f. In particular, the set B described in the proof is a subset of $\mathcal{P}(A)$, and therefore it is in the range of f. We then use a logically valid argument to derive the obviously false statement "$a \in B$ if and only if $a \notin B$."

By Theorem 4.12, $\mathbb{N}$ is not equinumerous with $\mathcal{P}(\mathbb{N})$. Which of these two sets is the "bigger" one? Let's consider the function $f: \mathbb{N} \to \mathcal{P}(\mathbb{N})$ defined by $f(n) = \{n\}$. This function looks like this:

$$0 \quad 1 \quad 2 \quad 3 \quad 4 \ldots$$
$$\{0\} \ \{1\} \ \{2\} \ \{3\} \ \{4\} \ldots$$

Observe that we are matching up each natural number with a subset of natural numbers (a very simple subset consisting of just one natural number) in a way so that different natural numbers get matched with different subsets. In other words, we defined an injective function from $\mathbb{N}$ to $\mathcal{P}(\mathbb{N})$. It seems like there are lots of subsets of $\mathbb{N}$ that didn't get mapped to (for example, all infinite subsets of $\mathbb{N}$). So, it seems that $\mathbb{N}$ is a "smaller" set than $\mathcal{P}(\mathbb{N})$.

We use the notation $A \preccurlyeq B$ if there is an injective function from A to B.

$$A \preccurlyeq B \text{ if and only if } \exists f(f: A \hookrightarrow B)$$

Recall: The symbol $\exists$ is called an **existential quantifier**, and it is pronounced "There exists" or "There is." The expression $\exists f(f: A \hookrightarrow B)$ can be translated into English as "There exists an f such that f is an injective function from A to B."

We write $A \prec B$ if $A \preccurlyeq B$ and $A \not\sim B$.

So, for example, $\mathbb{N} \prec \mathcal{P}(\mathbb{N})$.

Theorem 4.13: If A is any set, then $A \prec \mathcal{P}(A)$.

Proof: The function $f: A \to P(A)$ defined by $f(a) = \{a\}$ is injective. So, $A \preccurlyeq P(A)$. By Theorem 4.12, $A \nsim P(A)$. It follows that $A \prec P(A)$. $\qquad\qquad\qquad\qquad\qquad\qquad\qquad\qquad\qquad$ $\square$

Example 4.14: If we let $A = P(\mathbb{N})$, we can apply Theorem 4.13 to this set A to see that $P(\mathbb{N}) \prec P\big(P(\mathbb{N})\big)$. Continuing in this fashion, we get a sequence of increasingly larger sets.

$$\mathbb{N} \prec P(\mathbb{N}) \prec P\big(P(\mathbb{N})\big) \prec P\left(P\big(P(\mathbb{N})\big)\right) \prec \cdots$$

If A and B are arbitrary sets, in general it can be difficult to determine if A and B are equinumerous by producing a bijection. Luckily, the next theorem provides an easier way.

Theorem 4.15 (The Cantor-Schroeder-Bernstein Theorem): If A and B are sets such that $A \preccurlyeq B$ and $B \preccurlyeq A$, then $A \sim B$.

Note: At first glance, many students think that Theorem 4.15 is obvious and that the proof must be trivial. This is not true. The theorem says that if there is an injective function from A to B and another injective function from B to A, then there is a bijective function from A to B. This is a deep result, which is far from obvious. Constructing a bijection from two arbitrary injections is not an easy thing to do. I suggest that the reader takes a few minutes to try to do it, if for no other reason than to convince themselves that the proof is difficult. I leave the proof itself as an exercise (see Problem 18 below).

Example 4.16: Let's use Theorem 4.15 to prove that the open interval of real numbers $(0, 1)$ is equinumerous to the closed interval of real numbers $[0, 1]$.

Analysis: Since $(0, 1) \subseteq [0, 1]$, there is an obvious injective function $f: (0, 1) \to [0, 1]$ (just send each element to itself).

The harder direction is finding an injective function g from $[0, 1]$ into $(0, 1)$. We will do this by drawing a line segment with endpoints $\left(0, \frac{1}{4}\right)$ and $\left(1, \frac{3}{4}\right)$. This will give us a bijection from $[0, 1]$ to $\left[\frac{1}{4}, \frac{3}{4}\right]$. We can visualize this bijection using the graph to the right. We will write an equation for this line segment in the slope-intercept form $y = mx + b$. Here m is the slope of the line and b is the y-intercept of the line. We can use the graph to see that $b = \frac{1}{4}$ and $m = \frac{\text{rise}}{\text{run}} = \frac{\frac{3}{4} - \frac{1}{4}}{1 - 0} = \frac{\frac{2}{4}}{1} = \frac{1}{2}$. So, we define $g: [0, 1] \to (0, 1)$ by $g(x) = \frac{1}{2}x + \frac{1}{4}$.

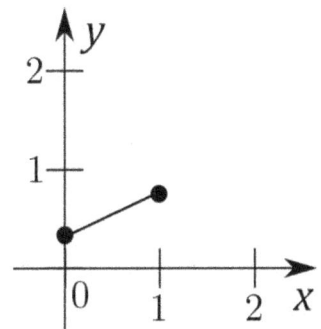

Let's write out the details of the proof.

Proof: Let $f: (0, 1) \to [0, 1]$ be defined by $f(x) = x$. Clearly, f is injective, so that $(0, 1) \preccurlyeq [0, 1]$.

Next, we define $g: [0, 1] \to \mathbb{R}$ by $g(x) = \frac{1}{2}x + \frac{1}{4}$. If $0 \le x \le 1$, then $0 \le \frac{1}{2}x \le \frac{1}{2}$, and therefore, $\frac{1}{4} \le \frac{1}{2}x + \frac{1}{4} \le \frac{3}{4}$. Since $0 < \frac{1}{4}$ and $\frac{3}{4} < 1$, we have $0 < g(x) < 1$. Therefore, $g: [0, 1] \to (0, 1)$. If $x \ne x'$, then $\frac{1}{2}x \ne \frac{1}{2}x'$, and so, $g(x) = \frac{1}{2}x + \frac{1}{4} \ne \frac{1}{2}x' + \frac{1}{4} = g(x')$. This shows that g is injective. It follows that $[0, 1] \preccurlyeq (0, 1)$.

Since $(0, 1) \preccurlyeq [0, 1]$ and $[0, 1] \preccurlyeq (0, 1)$, it follows from the Cantor-Schroeder-Bernstein Theorem that $(0, 1) \sim [0, 1]$. □

Notes: (1) If $A \subseteq B$, then the function $f: A \to B$ defined by $f(a) = a$ for all $a \in A$ is always injective. It is called the **inclusion map**.

(2) It is unfortunate that the same notation is used for points (ordered pairs of real numbers) and open intervals. Normally this isn't an issue, but in this particular example both usages of this notation appear. Take another look at the analysis above and make sure you can see when the notation (a, b) is being used for a point and when it is being used for an open interval.

(3) We could have used any closed interval $[a, b]$ with $0 < a < b < 1$ in place of $\left[\frac{1}{4}, \frac{3}{4}\right]$.

Example 4.17: Let's use Theorem 4.15 to prove that the half-open interval of real numbers $[0, 1)$ is equinumerous to $\mathcal{P}(\mathbb{N})$.

Proof: Let $f: \mathcal{P}(\mathbb{N}) \to [0, 1)$ be defined by $f(A) = 0. a_0 a_1 \ldots$, where $a_n = \begin{cases} 0 & \text{if } n \notin A. \\ 1 & \text{if } n \in A. \end{cases}$

If $A, B \in \mathcal{P}(\mathbb{N})$ with $A \neq B$, then there is $n \in \mathbb{N}$ with $n \in A \setminus B$ or $n \in B \setminus A$. **Without loss of generality**, assume that $n \in A \setminus B$. Assuming that $f(A) = 0. a_0 a_1 \ldots$ and $f(B) = 0. b_0 b_1 \ldots$, we have $a_n = 1$ and $b_n = 0$. Therefore, $f(A) \neq f(B)$. This shows that f is injective. It follows that $\mathcal{P}(\mathbb{N}) \preccurlyeq [0, 1)$.

Next, we define $g: [0, 1) \to \mathcal{P}(\mathbb{N})$. Let $a \in [0, 1)$. We define $A_0 = \left[0, \frac{1}{2}\right)$ and $A_1 = \left[\frac{1}{2}, 1\right)$. Then $\{A_0, A_1\}$ is a partition of $[0, 1)$, and so, it follows that $a \in A_0$ or $a \in A_1$, but not both. Similarly, we define $A_{00} = \left[0, \frac{1}{4}\right)$, $A_{01} = \left[\frac{1}{4}, \frac{1}{2}\right)$, $A_{10} = \left[\frac{1}{2}, \frac{3}{4}\right)$, and $A_{11} = \left[\frac{3}{4}, 1\right)$. Then $\{A_{00}, A_{01}, A_{10}, A_{11}\}$ is a partition of $[0, 1)$, and so, it follows that a is in exactly one of the four sets A_{00}, A_{01}, A_{10}, or A_{11}. Note that $\{A_{00}, A_{01}\}$ is a partition of A_0 and $\{A_{10}, A_{11}\}$ is a partition of A_1. It follows that if $a \in A_0$, then a must be in either A_{00} or A_{01}, and similarly if $a \in A_1$. Continuing in this fashion, we get a sequence of subscripts $a_0, \; a_0 a_1, \; a_0 a_1 a_2, \ldots$, where each subscript a_n is either 0 or 1. We let $g(a) = \{n \in \mathbb{N} \mid a_n = 1\}$. Now, suppose that $a \neq b$. Without loss of generality assume that $a < b$. If $a \in A_0$ and $b \in A_1$, then $0 \notin g(a)$, $0 \in g(b)$. Otherwise, there must be n such that $a, b \in A_{a_0 a_1 a_2 \cdots a_n}$, $a \in A_{a_0 a_1 a_2 \cdots a_n 0}$, and $b \in A_{a_0 a_1 a_2 \cdots a_n 1}$. It follows that $n + 1 \notin g(a)$ and $n + 1 \in g(b)$. Therefore, $g(a) \neq g(b)$. This shows that g is injective. It follows that $[0, 1) \preccurlyeq \mathcal{P}(\mathbb{N})$.

Since $\mathcal{P}(\mathbb{N}) \preccurlyeq [0, 1)$ and $[0, 1) \preccurlyeq \mathcal{P}(\mathbb{N})$, it follows from the Cantor-Schroeder-Bernstein Theorem that $\mathcal{P}(\mathbb{N}) \sim [0, 1)$. □

Notes: (1) In the second paragraph of the proof above, we used the expression "Without loss of generality." Recall from Note 1 following Example 4.4 that this expression can be used when an argument can be split up into 2 or more cases, and the proof of each of the cases is nearly identical.

For the proof above, the two cases are (i) $n \in A \setminus B$ and (ii) $n \in B \setminus A$. The argument for case (ii) is the same as the argument for case (i) except that the roles of A and B (and the roles of a_n and b_n for each n) are reversed.

(2) Given $a \in [0, 1)$, the number $. a_0 a_1 a_2 \ldots$ is called the **binary expansion** of a.

As an example, let's find the binary expansion of $\frac{1}{8}$.

Since $0 \le \frac{1}{8} < \frac{1}{2}$, we have $\frac{1}{8} \in \left[0, \frac{1}{2}\right) = A_0$. So, $a_0 = 0$. (Here we are in the "leftmost" set in the partition of $[0,1) = \left[0, \frac{1}{2}\right) \cup \left[\frac{1}{2}, 1\right)$.)

Next, since $0 \le \frac{1}{8} < \frac{1}{4}$, we have $\frac{1}{8} \in \left[0, \frac{1}{4}\right) = A_{00}$. So, $a_1 = 0$. (Here we are once again in the "leftmost" set in the partition of $\left[0, \frac{1}{2}\right) = \left[0, \frac{1}{4}\right) \cup \left[\frac{1}{4}, \frac{1}{2}\right)$. It may help to rewrite this using only denominators of 4 as follows: $\left[0, \frac{2}{4}\right) = \left[0, \frac{1}{4}\right) \cup \left[\frac{1}{4}, \frac{2}{4}\right)$.)

Now, since $\frac{1}{8} \le \frac{1}{8} < \frac{1}{4}$, we have $\frac{1}{8} \in \left[\frac{1}{8}, \frac{1}{4}\right) = A_{001}$. So, $a_2 = 1$. (This time we are in the "rightmost" set in the partition of $\left[0, \frac{1}{4}\right) = \left[0, \frac{1}{8}\right) \cup \left[\frac{1}{8}, \frac{1}{4}\right)$. Once again, it may help to rewrite this using only denominators of 8 as follows: $\left[0, \frac{2}{8}\right) = \left[0, \frac{1}{8}\right) \cup \left[\frac{1}{8}, \frac{2}{8}\right)$.)

Since $\frac{1}{8} \le \frac{1}{8} < \frac{3}{16}$, we have $\frac{1}{8} \in \left[\frac{1}{8}, \frac{3}{16}\right) = A_{0010}$. So, $a_3 = 0$. (This time we are back in the "leftmost" set in the partition of $\left[\frac{1}{8}, \frac{1}{4}\right) = \left[\frac{1}{8}, \frac{3}{16}\right) \cup \left[\frac{3}{16}, \frac{1}{4}\right)$. Again, it may help to rewrite this using only denominators of 16 as follows: $\left[\frac{2}{16}, \frac{4}{16}\right) = \left[\frac{2}{16}, \frac{3}{16}\right) \cup \left[\frac{3}{16}, \frac{4}{16}\right)$.)

From this point on, it should be easy to see that we will always be choosing the "leftmost" set in the partition. So, the remaining digits will all be 0. It follows that the binary expansion of $\frac{1}{8}$ is 0.001.

(3) Using this method of writing a binary expansion, we will never get a tail of 1's. Therefore, using this method, each real number corresponds to a unique binary expansion without a tail of 1's. See part 2 of Example 1.6 for the definition of a "tail."

(4) We can use this same method to form the **decimal expansion** of a real number as well. In this case, we partition $[0, 1)$ into $A_0 = \left[0, \frac{1}{10}\right)$, $A_1 = \left[\frac{1}{10}, \frac{2}{10}\right)$, $A_2 = \left[\frac{2}{10}, \frac{3}{10}\right)$,..., and $A_9 = \left[\frac{9}{10}, 1\right)$. Using $\frac{1}{8}$ as an example again, we have $\frac{1}{8} \in \left[\frac{1}{10}, \frac{2}{10}\right) = A_1$. So, $a_0 = 1$. We then partition $\left[\frac{1}{10}, \frac{2}{10}\right) = \left[\frac{10}{100}, \frac{20}{100}\right)$ into $A_{10} = \left[\frac{10}{100}, \frac{11}{100}\right)$, $A_{11} = \left[\frac{11}{100}, \frac{12}{100}\right)$, $A_{12} = \left[\frac{12}{100}, \frac{13}{100}\right)$,..., and $A_{19} = \left[\frac{19}{100}, \frac{20}{100}\right)$. Now, we have $\frac{1}{8} \in \left[\frac{12}{100}, \frac{13}{100}\right) = A_{12}$, and so, $a_1 = 2$. You should now check that $\frac{1}{8} \in \left[\frac{125}{1000}, \frac{126}{1000}\right) = A_{125}$, and so, $a_2 = 5$. It is then not too hard to check that the remaining digits will be 0. It follows that the decimal expansion of $\frac{1}{8}$ is 0.125.

(5) We can use this same method to form n-ary expansions for any integer $n \ge 2$. If $n = 2$, we get the binary expansion, as defined in Note 1. If $n = 10$, we get the decimal expansion, as defined in Note 3. If $n = 3$, we get the **ternary expansion**. The reader may want to check that the ternary expansion of $\frac{1}{8}$ is 0.010101010101 ...

Problem Set 4

Full solutions to these problems are available for free download here:

www.SATPrepGet800.com/ TFBZLF

LEVEL 1

1. Determine if each of the following relations are functions. For each such function, determine if it is injective. State the domain and range of each function.

    (i) $R = \{(a, b), (b, b), (c, d), (e, a)\}$

    (ii) $S = \{(a, a), (a, b), (b, a)\}$

    (iii) $T = \{(a, b) \mid a, b \in \mathbb{R} \land b < 0 \land a^2 + b^2 = 9\}$

2. Define $f: \mathbb{Z} \to \mathbb{Z}$ by $f(n) = n^2$. Let $A = \{0, 1, 2, 3, 4\}$, $B = \mathbb{N}$, and $C = \{-2n \mid n \in \mathbb{N}\}$. Evaluate each of the following:

    (i) $f[A]$

    (ii) $f^{-1}[A]$

    (iii) $f^{-1}[B]$

    (iv) $f[B \cup C]$

3. Let A, B, and C be sets. Prove the following:

    (i) $\preccurlyeq$ is transitive.

    (ii) $\prec$ is transitive.

    (iii) If $A \preccurlyeq B$ and $B \prec C$, then $A \prec C$.

    (iv) If $A \prec B$ and $B \preccurlyeq C$, then $A \prec C$.

LEVEL 2

4. Find sets A and B and a function f such that $f[A \cap B] \neq f[A] \cap f[B]$.

5. Let $f: A \to B$ and let $V \subseteq B$. Prove that $f[f^{-1}[V]] \subseteq V$.

6. Define $\mathcal{P}_k(\mathbb{N})$ for each $k \in \mathbb{N}$ by $\mathcal{P}_0(\mathbb{N}) = \mathbb{N}$ and $\mathcal{P}_{k+1}(\mathbb{N}) = \mathcal{P}(\mathcal{P}_k(\mathbb{N}))$ for $k > 0$. Find a set B such that for all $k \in \mathbb{N}$, $\mathcal{P}_k(\mathbb{N}) \prec B$.

7. Prove that if $A \sim B$ and $C \sim D$, then $A \times C \sim B \times D$.

LEVEL 3

8. For $f, g \in {}^{\mathbb{R}}\mathbb{R}$, define $f \preccurlyeq g$ if and only if for all $x \in \mathbb{R}$, $f(x) \leq g(x)$. Is $({}^{\mathbb{R}}\mathbb{R}, \preccurlyeq)$ a poset? Is it a linearly ordered set? What if we replace $\preccurlyeq$ by $\preccurlyeq^*$, where $f \preccurlyeq^* g$ if and only if there is an $x \in \mathbb{R}$ such that $f(x) \leq g(x)$?

9. Prove that the function $f: \mathbb{N} \to \mathbb{Z}$ defined by $f(n) = \begin{cases} \frac{n}{2} & \text{if } n \text{ is even} \\ -\frac{n+1}{2} & \text{if } n \text{ is odd} \end{cases}$ is a bijection.

10. Define a partition P of $\mathbb{N}$ such that $P \sim \mathbb{N}$ and for each $X \in P$, $X \sim \mathbb{N}$.

11. Prove that a countable union of countable sets is countable.

12. Let A and B be sets such that $A \sim B$. Prove that $\mathcal{P}(A) \sim \mathcal{P}(B)$.

LEVEL 4

13. Prove the following:

    (i) $\mathbb{N} \times \mathbb{N} \sim \mathbb{N}$.

    (ii) $\mathbb{Q} \sim \mathbb{N}$.

    (iii) Any two intervals of real numbers are equinumerous (including $\mathbb{R}$ itself).

    (iv) $^{\mathbb{N}}\mathbb{N} \sim \mathcal{P}(\mathbb{N})$.

14. Prove that if $A \sim B$ and $C \sim D$, then $^{A}C \sim {}^{B}D$.

LEVEL 5

15. Let X be a nonempty set of sets and let f be a function such that $\cup X \subseteq \text{dom } f$. Prove each of the following:

    (i) $f[\cup X] = \cup\{f[A] \mid A \in X\}$

    (ii) $f[\cap X] \subseteq \cap\{f[A] \mid A \in X\}$

    (iii) $f^{-1}[\cup X] = \cup\{f^{-1}[A] \mid A \in X\}$

    (iv) $f^{-1}[\cap X] = \cap\{f^{-1}[A] \mid A \in X\}$

16. Prove that for any sets A, B, and C, $^{B \times C}A \sim {}^{C}(^{B}A)$.

17. Prove the following:

    (i) $\mathcal{P}(\mathbb{N}) \sim \{f \in {}^{\mathbb{N}}\mathbb{N} \mid f \text{ is a bijection}\}$.

    (ii) $^{\mathbb{N}}\mathbb{R} \nsim {}^{\mathbb{R}}\mathbb{N}$.

CHALLENGE PROBLEM

18. Prove the Cantor-Schroeder-Bernstein Theorem.

LESSON 5
NUMBER SYSTEMS AND INDUCTION

The Natural Numbers

At this point, let's provide more formal definitions of the number systems we use most frequently, beginning with the set of natural numbers.

We define the following:

$$0 = \emptyset$$
$$1 = \{\emptyset\} = \{0\}$$
$$2 = \{\emptyset, \{\emptyset\}\} = \{0, 1\}$$
$$3 = \left\{\emptyset, \{\emptyset\}, \{\emptyset, \{\emptyset\}\}\right\} = \{0, 1, 2\}$$

In general, we let $n = \{0, 1, 2, \dots, n-1\}$ and we define the **natural numbers** to be the set

$$\mathbb{N} = \{0, 1, 2, 3, 4, \dots\}$$

If n is a natural number, we define the **successor** of n, written n^+, to be the natural number $n^+ = n \cup \{n\}$. Note that $n^+ = \{0, 1, 2, \dots, n-1, n\}$.

Example 5.1:

1. $0^+ = 0 \cup \{0\} = \{0\} = 1$.
2. $1^+ = 1 \cup \{1\} = \{0\} \cup \{1\} = \{0, 1\} = 2$.
3. $2^+ = 2 \cup \{2\} = \{0, 1\} \cup \{2\} = \{0, 1, 2\} = 3$.

If n is a natural number such that $n \neq 0$, we define the **predecessor** of n, written n^-, to be the natural number k such that $n = k^+$ (by Problem 4, part (vii) below, k is unique). Thus, $n = n^- \cup \{n^-\}$.

Note that for all $n \in \mathbb{N}$, $(n^+)^- = n$ and for all $n \in \mathbb{N}$ with $n \neq 0$, $(n^-)^+ = n$.

Example 5.2:

1. Since $1 = 0^+$, $1^- = 0$.
2. Since $2 = 1^+$, $2^- = 1$.
3. Since $3 = 2^+$, $3^- = 2$.

We define the ordering $<_\mathbb{N}$ on $\mathbb{N}$ by $n <_\mathbb{N} m$ if and only if $n \in m$. We will usually abbreviate $<_\mathbb{N}$ simply by $<$, especially if it is already clear that we are working with the natural numbers.

With this definition, $<_\mathbb{N}$ is a strict linear ordering on $\mathbb{N}$. You will be asked to prove this as part of Problem 28 below.

Example 5.3:

1. $0 \in \{0\}$. Therefore, $0 \in 0 \cup \{0\} = 1$. So, $0 < 1$.

2. $1 = \{0\} \notin \emptyset = 0$. So, $1 \nless 0$.

3. $1 = \{0\} \notin 0$ and $1 = \{0\} \notin \{0\}$. Therefore, $1 = \{0\} \notin 0 \cup \{0\} = 1$. So, $1 \nless 1$.

4. If $n \neq 0$, then $n^- \in \{n^-\}$. Therefore, $n^- \in n^- \cup \{n^-\} = n$. So, $n^- < n$.

We add the natural numbers m and n as follows:

    (i) If $n = 0$, then $m + n = m$.
    (ii) If $n = k^+$, then $m + n = (m + k)^+$.

We multiply the natural numbers m and n as follows:

    (i) If $n = 0$, then $mn = 0$.
    (ii) If $n = k^+$, then $mn = (mk) + m$.

Notes: (1) With the definition of addition given above, we see that for all natural numbers m, we have $m + 1 = (m + 0)^+ = m^+$. So, from now on, we can write $m + 1$ instead of m^+. In particular, clause (ii) in the definition of the sum of two natural numbers can be written as follows:

    (ii) If $n = k + 1$, then $m + n = (m + k) + 1$.

(2) Whenever it will not cause confusion, we will use the usual order of operations that we learned in elementary school. So, for example, we can abbreviate $(mk) + m$ as $mk + m$.

(3) Addition and multiplication of natural numbers are examples of **recursive definitions**. A full exploration of recursive definitions lies outside the scope of this book. Here we will just accept the existence of recursively defined functions without further mention.

Example 5.4:

1. $0 + 0 = 0$ by the first part of the definition of addition of natural numbers.

2. $0 + 1 = 0 + 0^+ = (0 + 0)^+ = 0^+ = 0 \cup \{0\} = \emptyset \cup \{\emptyset\} = \{\emptyset\} = 1$.

3. $0 + 2 = 0 + 1^+ = (0 + 1)^+ = 1^+ = 1 \cup \{1\} = \{0\} \cup \{1\} = \{0, 1\} = 2$.

4. $2 + 0 = 2$ by the first part of the definition of addition of natural numbers.

5. $2 + 1 = 2 + 0^+ = (2 + 0)^+ = 2^+ = 2 \cup \{2\} = \{0, 1\} \cup \{2\} = \{0, 1, 2\} = 3$.

6. $2 + 2 = 2 + 1^+ = (2 + 1)^+ = 3^+ = 3 \cup \{3\} = \{0, 1, 2\} \cup \{3\} = \{0, 1, 2, 3\} = 4$.

7. $2 \cdot 0 = 0$ by the first part of the definition of multiplication of natural numbers.

8. $2 \cdot 1 = 2 \cdot 0^+ = 2 \cdot 0 + 2 = 0 + 2 = 2$ (by part 3 above).

9. $2 \cdot 2 = 2 \cdot 1^+ = 2 \cdot 1 + 2 = 2 + 2 = 4$ (by part 6 above).

Well Ordering and the Principle of Mathematical Induction

If A is a subset of natural numbers, then we say that a is the **least** element of A if a is less than every other element of A. Symbolically, $\forall b \in A(b \neq a \rightarrow a < b)$.

Example 5.5: The least element of $\mathbb{N}$ itself is 0.

We now describe two important principles that are equivalent to each other.

The **Well Ordering Principle** (abbreviated **WOP**) is the following statement: Every nonempty subset of natural numbers has a least element.

The **Principle of Mathematical Induction** (abbreviated **POMI**) is the following statement: Let S be a set of natural numbers such that (i) $0 \in S$ and (ii) for all $k \in \mathbb{N}$, $k \in S \rightarrow k^+ \in S$. Then $S = \mathbb{N}$.

Notes: (1) The Principle of Mathematical Induction works like a chain reaction. We know that $0 \in S$ (this is condition (i)). Substituting 0 in for k in the expression "$k \in S \rightarrow k^+ \in S$" (condition (ii)) gives us $0 \in S \rightarrow 1 \in S$. So, we have that 0 is in the set S, and "if 0 is in the set S, then 1 is in the set S." So, $1 \in S$ must also be true.

(2) In terms of logic, if we let p be the statement $0 \in S$ and q the statement $1 \in S$, then we are given that $p \wedge (p \rightarrow q)$ is true. Observe that the only way that this statement can be true is if q is also true. Indeed, we must have both $p \equiv \text{T}$ and $p \rightarrow q \equiv \text{T}$. If q were false, then we would have $p \rightarrow q \equiv \text{T} \rightarrow \text{F} \equiv \text{F}$. So, we must have $q \equiv \text{T}$.

(3) Now that we showed $1 \in S$ is true (from Note 1 above), we can substitute 1 for k in the expression "$k \in S \rightarrow k^+ \in S$" (condition (ii)) to get $1 \in S \rightarrow 2 \in S$. So, we have $1 \in S \wedge (1 \in S \rightarrow 2 \in S)$ is true. So, $2 \in S$ must also be true.

(4) In general, we get the following chain reaction:

$$0 \in S \rightarrow 1 \in S \rightarrow 2 \in S \rightarrow 3 \in S \rightarrow \cdots$$

I hope that the "argument" presented in Notes 1 through 4 above convinces you that the Principle of Mathematical Induction is a reasonable one.

Now let's show that the Principle of Mathematical Induction follows from the Well Ordering Principle (In Problem 11 below, you will be asked to show that the Well Ordering Principle follows from the Principle of Mathematical Induction, thus showing that the two statements are equivalent). Proofs involving the Well Ordering Principle are generally done by contradiction.

Theorem 5.6: WOP $\rightarrow$ POMI.

Proof: Assume WOP and let S be a set of natural numbers such that $0 \in S$ (condition (i)), and such that whenever $k \in S$, $k^+ \in S$ (condition (ii)). Assume toward contradiction that $S \neq \mathbb{N}$. Let $A = \{k \in \mathbb{N} \mid k \notin S\}$ (so, A is the set of natural numbers **not** in S). Since $S \neq \mathbb{N}$, A is nonempty. So, by the Well Ordering Principle, A has a least element, let's call it a. $a \neq 0$ because $0 \in S$ and $a \notin S$. So, $a^- \in \mathbb{N}$. Letting $k = a^-$, we have $a^- \in S \rightarrow k \in S \rightarrow k^+ \in S \rightarrow (a^-)^+ \in S \rightarrow a \in S$. But $a \in A$, which means that $a \notin S$. This is a contradiction, and so, $S = \mathbb{N}$. $\square$

Recall: A proof by contradiction works as follows:

1. We assume the negation of what we are trying to prove.

2. We use a logically valid argument to derive a statement which is false.

3. Since the argument was logically valid, the only possible error is our original assumption. Therefore, the negation of our original assumption must be true.

In this problem we are trying to prove that $S = \mathbb{N}$. The negation of this statement is that $S \neq \mathbb{N}$, and so that is what we assume.

We then define a set A which contains elements of $\mathbb{N}$ that are not in S. In reality, this set is empty (because the conclusion of the theorem is $S = \mathbb{N}$). However, our (wrong!) assumption that $S \neq \mathbb{N}$ tells us that this set A actually has something in it. Saying that A has something in it is an example of a false statement that was derived from a logically valid argument. This false statement occurred not because of an error in our logic, but because we started with an incorrect assumption ($S \neq \mathbb{N}$).

The Well Ordering Principle then allows us to pick out the least element of this set A. Note that we can do this because A is a subset of $\mathbb{N}$. This wouldn't work if we knew only that A was a subset of $\mathbb{Z}$, as $\mathbb{Z}$ does **not** satisfy the Well Ordering Principle (for example, $\mathbb{Z}$ itself has no least element).

Again, although the argument that A has a least element is logically valid, A does not actually have any elements at all. We are working from the (wrong!) assumption that $S \neq \mathbb{N}$.

Once we have our hands on this least element a, we can get our contradiction. What can this least element a be? Well a was chosen to **not** be in S, so a cannot be 0 (because 0 **is** in S). Also, we know that $a^- \in S$ (because a is the **least** element not in S and $a^- < a$). But condition (ii) then forces a to be in S (because $a = (a^-)^+$).

So, we wind up with $a \in S$, contradicting the fact that a is the least element **not** in S.

The Principle of Mathematical Induction is often written in the following way:

($\star$) Let $P(n)$ be a statement and suppose that (i) $P(0)$ is true and (ii) for all $k \in \mathbb{N}$, $P(k) \to P(k^+)$. Then $P(n)$ is true for all $n \in \mathbb{N}$.

In Problem 14 below, you will be asked to show that statement ($\star$) is equivalent to POMI.

There are essentially two steps involved in a proof by mathematical induction. The first step is to prove that $P(0)$ is true (this is called the **base case**), and the second step is to assume that $P(k)$ is true, and use this to show that $P(k^+)$ is true (this is called the **inductive step**). While doing the inductive step, the statement "$P(k)$ is true" is often referred to as the **inductive hypothesis**.

Theorem 5.7: The sum of two natural numbers is a natural number.

Note: Consider the sum $m + n$ of the natural numbers m and n. We will prove this theorem by induction on n (and **not** by induction on m). We will start by letting m be an arbitrary natural number.

The base case will be to show that $m + 0$ is a natural number.

The inductive step will be to assume that $m + k$ is a natural number and then to prove that $m + k^+$ (or equivalently, $m + (k + 1)$) is a natural number.

Let's begin:

Proof: Assume that m is a natural number.

Base Case $(k = 0)$: $m + 0 = m$ (by clause (i) in the definition of the sum of two natural numbers), which we assumed was a natural number. Thus, we have shown that $m + 0$ is a natural number.

Inductive Step: Let k be a natural number and assume that $m + k$ is also a natural number. Then $m + k^+ = (m + k)^+$ (or equivalently, $m + (k + 1) = (m + k) + 1$), which is also a natural number.

Here we used the fact that the successor of a natural number is a natural number.

By the Principle of Mathematical Induction, $m + n$ is a natural number for all natural numbers n.

Since m was an arbitrary natural number, we have shown that the sum of any two natural numbers is a natural number. $\square$

Note: Since the sum of two natural numbers is always a natural number (by Theorem 5.7), we say that $\mathbb{N}$ is **closed** under addition. We may also say that $+$ is a **binary operation** on $\mathbb{N}$.

Theorem 5.8: For all natural numbers n, $0 + n = n$.

Proof: Base Case $(k = 0)$: $0 + 0 = 0$ by the definition of addition of natural numbers.

Inductive Step: Let $k \in \mathbb{N}$ and assume that $0 + k = k$. Then $0 + (k + 1) = (0 + k) + 1 = k + 1$, as desired.

For the first equality, we used the definition of addition of natural numbers. For the second equality, we used the inductive hypothesis.

By the Principle of Mathematical Induction, for all natural numbers n, $0 + n = n$. $\square$

Notes: (1) It's also true that for all natural numbers n, $n + 0 = n$. This follows right from the definition of addition of natural numbers.

(2) Since for all natural numbers n, we have $0 + n = n + 0 = 0$, we say that 0 is an **identity** with respect to addition, or that 0 is an **additive identity**.

Theorem 5.9: For all natural numbers m, n, and t, $(m + n) + t = m + (n + t)$.

Proof: Let m and n be natural numbers.

Base Case $(k = 0)$: $(m + n) + 0 = m + n = m + (n + 0)$ by the definition of addition in $\mathbb{N}$.

Inductive Step: Let $k \in \mathbb{N}$ and assume that $(m + n) + k = m + (n + k)$. Then we have

$$(m + n) + (k + 1) = \big((m + n) + k\big) + 1 = \big(m + (n + k)\big) + 1$$
$$= m + \big((n + k) + 1\big) = m + \big(n + (k + 1)\big).$$

For the first, third, and fourth equalities, we used the definition of addition of natural numbers. For the second equality, we used the inductive hypothesis.

By the Principle of Mathematical Induction, for all natural numbers t, $(m + n) + t = m + (n + t)$.

Since m and n were arbitrary natural numbers, we have shown that for all natural numbers m, n, and t, $(m + n) + t = m + (n + t)$. □

Notes: (i) Since for all natural numbers m, n, and t, we have $(m + n) + t = m + (n + t)$, we say that addition is **associative** in $\mathbb{N}$.

(ii) A **monoid** is a pair $(S, \star)$, where S is a set, $\star$ is an associative binary operation on S, and there is an identity $e \in S$ with respect to the operation $\ast$. Theorems 5.7, 5.8, and 5.9 prove that $(\mathbb{N}, +)$ is a monoid.

(iii) Addition is also **commutative** in $\mathbb{N}$. That is, for all natural numbers m and n, $m + n = n + m$. You will be asked to prove this in part (i) of Problem 4 below. We say that $(\mathbb{N}, +)$ is a **commutative monoid**.

Just like the sum operation, the product operation on the set of natural numbers has many nice algebraic properties such as

1. **Closure:** For all natural numbers m and n, mn is a natural number.
2. **Identity:** For all natural numbers n, $1 \cdot n = n \cdot 1 = n$.
3. **Associativity:** For all natural numbers m, n, and t, $(mn)t = m(nt)$.
4. **Commutativity:** For all natural numbers m and n, $mn = nm$.
5. **Distributivity:** For all natural numbers m, n, and t, $m(n + t) = mn + mt$.

The proofs that these properties hold are very similar to the proofs already given for addition. You will be asked to provide detailed proofs in Problem 4 below.

If $n \in \mathbb{N}$, we define n^2 to be the product of n with itself. So, $n^2 = n \cdot n$.

We define the **difference** between two natural numbers as follows: we say that $m - n = t$ if and only if $m = t + n$. For example, $7 - 2 = 5$ because $7 = 5 + 2$.

A natural number n is called **even** if there is another natural number b such that $n = 2b$.

Example 5.10:

1. 6 is even because $6 = 2 \cdot 3$.
2. 14 is even because $14 = 2 \cdot 7$.
3. We can write $1 = 2 \cdot \frac{1}{2}$, but this does **not** show that 1 is even (and as we all know, it is not). In the definition of even, it is very important that b is a natural number. The problem here is that $\frac{1}{2}$ is not a natural number, and so, it cannot be used as a value for b in the definition of even.

Let's use the Principle of Mathematical Induction to prove a theorem about natural numbers using these definitions.

Theorem 5.11: For all natural numbers n, $n^2 - n$ is an even natural number.

The proof of this result will require two very simple technical theorems about differences between natural numbers. A **lemma** is a theorem whose primary purpose is to prove a more important theorem. Let's prove these two preliminary lemmas first before we prove the main theorem.

Lemma 5.12: For all natural numbers m, n, and t, if $n - t \in \mathbb{N}$, then $m(n - t) = mn - mt$.

Proof: Suppose that $n - t \in \mathbb{N}$, say $n - t = a$. Then $n = a + t$. So, $mn = m(a + t) = ma + mt$ (using the distributivity of multiplication over addition in $\mathbb{N}$). Therefore, $mn - mt = ma = m(n - t)$. $\quad\square$

Lemma 5.13: For all natural numbers m, n, and t, $(m + n) - n = m$.

Proof: The equation $m + n = m + n$ is equivalent to $(m + n) - n = m$. $\quad\square$

Proof of Theorem 5.11: We will prove this theorem by induction on n.

Base Case $(k = 0)$: $0^2 - 0 = 2 \cdot 0$ because $0^2 = 2 \cdot 0 + 0$. So, $0^2 - 0$ is even.

Inductive Step: Let $k \in \mathbb{N}$ and assume that $k^2 - k$ is an even natural number. Then $k^2 - k = 2b$ for some natural number b. Now,

$$(k + 1)^2 - (k + 1) = (k + 1)(k + 1) - (k + 1) \cdot 1 = (k + 1)[(k + 1) - 1] = (k + 1) \cdot k$$
$$= k(k + 1) = k \cdot k + k \cdot 1 = k^2 + k = (k^2 - k) + 2k = 2b + 2k = 2(b + k).$$

Since $\mathbb{N}$ is closed under addition, $b + k \in \mathbb{N}$. Therefore, $(k + 1)^2 - (k + 1)$ is an even natural number. By the Principle of Mathematical Induction, $n^2 - n$ is an even natural number for all $n \in \mathbb{N}$. $\quad\square$

Notes: (1) For the first equality, we used the definition of $(k + 1)^2$ and the multiplicative identity property in $\mathbb{N}$.

(2) For the second equality, we used Lemma 5.12.

(3) For the third equality, we used Lemma 5.13.

(4) For the fourth equality, we used the commutativity of multiplication in $\mathbb{N}$.

(5) For the fifth and ninth equalities, we used the distributivity of multiplication over addition in $\mathbb{N}$.

(6) For the sixth equality, we used the definition of k^2 and the multiplicative identity property in $\mathbb{N}$.

(7) For the seventh equality, we used what I call the "**Standard Advanced Calculus Trick**." I sometimes abbreviate this as **SACT**. The trick is simple. If you need something to appear, just put it in. Then correct it by performing the opposite of what you just did.

In this case, in order to use the inductive hypothesis, we need $k^2 - k$ to appear, but unfortunately, we have $k^2 + k$ instead. Using SACT, I do the following:

- I simply put in what I need (and exactly where I need it): $k^2 - \boldsymbol{k} + k$

- Now, I undo the damage by performing the reverse operation: $k^2 - k + \boldsymbol{k} + k$

- Finally, I leave the part I need as is, and simplify the rest: $(k^2 - k) + 2k$

(8) For the eighth equality, we simply replaced $k^2 - k$ by $2b$. We established that these two quantities were equal in the second sentence of the inductive step.

Sometimes a statement involving the natural numbers may be false for 0, but true from some natural number on. In this case, we can still use induction. We just need to adjust the base case.

Theorem 5.14: $n^2 > 2n + 1$ for all natural numbers $n \geq 3$.

Proof: Base Case ($k = 3$): $3^2 = 9$ and $2 \cdot 3 + 1 = 6 + 1 = 7$. So, $3^2 > 2 \cdot 3 + 1$.

Inductive Step: Let $k \in \mathbb{N}$ with $k \geq 3$ and assume that $k^2 > 2k + 1$. Then we have

$$(k + 1)^2 = (k + 1)(k + 1) = (k + 1)k + (k + 1)(1) = k^2 + k + k + 1 > (2k + 1) + k + k + 1$$
$$= 2k + 2 + k + k = 2(k + 1) + k + k \geq 2(k + 1) + 1 \text{ (because } k + k \geq 3 + 3 = 6 \geq 1\text{)}.$$

By the Principle of Mathematical Induction, $n^2 > 2n + 1$ for all $n \in \mathbb{N}$ with $n \geq 3$. $\square$

Notes: (1) If we have a sequence of equations and inequalities of the form $=$, $\geq$, and $>$ (with at least one inequality symbol appearing), beginning with a and ending with b, then the final result is $a > b$ if $>$ appears at least once and $a \geq b$ otherwise.

For example, if $a = j = h = m > n = p = q \geq b$, then $a > b$. The sequence that appears in the solution above has this form.

(2) By definition, $x^2 = x \cdot x$. We used this in the first equality in the inductive step to write $(k + 1)^2$ as $(k + 1)(k + 1)$.

(3) For the second equality in the inductive step, we used distributivity to write $(k + 1)(k + 1)$ as $(k + 1)k + (k + 1)(1)$.

(4) For the third equality in the inductive step, we used commutativity, distributivity, and the multiplicative identity property to write $(k + 1)k$ as $k(k + 1) = k \cdot k + k \cdot 1 = k^2 + k$. We also used the multiplicative identity property to write $(k + 1)(1) = k + 1$.

(5) Associativity of addition is being used when we write the expression $k^2 + k + k + 1$. Notice the lack of parentheses. Technically speaking, we should have written $(k^2 + k) + (k + 1)$ and then taken another step to rewrite this as $k^2 + (k + (k + 1))$. However, since we have associativity, we can simply drop all those parentheses.

(6) The inequality "$k^2 + k + k + 1 > (2k + 1) + k + k + 1$" was attained by using the inductive hypothesis "$k^2 > 2k + 1$" together with part (x) from Problem 4 below.

71

(7) The dedicated reader should verify that the remaining equalities and inequalities in the proof are valid by determining which properties were used at each step.

Theorem 5.15: For every natural number n, there is a natural number j such that $n = 2j$ or $n = 2j + 1$.

Proof: Base Case $(k = 0)$: $0 = 2 \cdot 0$ by definition of mutiplication in $\mathbb{N}$.

Inductive Step: Suppose that $k \in \mathbb{N}$ and there is $j \in \mathbb{N}$ such that $k = 2j$ or $k = 2j + 1$. If $k = 2j$, then $k + 1 = 2j + 1$. If $k = 2j + 1$, then $k + 1 = (2j + 1) + 1 = 2j + (1 + 1) = 2j + 2 = 2(j + 1)$. Here we used associativity of addition in $\mathbb{N}$ and distributivity of multiplication over addition in $\mathbb{N}$. Since $\mathbb{N}$ is closed under addition, $j + 1 \in \mathbb{N}$.

By the Principle of Mathematical Induction, for every natural number n, there is a natural number j such that $n = 2j$ or $n = 2j + 1$. $\qquad\qquad\qquad\qquad\qquad\qquad\qquad\qquad\qquad\qquad\qquad$ □

The Integers

To motivate the definition of the integers, note that we can think of every integer as a difference of two natural numbers. For example, the integer -3 can be thought of as $1 - 4$. However, -3 can also be thought of as $2 - 5$. So, we must insist that $1 - 4 = 2 - 5$, or equivalently, $1 + 5 = 2 + 4$.

We define a relation R on $\mathbb{N} \times \mathbb{N}$ by $R = \{((a, b), (c, d)) \in (\mathbb{N} \times \mathbb{N})^2 \mid a + d = b + c\}$. In part 5 of Example 3.17, we showed that this relation is an equivalence relation. We can now define the set of integers to be the set of equivalence classes for this equivalence relation. That is, we define the set of integers to be $\mathbb{Z} = \{[(a, b)] \mid (a, b) \in \mathbb{N} \times \mathbb{N}\}$.

We identify the integer $[(n, 0)]$ with the natural number n. In this way, we have $\mathbb{N} \subseteq \mathbb{Z}$.

We define the ordering $<_{\mathbb{Z}}$ on $\mathbb{Z}$ by $[(a, b)] <_{\mathbb{Z}} [(c, d)]$ if and only if $a + d <_{\mathbb{N}} b + c$, where $<_{\mathbb{N}}$ is the usual ordering on $\mathbb{N}$ ($n <_{\mathbb{N}} m$ if and only if $n \in m$).

In Problem 12 below, you will be asked to show that $<_{\mathbb{Z}}$ is a well-defined strict linear ordering on $\mathbb{Z}$.

We add and multiply two integers using the following rules:

$$[(a, b)] + [(c, d)] = [(a + c, b + d)]$$
$$[(a, b)] \cdot [(c, d)] = [(ac + bd, ad + bc)]$$

In Problems 15 and 26 below, you will be asked to show that these two operations are well-defined.

Notes: (1) We will usually abbreviate $<_{\mathbb{Z}}$ simply by $<$, especially if it is already clear that we are working with the integers.

(2) If $a, b \in \mathbb{N}$ with $a \geq b$, then $[(a, b)] = [(a - b, 0)]$. If $a < b$, then $[(a, b)] = [(0, b - a)]$. In this way, we see that every integer can be written in the form $[(n, 0)]$ or $[(0, n)]$ for some $n \in \mathbb{N}$. We abbreviate $[(n, 0)]$ by n and we abbreviate $[(0, n)]$ by $-n$. For example, $[(2, 7)] = [(0, 5)] = -5$.

Example 5.16:

1. $[(k, k)] = [(0, 0)]$ for all $k \in \mathbb{N}$ because $k + 0 = k + 0$.

2. $[(5, 0)] = [(6, 1)]$ because $5 + 1 = 0 + 6$. Similarly, we have $[(5, 0)] = [(7, 2)] = [(8, 3)]$, and in general $[(5, 0)] = [(5 + k, k)]$ for any natural number k. $[(5, 0)]$ is the most "natural" way to express the natural number 5 as an integer. More generally, the natural number n can be expressed as an integer as $[(n, 0)]$.

3. We usually abbreviate the integer $[(0, k)]$ as $-k$. For example, -3 is an abbreviation for $[(0, 3)]$. We can also write -3 as $[(1, 4)]$ because $1 + 3 = 4 + 0$.

4. $[(0, 0)] < [(4, 0)]$ because $0 + 0 < 0 + 4$. More generally, for any natural number $k \neq 0$, we have $[(0, 0)] < [(k, 0)]$ because $0 + 0 < 0 + k$. This shows that for any natural number $k \neq 0$, the natural number k satisfies $0 < k$.

5. $[(0, 4)] < [(0, 0)]$ because $0 + 0 < 4 + 0$. More generally, for any natural number $k \neq 0$, we have $[(0, k)] < [(0, 0)]$ because $0 + 0 < k + 0$. This shows that for any natural number $k \neq 0$, the integer $-k$ satisfies $-k < 0$.

6. $7 + (-2) = [(7, 0)] + [(0, 2)] = [(7, 2)] = [(5, 0)] = 5$.

7. $-3 \cdot 5 = [(0, 3)] \cdot [(5, 0)] = [(0 \cdot 5 + 3 \cdot 0, 0 \cdot 0 + 3 \cdot 5)] = [(0, 15)] = -15$.

The Rational Numbers

In part 6 of Example 3.17, we showed that $R = \{((a, b), (c, d)) \in (\mathbb{Z} \times \mathbb{Z}^*)^2 \mid ad = bc\}$ is an equivalence relation on $\mathbb{Z} \times \mathbb{Z}^*$. For each $a \in \mathbb{Z}$ and $b \in \mathbb{Z}^*$, we define the **rational number** $\frac{a}{b}$ to be the equivalence class of (a, b). So, $\frac{a}{b} = [(a, b)]$, and we have $\frac{a}{b} = \frac{c}{d}$ if and only if $(a, b)R(c, d)$ if and only if $ad = bc$. The set of rational numbers is $\mathbb{Q} = \{\frac{a}{b} \mid a \in \mathbb{Z} \wedge b \in \mathbb{Z}^*\}$. In words, $\mathbb{Q}$ is "the set of quotients a over b such that a and b are integers and b is not zero."

We identify the rational number $\frac{a}{1}$ with the integer a. In this way, we have $\mathbb{Z} \subseteq \mathbb{Q}$.

We define $<_\mathbb{Q}$ on $\mathbb{Q}$ by $\frac{a}{b} <_\mathbb{Q} \frac{c}{d}$ if and only if $ad <_\mathbb{Z} bc$, where $<_\mathbb{Z}$ is the usual ordering on $\mathbb{Z}$.

In Problem 28 below, you will be asked to show that $<_\mathbb{Q}$ is a well-defined strict linear ordering on $\mathbb{Q}$.

We add and multiply two rational numbers using the following rules:

$$\frac{a}{b} + \frac{c}{d} = \frac{a \cdot d + b \cdot c}{b \cdot d} \qquad \frac{a}{b} \cdot \frac{c}{d} = \frac{a \cdot c}{b \cdot d}$$

In Problem 16 below, you will be asked to show that these two operations are well-defined.

Example 5.17:

1. $\frac{2}{3} < \frac{5}{4}$ because $2 \cdot 4 < 3 \cdot 5$. Also, $\frac{2}{3} + \frac{5}{4} = \frac{2 \cdot 4 + 3 \cdot 5}{3 \cdot 4} = \frac{23}{12}$ and $\frac{2}{3} \cdot \frac{5}{4} = \frac{2 \cdot 5}{3 \cdot 4} = \frac{10}{12} = \frac{5}{6}$ (because $10 \cdot 6 = 12 \cdot 5$).

2. $\frac{-3}{4} < \frac{-5}{7}$ because $-3 \cdot 7 < 4(-5)$, $\frac{-3}{4} + \frac{-5}{7} = \frac{-3 \cdot 7 + 4(-5)}{4 \cdot 7} = \frac{-41}{28}$, and $\frac{-3}{4} \cdot \frac{-5}{7} = \frac{-3(-5)}{4 \cdot 7} = \frac{15}{28}$.

The Real Numbers

There are several equivalent ways to define the set of real numbers. We will define this set here as equivalence classes of **Cauchy sequences** (see Problem 24 from Problem Set 6 for another approach). Informally, a Cauchy sequence is a rational-valued sequence whose values get "closer and closer to each other" as we go further out into the sequence.

We can recognize a Cauchy sequence as one that seems to be "converging" to a fixed value.

For example, it looks like the sequence $(x_n) = \left(\frac{1}{n+1}\right) = \left(1, \frac{1}{2}, \frac{1}{3}, \frac{1}{4}, \frac{1}{5}, \ldots\right)$ is a Cauchy sequence, as it seems to be converging to 0. The sequence $(x_n) = (n) = (0, 1, 2, 3, 4, 5, \ldots)$ is not a Cauchy sequence, as it does not seem to be converging to a fixed number. See Example 5.18 below for more details.

Let's now define Cauchy sequence more formally. We first make a few preliminary definitions.

If $n = [(k, m)]$ is an integer, we define $-n$ to be the integer $[(m, k)]$. With this definition, we have $-(-n) = n$. Indeed, if $n = [(k, m)]$, then $-(-n) = -(-[(k, m)]) = -[(m, k)] = [(k, m)] = n$.

If $x = \frac{a}{b}$ is a rational number, then we define $-x$ to be $\frac{-a}{b}$.

The **absolute value** of the rational number x is then defined by $|x| = \begin{cases} x & \text{if } x \geq 0. \\ -x & \text{if } x < 0. \end{cases}$

For example, we have $\left|\frac{2}{3}\right| = \frac{2}{3}$, $|0| = 0$, and $\left|\frac{-5}{7}\right| = -\left(\frac{-5}{7}\right) = \frac{-(-5)}{7} = \frac{5}{7}$.

The **distance** between rational numbers x and y is $|x - y|$. For example, the distance between 3 and 5 is $|3 - 5| = |-2| = 2$ and the distance between $\frac{1}{5}$ and $\frac{1}{7}$ is $\left|\frac{1}{5} - \frac{1}{7}\right| = \left|\frac{7-5}{5 \cdot 7}\right| = \left|\frac{2}{35}\right| = \frac{2}{35}$.

Notice that the distance between 5 and 3 is the same as the distance between 3 and 5. Indeed, we also have $|5 - 3| = |2| = 2$. In general, for rational numbers x and y, we have $|x - y| = |y - x|$.

Let $f = (x_n)$ be a rational-valued sequence. We say that f is a **Cauchy sequence** if

$$\text{for every } k \in \mathbb{N}^+, \text{ there is } K \in \mathbb{N} \text{ such that } m \geq n > K \text{ implies } |x_m - x_n| < \frac{1}{k}.$$

The idea is that we can make the distance between any two terms of the sequence as small as we choose by deleting a finite portion of the beginning of the sequence. If we wish to make the distance between any two terms less than $\frac{1}{k}$, we delete the first $K + 1$ terms of the sequence.

Example 5.18:

1. The sequence $(x_n) = \left(\frac{1}{n+1}\right)$ is a Cauchy sequence. To see this, let $k \in \mathbb{N}$, let $K = k$, and let $m \geq n > K$. Then

    $$|x_m - x_n| = \left|\frac{1}{m+1} - \frac{1}{n+1}\right| = \left|\frac{n-m}{(m+1)(n+1)}\right| \leq \left|\frac{m}{(m+1)(n+1)}\right| = \left|\frac{m}{mn+m+n+1}\right| \leq \left|\frac{m}{mn}\right| = \frac{1}{n} < \frac{1}{K} = \frac{1}{k}.$$

2. The sequence $(x_n) = ((-1)^n n)$ is **not** a Cauchy sequence. To see this, let $k = 1$ and let $K \in \mathbb{N}$. Then $K + 2 \geq K + 1 > K$, and

$$|x_{K+2} - x_{K+1}| = |(-1)^{K+2}(K+2) - (-1)^{K+1}(K+1)| = |(K+2) + (K+1)| = 2K + 3 \geq 3$$

(See the Note below). However, $\frac{1}{k} = \frac{1}{1} = 1$ and $3 \not< 1$.

3. For each $q \in \mathbb{Q}$, the **constant sequence** $(x_n) = (q)$ is a Cauchy sequence. To see this, let $k \in \mathbb{N}$, let $K = 0$, and let $m \geq n > 0$. Then

$$|x_m - x_n| = |q - q| = |0| = 0 < \frac{1}{k}.$$

Note: If K is even, then $|(-1)^{K+2}(K+2) - (-1)^{K+1}(K+1)| = (K+2) + (K+1) = 2K+3$, whereas, if K is odd, then $(-1)^{K+2}(K+2) - (-1)^{K+1}(K+1) = -(K+2) - (K+1) = -(2K+3)$.

Next, we would like to identify Cauchy sequences that seem to be converging to the same value. For example, we will identify the Cauchy sequences $(x_n) = \left(\frac{1}{n+1}\right)$ and $(y_n) = (0)$.

An equivalent way of saying that (x_n) and (y_n) converge to the same value is to say that $(x_n - y_n)$ converges to 0.

Let $A = \{(x_n) \mid (x_n) \text{ is a Cauchy sequence of rational numbers}\}$. We define a relation R on A as follows:

$(x_n)R(y_n)$ if and only if for every $k \in \mathbb{N}^+$, there is $K \in \mathbb{N}$ such that $n > K$ implies $|x_n - y_n| < \frac{1}{k}$.

In Problem 17 below, you will be asked to show that R is an equivalence relation on A.

Note: To show that R is transitive, you will need to use the **Triangle Inequality**. The Triangle Inequality says that if $a, b \in \mathbb{Q}$, then $|a + b| \leq |a| + |b|$.

One way to prove the Triangle Inequality is to analyze several cases separately. As an example of one such case, suppose that a and b are both nonnegative. Then $|a| = a$, $|b| = b$, and $|a + b| = a + b$. So, we get $|a + b| = a + b = |a| + |b|$ (in this case we get equality). I leave it to the reader to describe the other cases and to prove that the Triangle Inequality is true for each of these cases. You will be asked to prove the Triangle Inequality in a more general setting in Problem 6 in Problem Set 7.

We can now define the set of real numbers as follows:

$$\mathbb{R} = \{[(x_n)] \mid (x_n) \text{ is a Cauchy sequence of rational numbers}\}.$$

We identify the real number $[(q)]$ with the rational number q. In this way, we have $\mathbb{Q} \subseteq \mathbb{R}$.

We define the ordering $\leq_\mathbb{R}$ on $\mathbb{R}$ by

$[(x_n)] \leq_\mathbb{R} [(y_n)]$ if and only if there is $K \in \mathbb{N}$ such that $n > K$ implies $x_n \leq y_n$.

We can then define $<_\mathbb{R}$ on $\mathbb{R}$ by $[(x_n)] <_\mathbb{R} [(y_n)]$ if and only if $[(x_n)] \leq_\mathbb{R} [(y_n)]$ and $[(x_n)] \neq [(y_n)]$.

In Problem 28 below, you will be asked to show that $<_\mathbb{R}$ is a well-defined strict linear ordering on $\mathbb{R}$.

We add and multiply two real numbers using the following rules:

$$[(x_n)] + [(y_n)] = [(x_n + y_n)]$$
$$[(x_n)] \cdot [(y_n)] = [(x_n \cdot y_n)]$$

In Problems 25 and 27 below, you will be asked to show that $+$ and $\cdot$ are well-defined operations on the real numbers and that the sum and product of two real numbers are real numbers.

Note: We will generally drop the subscript from $<_{\mathbb{R}}$ and simply write $<$.

The Complex Numbers

The set of complex numbers is $\mathbb{C} = \{a + bi \mid a, b \in \mathbb{R}\}$.

In addition to visualizing complex numbers as points in the Complex Plane, as we did in part 3 of Example 1.6, we can also visualize the complex number $a + bi$ as a directed line segment (or **vector**) starting at the origin and ending at the point (a, b). Three examples are shown to the right.

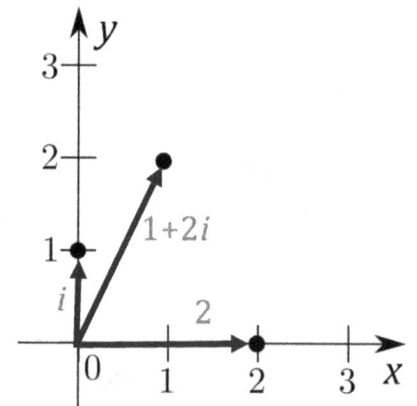

If $z = a + bi$ is a complex number, we call a the **real part** of z and b the **imaginary part** of z, and we write $a = \operatorname{Re} z$ and $b = \operatorname{Im} z$.

Two complex numbers are **equal** if and only if they have the same real part and the same imaginary part. In other words,

$$a + bi = c + di \text{ if and only if } a = c \text{ and } b = d.$$

We add two complex numbers by simply adding their real parts and adding their imaginary parts. So,

$$(a + bi) + (c + di) = (a + c) + (b + d)i.$$

As a point, this sum is $(a + c, b + d)$. We can visualize this sum as the vector starting at the origin that is the diagonal of the parallelogram formed from the vectors $a + bi$ and $c + di$. Here is an example showing that $(1 + 2i) + (-3 + i) = -2 + 3i$.

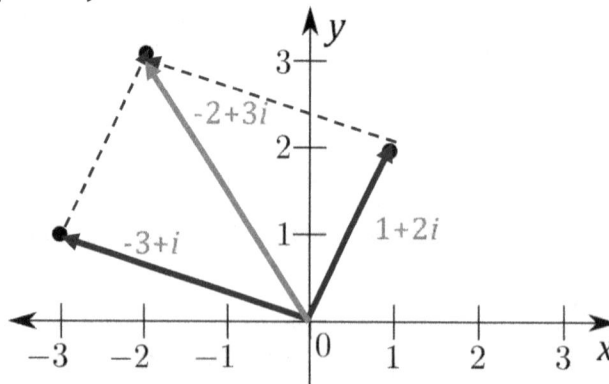

The definition for multiplying two complex numbers is a bit more complicated:

$$(a + bi)(c + di) = (ac - bd) + (ad + bc)i.$$

Notes: (1) If $b = 0$, then we call $a + bi = a + 0i = a$ a **real number**. Note that when we add or multiply two real numbers, we always get another real number.

$$(a + 0i) + (b + 0i) = (a + b) + (0 + 0)i = (a + b) + 0i = a + b.$$

$$(a + 0i)(b + 0i) = (ab - 0 \cdot 0) + (a \cdot 0 + 0b)i = (ab - 0) + (0 + 0)i = ab + 0i = ab.$$

In this way, we have $\mathbb{R} \subseteq \mathbb{C}$.

(2) If $a = 0$, then we call $a + bi = 0 + bi = bi$ a **pure imaginary number**.

(3) $i^2 = -1$. To see this, note that $i^2 = i \cdot i = (0 + 1i)(0 + 1i)$, and we have

$$(0 + 1i)(0 + 1i) = (0 \cdot 0 - 1 \cdot 1) + (0 \cdot 1 + 1 \cdot 0)i = (0 - 1) + (0 + 0)i = -1 + 0i = -1.$$

(4) The definition of the product of two complex numbers is motivated by how we expect multiplication should behave, together with replacing i^2 by -1. If we were to naïvely multiply the two complex numbers, we would have

$$(a + bi)(c + di) = (a + bi)c + (a + bi)(di) = ac + bci + adi + bdi^2$$
$$= ac + bci + adi + bd(-1) = ac + (bc + ad)i - bd = (ac - bd) + (ad + bc)i.$$

Those familiar with the mnemonic FOIL may notice that "FOILing" will always work to produce the product of two complex numbers, provided we replace i^2 by -1 and simplify.

Example 5.19: Let $z = 2 - 3i$ and $w = -1 + 5i$. Then

$$z + w = (2 - 3i) + (-1 + 5i) = (2 + (-1)) + (-3 + 5)i = \mathbf{1 + 2i}.$$

$$zw = (2 - 3i)(-1 + 5i) = \big(2(-1) - (-3)(5)\big) + \big(2 \cdot 5 + (-3)(-1)\big)i$$
$$= (-2 + 15) + (10 + 3)i = \mathbf{13 + 13i}.$$

Exponential Form of a Complex Number

* The material in this section will not be used until Lesson 16. The reader is welcome to skip this section and come back to it after completing Lesson 15. If you skip this section, then you will also want to skip problems 7, 8, 9, 19, 20, 21, and 22 in the Problem Set below.

A **circle** in the Complex Plane is the set of all points that are at a fixed distance (called the **radius** of the circle) from a fixed point (called the **center** of the circle).

The **circumference** of a circle is the distance around the circle.

If C and C' are the circumferences of two circles with radii r and r', respectively, then it turns out that $\frac{C}{2r} = \frac{C'}{2r'}$. In other words, the value of the ratio $\frac{\text{Circumference}}{2(\text{radius})}$ is independent of the circle that we use to form this ratio. We leave the proof of this fact for the interested reader to investigate themselves. We call the common value of this ratio π (pronounced "pi"). So, we have $\frac{C}{2r} = \pi$, or equivalently, $C = 2\pi r$.

Example 5.20: The **unit circle** is the circle with radius 1 and center $(0,0)$. The equation of this circle is $|z| = 1$. If we write z in the standard form $z = x + yi$, we see that $|z| = \sqrt{x^2 + y^2}$, and so, the equation of the unit circle can also be written $x^2 + y^2 = 1$. To the right is a picture of the unit circle in the Complex Plane.

The circumference of the unit circle is $2\pi \cdot 1 = \mathbf{2\pi}$.

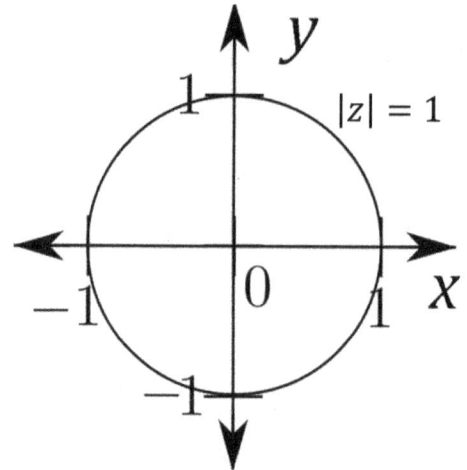

An **angle in standard position** consists of two **rays**, both of which have their initial point at the origin, and one of which is the positive x-axis. We call the positive x-axis the **initial ray** and we call the second ray the **terminal ray**. The **radian measure** of the angle is the part of the circumference of the unit circle beginning at the point $(1,0)$ on the positive x-axis and *eventually* ending at the point on the unit circle intercepted by the second ray. If the motion is in the counterclockwise direction, the radian measure is positive and if the motion is in the clockwise direction, the radian measure is negative.

Example 5.21: Let's draw a few angles where the terminal ray lies along the line $y = x$.

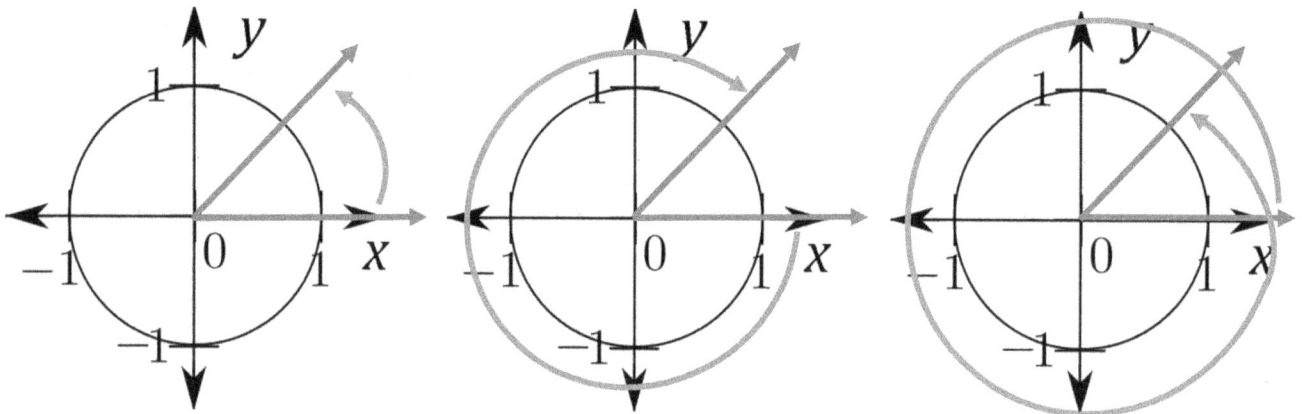

Observe that in the leftmost picture, the arc intercepted by the angle has a length that is one-eighth of the circumference of the circle. Since the circumference of the unit circle is 2π and the motion is in the counterclockwise direction, the angle has a radian measure of $\frac{2\pi}{8} = \frac{\pi}{4}$.

Similarly, in the center picture, the arc intercepted by the angle has a length that is seven-eighths of the circumference of the circle. This time the motion is in the clockwise direction, and so, the radian measure of the angle is $-\frac{7}{8} \cdot 2\pi = -\frac{7\pi}{4}$.

In the rightmost picture, the angle consists of a complete rotation, tracing out the entire circumference of the circle, followed by tracing out an additional length that is one-eighth the circumference of the circle. Since the motion is in the counterclockwise direction, the radian measure of the angle is $2\pi + \frac{2\pi}{8} = \frac{8\pi}{4} + \frac{\pi}{4} = \frac{9\pi}{4}$.

Let's find the point of intersection of the unit circle with the terminal ray of the angle $\frac{\pi}{4}$ that lies along the line with equation $y = x$ (as shown in the leftmost figure from Example 5.21 above). If we call this point (a, b), then we have $b = a$ (because (a, b) is on the line $y = x$) and $a^2 + b^2 = 1$ (because (a, b) is on the unit circle). Replacing b by a in the second equation gives us $a^2 + a^2 = 1$, or equivalently, $2a^2 = 1$. So, $a^2 = \frac{1}{2}$. The two solutions to this equation are $a = \pm\sqrt{\frac{1}{2}} = \pm\frac{\sqrt{1}}{\sqrt{2}} = \pm\frac{1}{\sqrt{2}}$. From the picture, it should be clear that we are looking for the positive solution, so that $a = \frac{1}{\sqrt{2}}$. Since $b = a$, we also have $b = \frac{1}{\sqrt{2}}$. Therefore, the point of intersection is $\left(\frac{1}{\sqrt{2}}, \frac{1}{\sqrt{2}}\right)$.

Notes: (1) The number $\frac{1}{\sqrt{2}}$ can also be written in the form $\frac{\sqrt{2}}{2}$. To see that these two numbers are equal, observe that we have

$$\frac{1}{\sqrt{2}} = \frac{1}{\sqrt{2}} \cdot 1 = \frac{1}{\sqrt{2}} \cdot \frac{\sqrt{2}}{\sqrt{2}} = \frac{1 \cdot \sqrt{2}}{\sqrt{2} \cdot \sqrt{2}} = \frac{\sqrt{2}}{2}.$$

(2) In the figure below on the left, we see a visual representation of the circle, the given angle, and the desired point of intersection.

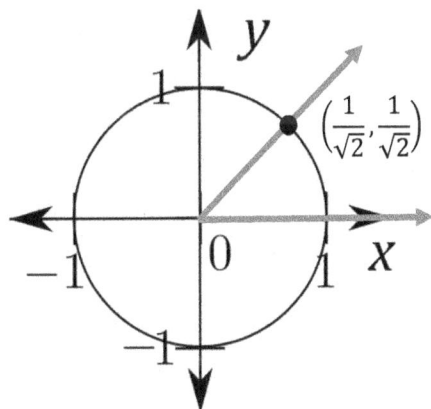

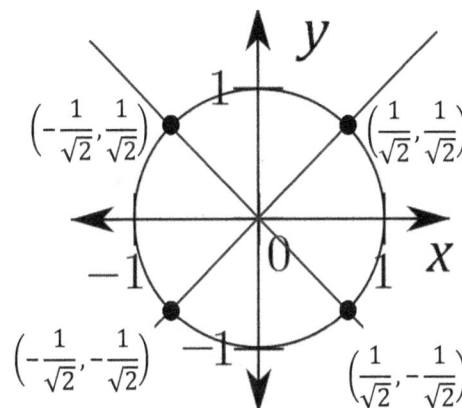

(3) In the figure above on the right, we have divided the Complex Plane into eight regions using the lines with equations $y = x$ and $y = -x$ (together with the x- and y-axes). We then used the symmetry of the circle to label the four points of intersection of the unit circle with each of these two lines.

If θ (pronounced "theta") is the radian measure of an angle in standard position such that the terminal ray intersects the unit circle at the point (x, y), then we will say that $W(\theta) = (x, y)$. This expression defines a function $W: \mathbb{R} \to \mathbb{R} \times \mathbb{R}$ called the **wrapping function**. Observe that the inputs of the wrapping function are real numbers, which we think of as the radian measure of angles in standard position. The outputs of the wrapping function are pairs of real numbers, which we think of as points in the Complex Plane. Also, observe that the range of the wrapping function is the unit circle.

We now define the cosine and sine of the angle θ by $\cos\theta = x$ and $\sin\theta = y$, where $W(\theta) = (x, y)$. For convenience, we also define the tangent of the angle by $\tan\theta = \frac{\sin\theta}{\cos\theta} = \frac{y}{x}$.

Notes: (1) The wrapping function is **not** one to one. For example, $W\left(\frac{\pi}{2}\right) = (0,1)$ and $W\left(\frac{5\pi}{2}\right) = (0,1)$. However, $\frac{\pi}{2} \neq \frac{5\pi}{2}$. There are actually infinitely many real numbers that map to $(0,1)$ under the wrapping function. Specifically, $W\left(\frac{\pi}{2} + 2k\pi\right) = (0,1)$ for every $k \in \mathbb{Z}$.

In general, each point on the unit circle is the image of infinitely many real numbers. Indeed, if $W(\theta) = (a,b)$, then $W(\theta + 2k\pi) = (a,b)$ for all $k \in \mathbb{Z}$.

(2) The wrapping function gives us a convenient way to associate an angle θ in standard position with the corresponding point (x, y) on the unit circle. It is mostly used only as a notational convenience. We will usually be more interested in the expressions $\cos\theta = x$ and $\sin\theta = y$.

Example 5.22: Using the rightmost figure above, we can make the following computations:

$$W\left(\frac{\pi}{4}\right) = \left(\frac{1}{\sqrt{2}}, \frac{1}{\sqrt{2}}\right) \quad W\left(\frac{3\pi}{4}\right) = \left(-\frac{1}{\sqrt{2}}, \frac{1}{\sqrt{2}}\right) \quad W\left(\frac{5\pi}{4}\right) = \left(-\frac{1}{\sqrt{2}}, -\frac{1}{\sqrt{2}}\right) \quad W\left(\frac{7\pi}{4}\right) = \left(\frac{1}{\sqrt{2}}, -\frac{1}{\sqrt{2}}\right)$$

$$\cos\frac{\pi}{4} = \frac{1}{\sqrt{2}} \qquad \sin\frac{\pi}{4} = \frac{1}{\sqrt{2}} \qquad \cos\frac{3\pi}{4} = -\frac{1}{\sqrt{2}} \qquad \sin\frac{3\pi}{4} = \frac{1}{\sqrt{2}}$$

$$\cos\frac{5\pi}{4} = -\frac{1}{\sqrt{2}} \qquad \sin\frac{5\pi}{4} = -\frac{1}{\sqrt{2}} \qquad \cos\frac{7\pi}{4} = \frac{1}{\sqrt{2}} \qquad \sin\frac{7\pi}{4} = -\frac{1}{\sqrt{2}}$$

It's also easy to compute the cosine and sine of the four **quadrantal angles** $0, \frac{\pi}{2}, \pi$, and $\frac{3\pi}{2}$. Here we use the fact that the points $(1,0), (0,1), (-1,0)$, and $(0,-1)$ lie on the unit circle.

$$W(0) = (1,0) \quad W\left(\frac{\pi}{2}\right) = (0,1) \quad W(\pi) = (-1,0) \quad W\left(\frac{3\pi}{2}\right) = (0,-1)$$

$$\cos 0 = 1 \qquad \sin 0 = 0 \qquad \cos\frac{\pi}{2} = 0 \qquad \sin\frac{\pi}{2} = 1$$

$$\cos\pi = -1 \qquad \sin\pi = 0 \qquad \cos\frac{3\pi}{2} = 0 \qquad \sin\frac{3\pi}{2} = -1$$

Also, if we add any integer multiple of 2π to an angle, the cosine and sine of the new angle have the same values as the old angle. For example, $\cos\frac{9\pi}{4} = \cos\left(\frac{\pi}{4} + \frac{8\pi}{4}\right) = \cos\left(\frac{\pi}{4} + 2\pi\right) = \cos\frac{\pi}{4} = \frac{1}{\sqrt{2}}$. This is a direct consequence of the fact that $W(\theta + 2k\pi) = W(\theta)$ for all $k \in \mathbb{Z}$.

We can also compute the tangent of each angle by dividing the sine of the angle by the cosine of the angle. For example, we have

$$\tan\frac{\pi}{4} = \frac{\sin\frac{\pi}{4}}{\cos\frac{\pi}{4}} = \frac{\frac{1}{\sqrt{2}}}{\frac{1}{\sqrt{2}}} = 1.$$

Similarly, we have

$$\tan\frac{3\pi}{4} = -1 \qquad \tan\frac{5\pi}{4} = 1 \qquad \tan\frac{7\pi}{4} = -1 \qquad \tan 0 = 0 \qquad \tan\pi = 0$$

When $\theta = \frac{\pi}{2}$ or $\frac{3\pi}{2}$, $\tan \theta$ is **undefined**.

Notes: (1) If $z = x + yi$ is any complex number, then the point (x, y) lies on a circle of radius r centered at the origin, where $r = |z| = \sqrt{x^2 + y^2}$. If θ is the radian measure of an angle in standard position such that the terminal ray intersects this circle at the point (x, y), then it can be proved that the cosine and sine of the angle are equal to $\cos \theta = \frac{x}{r}$ and $\sin \theta = \frac{y}{r}$.

(2) It is standard to use the abbreviations $\cos^2 \theta$ and $\sin^2 \theta$ for $(\cos \theta)^2$ and $(\sin \theta)^2$, respectively.

From the definition of cosine and sine, we have the following formula called the **Pythagorean Identity**:
$$\cos^2 \theta + \sin^2 \theta = 1$$

(3) Also, from the definition of cosine and sine, we have the following two formulas called the **Negative Identities**:
$$\cos(-\theta) = \cos \theta \qquad \qquad \sin(-\theta) = -\sin \theta.$$

Theorem 5.23: Let θ and ϕ be the radian measures of angles A and B, respectively. Then we have
$$\cos(\theta + \phi) = \cos \theta \cos \phi - \sin \theta \sin \phi$$
$$\sin(\theta + \phi) = \sin \theta \cos \phi + \cos \theta \sin \phi.$$

Notes: (1) The two formulas appearing in Theorem 5.23 are called the **Sum Identities**. You will be asked to prove Theorem 5.23 in Problem 22 below (parts (i) and (v)).

(2) Theorem 5.23 will be used to prove De Moivre's Theorem (Theorem 5.25) below. De Moivre's Theorem provides a fast method for performing exponentiation of complex numbers.

(3) θ and ϕ are Greek letters pronounced "theta" and "phi," respectively. These letters are often used to represent angle measures. We may sometimes also use the capital versions of these letters, Θ and Φ, especially when insisting that the radian measures of the given angles are between $-\pi$ and π.

The **standard form** (or **rectangular form**) of a complex number z is $z = x + yi$, where x and y are real numbers. Recall from part 3 of Example 1.6 that we can visualize the complex number $z = x + yi$ as the point (x, y) in the Complex Plane.

If for $z \neq 0$, we let $r = |z| = |x + yi| = \sqrt{x^2 + y^2}$ and we let θ be the radian measure of an angle in standard position such that the terminal ray passes through the point (x, y), then we see that r and θ determine this point. So, we can also write this point as (r, θ).

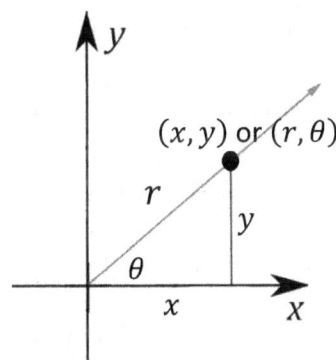

In Note 1 following Example 5.22, we saw that $\cos \theta = \frac{x}{r}$ and $\sin \theta = \frac{y}{r}$. By multiplying each side of these last two equations by r, we get $x = r \cos \theta$ and $y = r \sin \theta$. These equations allow us to rewrite the complex number $z = x + yi$ in the **polar form** $z = r \cos \theta + (r \sin \theta)i = r(\cos \theta + i \sin \theta)$.

If we also make the definition $e^{i\theta} = \cos\theta + i\sin\theta$, we can write the complex number $z = x + yi$ in the **exponential form** $z = re^{i\theta}$.

$r = |z|$ is called the **absolute value** or **modulus** of the complex number. We will call the angle θ an **argument** of the complex number and we may sometimes write $\theta = \arg z$.

Note that although $r = |z|$ and $\theta = \arg z$ uniquely determine a point (r, θ), there are infinitely many other values for $\arg z$ that represent the same point. Indeed, $(r, \theta + 2k\pi)$ represents the same point for each $k \in \mathbb{Z}$. However, there is a unique such value Θ for $\arg z$ such that $-\pi < \Theta \leq \pi$. We call this value Θ the **principal argument** of z, and we write $\Theta = \text{Arg } z$.

Notes: (1) The definition $e^{i\theta} = \cos\theta + i\sin\theta$ is known as **Euler's formula**.

(2) When written in exponential form, two complex numbers $z = re^{i\theta}$ and $w = se^{i\phi}$ are equal if and only if $r = s$ and $\phi = \theta + 2k\pi$ for some $k \in \mathbb{Z}$.

Example 5.24: Let's convert the complex number $z = 1 + i$ to exponential form. To do this, we need to find r and θ. We have $r = |z| = \sqrt{1^2 + 1^2} = \sqrt{1+1} = \sqrt{2}$. Next, we have $\tan\theta = \frac{1}{1} = 1$. It follows that $\theta = \frac{\pi}{4}$. So, in exponential form, we have $z = \sqrt{2}e^{\frac{\pi}{4}i}$.

Note: $\frac{\pi}{4}$ is the principal argument of $z = 1 + i$ because $-\pi < \frac{\pi}{4} \leq \pi$. When we write a complex number in exponential form, we will usually use the principle argument.

If $z \in \mathbb{C}$, we define z^2 to be the complex number $z \cdot z$. Similarly, $z^3 = z \cdot z \cdot z = z^2 \cdot z$. More generally, for $z \in \mathbb{C}$ and $n \in \mathbb{Z}$ we define z^n as follows:

- For $n = 0$, $z^n = z^0 = 1$.

- For $n \in \mathbb{Z}^+$, $z^{n+1} = z^n \cdot z$.

- For $n \in \mathbb{Z}^-$, $z^n = \frac{1}{z^{-n}}$.

Due to the following theorem, it's often easier to compute z^n when z is written in exponential form.

Theorem 5.25 (De Moivre's Theorem): For all $n \in \mathbb{Z}$, $\left(e^{i\theta}\right)^n = e^{i(n\theta)}$.

Proof: For $n = 0$, we have $\left(e^{i\theta}\right)^0 = (\cos\theta + i\sin\theta)^0 = 1 = e^0 = e^{i(0\theta)}$.

We prove De Moivre's Theorem for $n \in \mathbb{Z}^+$ by induction on n.

Base Case $(k = 1)$: $\left(e^{i\theta}\right)^1 = e^{i\theta} = e^{i(1\theta)}$.

Inductive Step: Assume that $k \geq 1$ and $\left(e^{i\theta}\right)^k = e^{i(k\theta)}$. We then have

$$\left(e^{i\theta}\right)^{k+1} = (\cos\theta + i\sin\theta)^{k+1} = (\cos\theta + i\sin\theta)^k(\cos\theta + i\sin\theta) = \left(e^{i\theta}\right)^k(\cos\theta + i\sin\theta)$$

$$= e^{i(k\theta)}(\cos\theta + i\sin\theta) = (\cos k\theta + i\sin k\theta)(\cos\theta + i\sin\theta)$$

$$= [(\cos k\theta)(\cos\theta) - (\sin k\theta)(\sin\theta)] + [(\sin k\theta)(\cos\theta) + (\cos k\theta)(\sin\theta)]i.$$

$$= \cos\big((k+1)\theta\big) + \sin\big((k+1)\theta\big)\,i \text{ (by Theorem 5.23)} = e^{i((k+1)\theta)}.$$

By the Principle of Mathematical Induction, $\left(e^{i\theta}\right)^n = e^{i(n\theta)}$ for all $n \in \mathbb{Z}^+$.

If $n < 0$, then

$$\left(e^{i\theta}\right)^n = \frac{1}{\left(e^{i\theta}\right)^{-n}} = \frac{1}{e^{i(-n\theta)}} = \frac{1}{\cos(-n\theta) + i\sin(-n\theta)}$$

$$= \frac{1}{\cos(n\theta) - i\sin(n\theta)} \text{ (by the Negative Identities)}$$

$$= \frac{1}{\cos(n\theta) - i\sin(n\theta)} \cdot \frac{\cos(n\theta) + i\sin(n\theta)}{\cos(n\theta) + i\sin(n\theta)} = \frac{\cos(n\theta) + i\sin(n\theta)}{\cos^2(n\theta) + \sin^2(n\theta)}$$

$$= \cos(n\theta) + i\sin(n\theta) \text{ (by the Pythagorean Identity)} = e^{i(n\theta)}. \qquad \square$$

Note: De Moivre's Theorem generalizes to all $n \in \mathbb{C}$ with a small "twist." In general, the expression $\left(e^{i\theta}\right)^n$ may have multiple values, whereas $e^{i(n\theta)}$ takes on just one value. However, for all $n \in \mathbb{C}$, $\left(e^{i\theta}\right)^n = e^{i(n\theta)}$ in the sense that $e^{i(n\theta)}$ is equal to one of the possible values of $\left(e^{i\theta}\right)^n$.

As a very simple example, let $\theta = 0$ and $n = \frac{1}{2}$. Then $e^{i(n\theta)} = e^0 = 1$ and $\left(e^{i\theta}\right)^n = 1^{\frac{1}{2}}$, which has two values: 1 and -1 (because $1^2 = 1$ and $(-1)^2 = 1$). Observe that $e^{i(n\theta)}$ is equal to one of the two possible values of $\left(e^{i\theta}\right)^n$.

We will not prove this more general result here.

Example 5.26: Let's compute $(2 - 2i)^6$. If we let $z = 2 - 2i$, we have $\tan\theta = \frac{-2}{2} = -1$, so that $\theta = \frac{7\pi}{4}$ (Why?). Also, $r = |z| = \sqrt{2^2 + (-2)^2} = \sqrt{2^2(1+1)} = \sqrt{2^2 \cdot 2} = \sqrt{2^2} \cdot \sqrt{2} = 2\sqrt{2}$. So, in exponential form, $z = 2\sqrt{2}e^{\frac{7\pi}{4}i}$, and therefore,

$$z^6 = \left(2\sqrt{2}e^{\frac{7\pi}{4}i}\right)^6 = 2^6\sqrt{2}^6\left(e^{\frac{7\pi}{4}i}\right)^6 = 64 \cdot 8e^{6\left(\frac{7\pi}{4}\right)i} = 512e^{\frac{21\pi}{2}i} = 512e^{\left(\frac{\pi}{2}+10\pi\right)i}$$

$$= 512e^{\frac{\pi}{2}i} = 512\left(\cos\frac{\pi}{2} + i\sin\frac{\pi}{2}\right) = 512(0 + i \cdot 1) = \mathbf{512i}.$$

Problem Set 5

Full solutions to these problems are available for free download here:

www.SATPrepGet800.com/TFBZLF

LEVEL 1

1. Use the Principle of Mathematical Induction to prove each of the following:

   (i) $2^n > n$ for all natural numbers $n \geq 1$.

   (ii) $0 + 1 + 2 + \cdots + n = \dfrac{n(n+1)}{2}$ for all natural numbers.

   (iii) $n! > 2^n$ for all natural numbers $n \geq 4$ (where $n! = 1 \cdot 2 \cdots n$ for all natural numbers $n \geq 1$).

   (iv) $2^n \geq n^2$ for all natural numbers $n \geq 4$.

2. A natural number n is **divisible** by a natural number k, written $k | n$, if there is another natural number b such that $n = kb$. Prove that $n^3 - n$ is divisible by 3 for all natural numbers n.

3. Let $z = -4 - i$ and $w = 3 - 5i$. Compute each of the following:

   (i) $z + w$

   (ii) zw

   (iii) $\text{Im } w$

LEVEL 2

4. Prove each of the following. (You may assume that $<$ is a strict linear ordering of $\mathbb{N}$.)

   (i)   Addition is commutative in $\mathbb{N}$.

   (ii)   The set of natural numbers is closed under multiplication.

   (iii) 1 is a multiplicative identity in $\mathbb{N}$.

   (iv)   Multiplication is distributive over addition in $\mathbb{N}$.

   (v)   Multiplication is associative in $\mathbb{N}$.

   (vi)   Multiplication is commutative in $\mathbb{N}$.

   (vii) For all natural numbers m, n, and k, if $m + k = n + k$, then $m = n$.

   (viii) For all natural numbers m, n, and k, if $mk = nk$, then $m = n$.

   (ix)   For all natural numbers m and n, $m < n$ if and only if there is a natural number $k > 0$ such that $n = m + k$.

   (x)   For all natural numbers m, n, and k, $m < n$ if and only if $m + k < n + k$.

   (xi)   For all natural numbers m and n, if $m > 0$ and $n > 0$, then $mn > 0$.

84

5. A set A is **transitive** if $\forall x (x \in A \to x \subseteq A)$ (in words, every element of A is also a subset of A). Prove that every natural number is transitive.

6. Determine if each of the following sequences are Cauchy sequences. Are any of the Cauchy sequences equivalent?

   (i) $(x_n) = \left(1 + \frac{1}{n+1}\right)$

   (ii) $(y_n) = (2^n)$

   (iii) $(z_n) = \left(1 - \frac{1}{2n+1}\right)$

7. Each of the following complex numbers is written in exponential form. Rewrite each complex number in standard form:

   (i) $e^{\pi i}$

   (ii) $e^{-\frac{5\pi}{2}i}$

   (iii) $3e^{\frac{\pi}{4}i}$

   (iv) $2e^{\frac{\pi}{3}i}$

   (v) $\sqrt{2}e^{\frac{7\pi}{6}i}$

   (vi) $\pi e^{-\frac{5\pi}{4}i}$

   (vii) $e^{\frac{19\pi}{12}}$

8. Each of the following complex numbers is written in standard form. Rewrite each complex number in exponential form:

   (i) $-1 - i$

   (ii) $\sqrt{3} + i$

   (iii) $1 - \sqrt{3}i$

   (iv) $\left(\frac{\sqrt{6}+\sqrt{2}}{4}\right) + \left(\frac{\sqrt{6}-\sqrt{2}}{4}\right)i$

9. Write the following complex numbers in standard form:

   (i) $\left(\frac{\sqrt{2}}{2} + \frac{\sqrt{2}}{2}i\right)^4$

   (ii) $\left(1 + \sqrt{3}i\right)^5$

LEVEL 3

10. Prove that if $n \in \mathbb{N}$ and A is a nonempty subset of n, then A has a least element.

11. Prove POMI $\to$ WOP.

12. Prove that $<_\mathbb{Z}$ is a well-defined strict linear ordering on $\mathbb{Z}$. You may use the fact that $<_\mathbb{N}$ is a well-defined strict linear ordering on $\mathbb{N}$.

LEVEL 4

13. Prove that $3^n - 1$ is even for all natural numbers n.

14. Show that the Principle of Mathematical Induction is equivalent to the following statement:

    ($\star$) Let $P(n)$ be a statement and suppose that (i) $P(0)$ is true and (ii) for all $k \in \mathbb{N}$, $P(k) \to P(k+1)$. Then $P(n)$ is true for all $n \in \mathbb{N}$.

15. Prove that addition of integers is well-defined.

16. Prove that addition and multiplication of rational numbers are well-defined.

17. Let $A = \{(x_n) \mid (x_n) \text{ is a Cauchy sequence of rational numbers}\}$ and define the relation R on A by $(x_n)R(y_n)$ if and only if for every $k \in \mathbb{N}^+$, there is $K \in \mathbb{N}$ such that $n > K$ implies $|x_n - y_n| < \frac{1}{k}$. Prove that R is an equivalence relation on A.

18. Prove that $\{A \in \mathcal{P}(\mathbb{N}) \mid A \text{ is finite}\}$ is countable and $\{A \in \mathcal{P}(\mathbb{N}) \mid A \text{ is infinite}\}$ is uncountable.

19. Consider triangle AOP, where $O = (0,0)$, $A = (1,0)$, and P is the point on the unit circle so that angle POA has radian measure $\frac{\pi}{3}$. Prove that triangle AOP is equilateral, and then use this to prove that $W\left(\frac{\pi}{3}\right) = \left(\frac{1}{2}, \frac{\sqrt{3}}{2}\right)$. You may use the following facts about triangles: (i) The interior angle measures of a triangle sum to π radians; (ii) Two sides of a triangle have the same length if and only if the interior angles of the triangle opposite these sides have the same measure; (iii) If two sides of a triangle have the same length, then the line segment beginning at the point of intersection of those two sides and terminating on the opposite base midway between the endpoints of that base is perpendicular to that base.

20. Prove that $W\left(\frac{\pi}{6}\right) = \left(\frac{\sqrt{3}}{2}, \frac{1}{2}\right)$. You can use facts (i), (ii), and (iii) described in Problem 19.

21. Let θ and ϕ be the radian measure of angles A and B, respectively. Prove the following identity:
    $$\cos(\theta - \phi) = \cos\theta\cos\phi + \sin\theta\sin\phi$$

22. Let θ and ϕ be the radian measure of angles A and B, respectively. Prove the following identities:
    (i) $\cos(\theta + \phi) = \cos\theta\cos\phi - \sin\theta\sin\phi$
    (ii) $\cos(\pi - \theta) = -\cos\theta$
    (iii) $\cos\left(\frac{\pi}{2} - \theta\right) = \sin\theta$
    (iv) $\sin\left(\frac{\pi}{2} - \theta\right) = \cos\theta$
    (v) $\sin(\theta + \phi) = \sin\theta\cos\phi + \cos\theta\sin\phi$
    (vi) $\sin(\pi - \theta) = -\sin\theta$

23. The Principle of Strong Induction is the following statement:

> (**) Let $P(n)$ be a statement and suppose that (i) $P(0)$ is true and (ii) for all $k \in \mathbb{N}$, $\forall j \leq k \left(P(j) \right) \to P(k + 1)$. Then $P(n)$ is true for all $n \in \mathbb{N}$.

Use the Principle of Mathematical Induction to prove the Principle of Strong Induction.

24. Use the Principle of Mathematical Induction to prove that for every $n \in \mathbb{N}$, if S is a set with $|S| = n$, then S has 2^n subsets. (Hint: Use Problem 16 from Problem Set 1.)

25. Prove that addition of real numbers is well-defined and that the sum of two real numbers is a real number.

CHALLENGE PROBLEMS

26. Prove that multiplication of integers is well-defined.

27. Prove that multiplication of real numbers is well-defined and that the product of two real numbers is a real number.

28. Prove that $<_\mathbb{N}$ is a strict linear ordering on $\mathbb{N}$, that $<_\mathbb{Q}$ is a well-defined strict linear ordering on $\mathbb{Q}$, and that $<_\mathbb{R}$ is a well-defined strict linear ordering on $\mathbb{R}$.

29. Define a set to be **selfish** if the number of elements it has is in the set. For example, the set $K_5 = \{1, 2, 3, 4, 5\}$ is selfish because it has 5 elements and 5 is in the set. A selfish set is **minimal** if none of its proper subsets is also selfish. For example, the set K_5 is not a minimal selfish set because $\{1\}$ is a selfish subset. Let $K_n = \{1, 2, 3, \dots n\}$. Determine with proof how many minimal selfish subsets K_n has in terms of n.

LESSON 6
ALGEBRAIC STRUCTURES AND COMPLETENESS

Binary Operations and Closure

A **binary operation** on a set is a rule that combines two elements of the set to produce another element of the set.

Example 6.1: Let $S = \{0, 1\}$. Multiplication on S is a binary operation, whereas addition on S is **not** a binary operation (here we are thinking of multiplication and addition in the "usual" sense, meaning the way we would think of them in elementary school or middle school).

To see that multiplication is a binary operation on S, observe that $0 \cdot 0 = 0$, $0 \cdot 1 = 0$, $1 \cdot 0 = 0$, and $1 \cdot 1 = 1$. Each of the four computations produces 0 or 1, both of which are in the set S.

To see that addition is not a binary operation on S, just note that $1 + 1 = 2$, and $2 \notin S$.

Formally, a **binary operation** $\star$ on a set S is a **function** $\star : S \times S \to S$. So, if $a, b \in S$, then we have $\star\,(a, b) \in S$. For easier readability, we usually write $\star\,(a, b)$ as $a \star b$.

When $\star$ is a binary operation on S, we say that S is **closed** under $\star$.

Example 6.2:

1. Let $S = \{u, v, w\}$ and define $\star$ using the following table:

$\star$	u	v	w
u	v	w	w
v	w	u	u
w	u	v	v

    The table given above is called a **multiplication table**. For $a, b \in S$, we evaluate $a \star b$ by taking the entry in the row given by a and the column given by b. For example, $v \star w = u$.

$\star$	u	v	w
u	v	w	w
v	w	u	u
w	u	v	v

    $\star$ is a binary operation on S because the only possible "outputs" are $u, v,$ and w.

2. The operation of addition on the set of natural numbers is a binary operation because whenever we add two natural numbers we get another natural number (we proved this in Theorem 5.7). Here, the set S is $\mathbb{N}$ and the operation $\star$ is $+$.

3. Similarly, the operation of multiplication on the set of natural numbers is a binary operation because whenever we multiply two natural numbers we get another natural number (you were asked to prove this in part (ii) of Problem 4 from Problem Set 5). Here, the set S is $\mathbb{N}$ and the operation $\star$ is $\cdot$.

4. The operation of addition on the set of integers is a binary operation because whenever we add two integers we get another integer. Here, the set S is $\mathbb{Z} = \{[(a,b)] \mid (a,b) \in \mathbb{N} \times \mathbb{N}\}$, where $(a,b) \sim (c,d)$ if and only if $a + d = b + c$. The operation $\star$ is $+$, which is defined by $[(a,b)] + [(c,d)] = [(a+c,b+d)]$. You were asked to prove that this operation is well-defined in Problem 15 from Problem Set 5. By 2 above, if $a, b \in \mathbb{N}$, then $a + c, b + d \in \mathbb{N}$. So, $[(a,b)], [(c,d)] \in \mathbb{Z}$ implies that $[(a,b)] + [(c,d)] = [(a+c,b+d)] \in \mathbb{Z}$, showing that $\mathbb{Z}$ is closed under addition.

5. Similarly, the operation of multiplication on the set of integers is also a binary operation. The operation $\star$ is $\cdot$, which is defined by $[(a,b)] \cdot [(c,d)] = [(ac + bd, ad + bc)]$. You were asked to prove that this operation is well-defined in Problem 26 from Problem Set 5. By 2 and 3 above, if $a, b, c, d \in \mathbb{N}$, then $ac + bd, ad + bc \in \mathbb{N}$. So, $[(a,b)], [(c,d)] \in \mathbb{Z}$ implies that $[(a,b)] \cdot [(c,d)] = [(ac + bd, ad + bc)] \in \mathbb{Z}$, showing that $\mathbb{Z}$ is closed under multiplication.

6. The operation of addition on the set of rational numbers is a binary operation because whenever we add two rational numbers we get another rational number. Here, the set S is $\mathbb{Q} = \{\frac{a}{b} \mid a \in \mathbb{Z} \wedge b \in \mathbb{Z}^*\}$, where $\frac{a}{b} = \frac{c}{d}$ if and only if $ad = bc$. The operation $\star$ is $+$, which is defined by $\frac{a}{b} + \frac{c}{d} = \frac{ad+bc}{bd}$. You were asked to prove that this operation is well-defined in Problem 16 from Problem Set 5. By 4 and 5 above, if $a, b, c, d \in \mathbb{Z}$, then $ad + bc, bd \in \mathbb{Z}$. We need to show that for integers $b, d \neq 0$, we have $bd \neq 0$. Let $b = [(m,n)]$ and $d = [(s,t)]$ and assume that $b \cdot d = [(m,n)] \cdot [(s,t)] = [(0,0)]$ and $b = [(m,n)] \neq [(0,0)]$. We must show that $d = [(s,t)] = [(0,0)]$. Since $b \cdot d = [(0,0)]$, we have $[(ms + nt, mt + ns)] = [(0,0)]$. So, $ms + nt = mt + ns$. Since $b \neq [(0,0)]$, we have $m \neq n$. Without loss of generality, assume that $m < n$, then by part (ix) of Problem 4 from Problem Set 5, there is a natural number $k > 0$ such that $n = m + k$. Using associativity of addition and distributivity of multiplication over addition in $\mathbb{N}$, we have

$$ms + mt + kt = ms + (m + k)t = ms + nt = mt + ns = mt + (m + k)s = mt + ms + ks.$$

By part (vii) of Problem 4 from Problem Set 5 and commutativity of addition in $\mathbb{N}$, $kt = ks$. By part (viii) of Problem 4 from Problem Set 5, $t = s$. So, $d = [(s,t)] = [(0,0)]$. Thus, $\mathbb{Q}$ is closed under addition.

7. Similarly, the operation of multiplication on the set of rational numbers is also a binary operation. The operation $\star$ is $\cdot$, which is defined by $\frac{a}{b} \cdot \frac{c}{d} = \frac{ac}{bd}$. You were asked to prove that this operation is well-defined in Problem 16 from Problem Set 5. By 5 above, if $a, b, c, d \in \mathbb{Z}$, then $ac, bd \in \mathbb{Z}$. Also, if $b, d \neq 0$, then $bd \neq 0$ by the same argument given in 6 above. Thus, $\mathbb{Q}$ is closed under multiplication.

8. The operation of addition on the set of real numbers is a binary operation because whenever we add two real numbers we get another real number. Here, the set S is $\mathbb{R} = \{[(x_n)] \mid (x_n) \text{ is a Cauchy sequence of rational numbers}\}$, where $(x_n) \sim (y_n)$ if and only if for every $k \in \mathbb{N}^+$, there is $K \in \mathbb{N}$ such that $n > K$ implies $|x_n - y_n| < \frac{1}{k}$. The operation $\star$ is $+$, which is defined by $[(x_n)] + [(y_n)] = [(x_n + y_n)]$. You were asked to prove that this operation is well-defined and that the sum of two real numbers is a real number in Problem 25 from Problem Set 5. Thus, $\mathbb{R}$ is closed under addition.

9. Similarly, the operation of multiplication on the set of real numbers is also a binary operation. The operation $\star$ is $\cdot$, which is defined by $[(x_n)] \cdot [(y_n)] = [(x_n \cdot y_n)]$. You were asked to prove that this operation is well-defined and that the product of two real numbers is a real number in Problem 27 from Problem Set 5. Thus, $\mathbb{R}$ is closed under multiplication.

10. Subtraction on the set of natural numbers is **not** a binary operation. To see this, we just need to provide a single **counterexample**. Recall from Lesson 5 that for natural numbers m, n, and t, we say that $m - n = t$ if and only if $m = t + n$. It is straightforward to prove by induction on $t \in \mathbb{N}$ that $1 \subset t + 2$ (1 is a *proper* subset of $t + 2$) for all $t \in \mathbb{N}$. In particular, for all $t \in \mathbb{N}$, $1 \neq t + 2$. Therefore, $1 - 2$ is not defined in $\mathbb{N}$.

Some authors refer to a binary operation $\star$ on a set S even when the binary operation is not defined on all pairs of elements $a, b \in S$. We will always refer to these "false operations" as **partial binary operations**.

We say that the set S is **closed** under the partial binary operation $\star$ if whenever $a, b \in S$, we have $a \star b \in S$.

In Example 6.2, part 10 above, we saw that subtraction is a partial binary operation on $\mathbb{N}$ that is not a binary operation. In other words, $\mathbb{N}$ is not **closed** under subtraction.

Groups

A **group** is a pair $(G, \star)$ consisting of a set G together with a binary operation $\star$ on G satisfying:

(1) **(Associativity)** For all $x, y, z \in G$, $(x \star y) \star z = x \star (y \star z)$.

(2) **(Identity)** There exists an element $e \in G$ such that for all $x \in G$, $e \star x = x \star e = x$.

(3) **(Inverse)** For each $x \in G$, there is $y \in G$ such that $x \star y = y \star x = e$.

Notes: (1) If $y \in G$ is an inverse of $x \in G$, we will usually write $y = x^{-1}$.

(2) Recall that the definition of a binary operation already implies closure. However, many books on groups will mention this property explicitly:

(**Closure**) For all $x, y \in G$, $x \star y \in G$.

(3) A group is **commutative** or **Abelian** if for all $x, y \in G$, $x \star y = y \star x$.

(4) The properties that define a group are called the **group axioms**. These are the statements that are **given** to be true in all groups. There are many other statements that are true in groups. However, any additional statements need to be **proved** using the axioms.

(5) If properties (1) and (2) hold, we get a **monoid**. So, a group is a monoid with the inverse property.

(6) If property (1) holds, we get a **semigroup**. So, a monoid is semigroup with an identity element.

Example 6.3:

1. Is $(\mathbb{N}, +)$ a group? By Theorem 5.7, $+$ is a binary operation on $\mathbb{N}$ (or equivalently, $\mathbb{N}$ is closed under $+$). By Theorem 5.9, $+$ is associative in $\mathbb{N}$. By Theorem 5.8, together with the definition of addition (see Note 1 following Theorem 5.8), 0 is an additive identity for $\mathbb{N}$. It follows that $(\mathbb{N}, +)$ is a monoid.

    However, $(\mathbb{N}, +)$ is **not** a group. The inverse property fails. For example, 1 has no inverse (and in fact, the only natural number with an additive inverse is 0). Indeed, suppose toward contradiction that $n \in \mathbb{N}$ with $n + 1 = 0$. Since $n + 1 = n^+ = n \cup \{n\}$, and $n \in n \cup \{n\}$, we must have $n \in 0 = \emptyset$. But the empty set has no elements. This contradiction proves that there is no natural number n such that $n + 1 = 0$. So, 1 has no additive inverse in $\mathbb{N}$. Therefore, the inverse property fails and $(\mathbb{N}, +)$ is **not** a group. It's worth mentioning that $(\mathbb{N}, +)$ is commutative. This follows from part (i) of Problem 4 from Problem Set 5. So, $(\mathbb{N}, +)$ is a commutative monoid that is **not** a group.

2. Similarly, $(\mathbb{N}, \cdot)$ is a commutative monoid with identity 1 that is not a group. I leave the details that $(\mathbb{N}, \cdot)$ is a commutative monoid to the reader. Let's prove that 2 has no multiplicative inverse in $\mathbb{N}$. We have $2 \cdot 0 = 0 \neq 1$. We now prove by induction that for all $n \geq 1$, $2n > 1$. The base case is $2 \cdot 1 = 2 > 1$. Assuming $2k > 1$, by parts (iv) and (x) of Problem 4 from Problem Set 5, we have $2(k + 1) = 2k + 2 > 1 + 2 > 1$. In particular, we showed that for all $n \in \mathbb{N}$, $2n \neq 1$. Therefore, 2 has no multiplicative inverse in $\mathbb{N}$.

3. $(\mathbb{Z}, +)$ is a commutative group with identity $0 = [(0, 0)]$. We showed that $+$ is a binary operation on $\mathbb{Z}$ in part 4 of Example 6.2 above. To see that $+$ is associative in $\mathbb{Z}$, observe that for $a, b, c, d, e, f \in \mathbb{N}$, we have

    $$([(a, b)] + [(c, d)]) + [(e, f)] = [(a + c, b + d)] + [(e, f)] = [((a + c) + e, (b + d) + f)]$$
    $$= [(a + (c + e), b + (d + f))] = [(a, b)] + [(c + e, d + f)] = [(a, b)] + ([(c, d)] + [(e, f)])$$

    For the first, second, fourth and fifth equalities, we simply used the definition of addition of integers. For the third equality, we used the associativity of addition in $\mathbb{N}$. I leave it to the reader to verify that $[(0, 0)]$ is an additive identity, that the inverse of $[(a, b)]$ is $[(b, a)]$, and that $+$ is commutative in $\mathbb{Z}$.

4. $(\mathbb{Z}, \cdot)$ is a commutative monoid with identity $1 = [(1, 0)]$ that is not a group. I leave it to the reader to verify that $(\mathbb{Z}, \cdot)$ is a commutative monoid. Let's prove that $2 = [(2, 0)]$ has no multiplicative inverse. If $[(a, b)]$ is a multiplicative inverse of 2, then we have $2a + 0b = 1$ and $2b + 0a = 0$. The first equation is equivalent to $2a = 1$. However, by 2 above, 2 has no multiplicative inverse in $\mathbb{N}$. Therefore, the equation $2a = 1$ has no solution, and so, 2 has no multiplicative inverse in $\mathbb{Z}$.

5. $(\mathbb{Q}, +)$ is a commutative group with identity $0 = \frac{0}{1}$. You will need to prove this as part of Problem 17 below.

6. $(\mathbb{R}, +)$ is a commutative group with identity $[(0)]$. We showed that $+$ is a binary operation on $\mathbb{R}$ in part 8 of Example 6.2 above. To see that $+$ is associative in $\mathbb{R}$, we use the associativity of $+$ in $\mathbb{Q}$. If $[(x_n)], [(y_n)], [(z_n)] \in \mathbb{R}$, then

$$([(x_n)] + [(y_n)]) + [(z_n)] = [(x_n + y_n)] + [(z_n)] = [((x_n + y_n) + z_n)]$$
$$= [(x_n + (y_n + z_n))] = [(x_n)] + [(y_n + z_n)] = [(x_n)] + ([(y_n)] + [(z_n)]).$$

To see that $[(0)]$ is the additive identity, using the fact that 0 is the additive identity in $\mathbb{Q}$, we have for $[(x_n)] \in \mathbb{R}$,

$$[(0)] + [(x_n)] = [(0 + x_n)] = [(x_n)] \text{ and } [(x_n)] + [(0)] = [(x_n + 0)] = [(x_n)].$$

The inverse of the real number $[(x_n)]$ is $[(-x_n)]$, where for each $n \in \mathbb{N}$, $-x_n$ is the additive inverse of x_n in $\mathbb{Q}$. To see this, simply observe that

$$[(x_n)] + [(-x_n)] = [(x_n + (-x_n))] = [(0)] \text{ and } [(-x_n)] + [(x_n)] = [(-x_n + x_n)] = [(0)].$$

Finally, to see that $+$ is commutative in $\mathbb{R}$, we use the commutativity of $+$ in $\mathbb{Q}$. If $[(x_n)], [(y_n)] \in \mathbb{R}$, then

$$[(x_n)] + [(y_n)] = [(x_n + y_n)] = [(y_n + x_n)] = [(y_n)] + [(x_n)].$$

7. $(\mathbb{Q}, \cdot)$ and $(\mathbb{R}, \cdot)$ fail to be groups, but only because 0 has no inverse (this follows immediately from part (iii) of Problem 7 below). However, $(\mathbb{Q}^*, \cdot)$ and $(\mathbb{R}^*, \cdot)$ are both commutative groups. You will need to prove this as part of Problem 17 below.

Fields

A **field** is a triple $(F, +, \cdot)$, where F is a set and $+$ and $\cdot$ are binary operations on F satisfying

(1) $(F, +)$ is a commutative group.

(2) $(F^*, \cdot)$ is a commutative group.

(3) $\cdot$ is **distributive** over $+$ in F. That is, for all $x, y, z \in F$, we have

$$x \cdot (y + z) = x \cdot y + x \cdot z \qquad \text{and} \qquad (y + z) \cdot x = y \cdot x + z \cdot x.$$

(4) $0 \neq 1$ (where 0 is the identity of $(F, +)$ and 1 is the identity of $(F^*, \cdot)$).

We will refer to the operation $+$ as addition, the operation $\cdot$ as multiplication, the additive identity as 0, the multiplicative identity as 1, the additive inverse of an element $x \in F$ as $-x$, and the multiplicative inverse of an element $x \in F^*$ as x^{-1}. We will often abbreviate $x \cdot y$ as xy.

Notes: (1) $(F, +)$ a commutative group means the following:

- **(Closure)** For all $x, y \in F$, $x + y \in F$.

- **(Associativity)** For all $x, y, z \in F$, $(x + y) + z = x + (y + z)$.

- **(Commutativity)** For all $x, y \in F$, $x + y = y + x$.

- **(Identity)** There exists an element $0 \in F$ such that for all $x \in F$, $0 + x = x + 0 = x$.

- **(Inverse)** For each $x \in F$, there is $-x \in F$ such that $x + (-x) = (-x) + x = 0$.

(2) Similarly, $(F^*, \cdot)$ a commutative group means the following:

- **(Closure)** For all $x, y \in F^*$, $xy \in F^*$.

- **(Associativity)** For all $x, y, z \in F^*$, $(xy)z = x(yz)$.

- **(Commutativity)** For all $x, y \in F^*$, $xy = yx$.

- **(Identity)** There exists an element $1 \in F^*$ such that for all $x \in F^*$, $1x = x \cdot 1 = x$.

- **(Inverse)** For each $x \in F^*$, there is $x^{-1} \in F^*$ such that $xx^{-1} = x^{-1}x = 1$.

(3) Recall that F^* is the set of nonzero elements of F. We can write $F^* = \{x \in F \mid x \neq 0\}$ (pronounced "the set of x in F such that x is not equal to 0") or $F^* = F \setminus \{0\}$ (pronounced "F with 0 removed").

(4) The properties that define a field are called the **field axioms**. These are the statements that are **given** to be true in all fields. There are many other statements that are true in fields. However, any additional statements need to be **proved** using the axioms.

(5) If we replace the condition that "$(F^*, \cdot)$ is a commutative group" by "$(F, \cdot)$ is a monoid," then the resulting structure is called a **ring**. The most well-known example of a ring is $\mathbb{Z}$, the ring of integers.

We also do not require 0 and 1 to be distinct in the definition of a ring. If $0 = 1$, we get the zero ring, which consists of just one element, namely 0 (Why?). The operations of addition and multiplication are defined by $0 + 0 = 0$ and $0 \cdot 0 = 0$. The reader may want to verify that the zero ring is in fact a ring.

The main difference between a ring and a field is that in a ring, there can be nonzero elements that do not have multiplicative inverses. For example, in $\mathbb{Z}$, 2 has no multiplicative inverse (see part 4 of Example 6.3 above). So, the equation $2x = 1$ has no solution.

(6) If we also replace "$(F, +)$ is a commutative group" by "$(F, +)$ is a commutative monoid," then the resulting structure is a **semiring**. The most well-known example of a semiring is $\mathbb{N}$, the semiring of natural numbers.

The main difference between a semiring and a ring is that in a semiring, there can be elements that do not have additive inverses. For example, in $\mathbb{N}$, 1 has no additive inverse (see part 1 of Example 6.3). Thus, the equation $x + 1 = 0$ has no solution.

Technical note: For a semiring, we include one additional axiom: For all $x \in F$, $0 \cdot x = x \cdot 0 = 0$. This statement follows from the ring axioms (see part (iii) of Problem 7 below), but **not** from the other semiring axioms. We need the additive inverse property to prove it.

(7) Every field is a commutative ring. Although this is not too hard to show (you will be asked to show this in Problem 12 below), it is worth observing that this is not completely obvious. For example, if $(F, +, \cdot)$ is a ring, then since $(F, \cdot)$ is a monoid with identity 1, it follows that $1 \cdot 0 = 0 \cdot 1 = 0$. However, in the definition of a field given above, this property of 0 is not given as an axiom. We **are** given that $(F^*, \cdot)$ is a commutative group, and so, it follows that 1 is an identity for F^*. But $0 \notin F^*$, and so, $1 \cdot 0 = 0 \cdot 1 = 0$ needs to be proved.

Similarly, in the definition of a field given above, 0 is excluded from associativity and commutativity. These need to be checked.

Example 6.4:

1. $(\mathbb{Z}, +, \cdot)$ is a ring that is **not** a field. In parts 3 and 4 of Example 6.3 above, we saw that $(\mathbb{Z}, +)$ is a commutative group and $(\mathbb{Z}, \cdot)$ is a monoid. All that is left to check is distributivity. We first check left distributivity. For $a, b, c, d, e, f \in \mathbb{N}$, we have

$$[(a, b)]([(c, d)] + [(e, f)]) = [(a, b)] \cdot [(c + e, d + f)]$$
$$= [(a(c + e) + b(d + f), a(d + f) + b(c + e))]$$
$$= [((ac + ae) + (bd + bf), (ad + af) + (bc + be))]$$
$$= [((ac + bd) + (ae + bf), (ad + bc) + (af + be))]$$
$$= [(ac + bd, ad + bc)] + [(ae + bf, af + be)] = [(a, b)] \cdot [(c, d)] + [(a, b)] \cdot [(e, f)].$$

   For the first, second, fifth and sixth equalities, we simply used the definitions of addition and multiplication of integers. For the third equality, we used the distributivity of multiplication over addition in $\mathbb{N}$. For the fourth equality, we used the associativity and commutativity of addition in $\mathbb{N}$. Since multiplication is commutative in $\mathbb{Z}$, right distributivity follows immediately from left distributivity.

$$([(c, d)] + [(e, f)])[(a, b)] = [(a, b)]([(c, d)] + [(e, f)])$$
$$= [(a, b)] \cdot [(c, d)] + [(a, b)] \cdot [(e, f)] = [(c, d)] \cdot [(a, b)] + [(e, f)] \cdot [(a, b)].$$

   Therefore, $(\mathbb{Z}, +, \cdot)$ is a ring. We already showed in part 4 of Example 6.3 that 2 has no multiplicative inverse in $\mathbb{Z}$. So, $(\mathbb{Z}, +, \cdot)$ is **not** a field.

2. $(\mathbb{Q}, +, \cdot)$ and $(\mathbb{R}, +, \cdot)$ are both fields. You will need to prove this as part of Problem 17 below.

3. $(\mathbb{C}, +, \cdot)$ is field. You will be asked to prove this in Problem 17 below. The proof is very straightforward and mostly uses the fact that $(\mathbb{R}, +, \cdot)$ is a field. For example, to verify that addition is commutative in $\mathbb{C}$, we have

$$(a + bi) + (c + di) = (a + c) + (b + d)i = (c + a) + (d + b)i = (c + di) + (a + bi).$$

   We have $a + c = c + a$ because $a, c \in \mathbb{R}$ and addition is commutative in $\mathbb{R}$. For the same reason, we have $b + d = d + b$.

   We note a few additional things of importance here: The identity for addition is $0 = 0 + 0i$. The identity for multiplication is $1 = 1 + 0i$. The additive inverse of $z = a + bi$ is $-z = -(a + bi) = -a - bi$. The multiplicative inverse of $z = a + bi$ is $z^{-1} = \frac{a}{a^2 + b^2} - \frac{b}{a^2 + b^2} i$.

Subtraction and Division: If $a, b \in F$, we define $a - b = a + (-b)$ and for $b \neq 0, \frac{a}{b} = ab^{-1}$.

Ordered Rings and Fields

An **ordered ring** is a quadruple $(R, +, \cdot, \leq)$, where $(R, +, \cdot)$ is a ring, and $(R, \leq)$ is a linearly ordered set such that:

(1) If $a, b, c \in R$, then $a \leq b \rightarrow a + c \leq b + c$.

(2) If $a, b \in R$ with $0 \leq a$ and $0 \leq b$, then $0 \leq ab$.

Example 6.5:

1. $(\mathbb{Z}, +, \cdot, \leq)$ is an ordered ring. We already showed that $(\mathbb{Z}, +, \cdot)$ is a ring in part 1 of Example 6.4.

   In Problem 12 from Problem Set 5, you were asked to show that $<$ is a well-defined strict linear ordering on $\mathbb{Z}$. $\leq$ is the corresponding linear ordering given by Theorem 3.13.

   Let's check that properties (1) and (2) above are satisfied.

   For (1), let $a, b, c, d, e, f \in \mathbb{N}$, and assume that $[(a, b)] \leq [(c, d)]$. Then $a + d \leq b + c$, and so, by part (x) of Problem 4 from Problem Set 5, $(a + d) + (e + f) \leq (b + c) + (e + f)$. We can then use associativity and commutativity of addition in $\mathbb{N}$, to get the inequality $(a + e) + (d + f) \leq (b + f) + (c + e)$. By definition, $[(a + e, b + f)] \leq [(c + e, d + f)]$. Therefore, $[(a, b)] + [(e, f)] \leq [(c, d)] + [(e, f)]$. For (2), let $a, b, c, d \in \mathbb{N}$, and assume that $[(0, 0)] \leq [(a, b)]$ and $[(0, 0)] \leq [(c, d)]$. Then $b \leq a$, and $d \leq c$. By part (ix) of Problem 4 from Problem Set 5, there are natural numbers $k > 0$ and $t > 0$ such that $a = b + k$ and $c = d + t$. Then we have

$$ad + bc = (b + k)d + b(d + t) = bd + kd + bd + bt.$$

$$ac + bd = (b + k)(d + t) + bd = (b + k)d + (b + k)t + bd = bd + kd + bt + kt + bd.$$

   It follows that $ac + bd = (ad + bc) + kt$. Therefore, again by part (ix) of Problem 4 from Problem Set 5, we have $ad + bc \leq ac + bd$. Thus, $[(0, 0)] \leq [(ac + bd, ad + bc)]$. Therefore, $[(0, 0)] \leq [(a, b)] \cdot [(c, d)]$.

2. $(\mathbb{Q}, +, \cdot, \leq)$ and $(\mathbb{R}, +, \cdot, \leq)$ are ordered fields. You will be asked to prove this in Problem 17 below.

3. $(\mathbb{C}, +, \cdot)$ **cannot** be ordered. To see this, assume toward contradiction that $\leq$ is an ordering of $(\mathbb{C}, +, \cdot)$. If $i \geq 0$, then $-1 = i^2 = i \cdot i \geq 0$ by property (2) of an ordered field. If $i \leq 0$, then by property (1) of an ordered field, $0 = -i + i \leq -i + 0 = -i$, and so, by property (2) of an ordered field, we have $-1 = i^2 = 1i^2 = (-1)(-1)i \cdot i = (-1i)(-1i) = (-i)(-i) \geq 0$ (here we have also used parts (vi) and (vii) of Problem 7 below together with commutativity and associativity of multiplication in $\mathbb{C}$, and the multiplicative identity property in $\mathbb{C}$). So, $-1 \geq 0$. By property (1) of an ordered field, we have $0 = -1 + 1 \geq 0 + 1 = 1$ and by property (2) of an ordered field, we have $1 = (-1)(-1) \geq 0$. Since $\leq$ is antisymmetric, $1 = 0$, a contradiction.

If $(R, +, \cdot, \leq)$ is an ordered ring, we will write $a < b$ if $a \leq b$ and $a \neq b$. By Theorem 3.12, $<$ is a strict linear ordering on R. We may also write $b \geq a$ in place of $a \leq b$ and $b > a$ in place of $a < b$.

In general, we may just use the name of the underlying set for a whole structure when there is no danger of confusion. For example, we may refer to the ring R or the ordered field F instead of the ring $(R, +, \cdot)$ or the ordered field $(F, +, \cdot, \leq)$.

Fields are particularly nice to work with because all the arithmetic and algebra we've learned through the years can be used in fields. For example, in the field of rational numbers, we can solve the equation $2x = 1$. The multiplicative inverse property allows us to do this. Indeed, the multiplicative inverse of 2 is $\frac{1}{2}$, and therefore, $x = \frac{1}{2}$ is a solution to the given equation. Compare this to the ring of integers. If we restrict ourselves to the integers, then the equation $2x = 1$ has no solution (see part 4 of Example 6.3).

Working with ordered fields is very nice as well. In the Problem Set below, you will be asked to derive some additional properties of fields and ordered fields that follow from the axioms. We will prove a few such properties now as examples.

Theorem 6.6: Let $(F, \leq)$ be an ordered field. Then for all $x \in F^*$, $x \cdot x > 0$.

Proof: There are two cases to consider:

Case 1: If $x > 0$, then $x \geq 0$ and $x \neq 0$. By property (2) of an ordered field, $x \cdot x \geq 0$. If $x \cdot x = 0$, then since x has an inverse, $x = 1 \cdot x = (x^{-1} \cdot x) \cdot x = x^{-1} \cdot (x \cdot x) = x^{-1} \cdot 0 = 0$ (by part (iii) of Problem 7 below), contrary to our assumption that $x \in F^*$. So, $x \cdot x \neq 0$, and therefore, $x \cdot x > 0$.

Case 2: If $x < 0$, then $x \leq 0$. By property (1) of an ordered field, $x + (-x) \leq 0 + (-x)$, and so, $0 \leq -x$. Therefore, $(-x)(-x) \geq 0$, by property (2) of an ordered field. Now, using Problem 7 (parts (vi) and (vii)) below, together with commutativity and associativity of multiplication in F, and the multiplicative identity property in F, we have

$$(-x)(-x) = (-1x)(-1x) = (-1)(-1)x \cdot x = 1(x \cdot x) = x \cdot x.$$

So, again we have $x \cdot x \geq 0$. The same argument used in case 1 can be used to rule out $x \cdot x = 0$. Therefore, $x \cdot x > 0$. $\qquad \square$

Theorem 6.7: Every ordered field $(F, +, \cdot, \leq)$ contains a copy of the natural numbers. Specifically, F contains a subset $\overline{\mathbb{N}} = \{\overline{n} \mid n \in \mathbb{N}\}$ such that for all $n, m \in \mathbb{N}$, we have $\overline{n + m} = \overline{n} + \overline{m}, \overline{n \cdot m} = \overline{n} \cdot \overline{m}$, and $n < m \leftrightarrow \overline{n} < \overline{m}$.

Proof: Let $(F, +, \cdot, \leq)$ be an ordered field. By the definition of a field, $0, 1 \in F$ and $0 \neq 1$.

We let $\overline{0} = 0$ and $\overline{n} = 1 + 1 + \cdots + 1$, where 1 appears n times. Let $\overline{\mathbb{N}} = \{\overline{n} \mid n \in \mathbb{N}\}$. Then $\overline{\mathbb{N}} \subseteq F$.

We first prove by induction on m that for all $n, m \in \mathbb{N}, \overline{n + m} = \overline{n} + \overline{m}$.

Base case ($k = 0$): $\overline{n + 0} = \overline{n} = \overline{n} + 0 = \overline{n} + \overline{0}$.

Inductive step: Suppose that $\overline{n + k} = \overline{n} + \overline{k}$. Then we have

$$\overline{n + (k + 1)} = \overline{(n + k) + 1} = \overline{n + k} + 1 = (\overline{n} + \overline{k}) + 1 = \overline{n} + (\overline{k} + 1) = \overline{n} + \overline{k + 1}.$$

By the Principle of Mathematical Induction, for all natural numbers $m, \overline{n+m} = \overline{n} + \overline{m}$.

Since $n \in \mathbb{N}$ was arbitrary, we have that for all $n, m \in \mathbb{N}, \overline{n+m} = \overline{n} + \overline{m}$.

Similarly, we now prove by induction on m that for all $n, m \in \mathbb{N}, \overline{n \cdot m} = \overline{n} \cdot \overline{m}$.

Base case $(k = 0)$: $\overline{n \cdot 0} = \overline{0} = \overline{n} \cdot \overline{0}$.

Inductive step: Suppose that $\overline{n \cdot k} = \overline{n} \cdot \overline{k}$. Then we have

$$\overline{n \cdot (k+1)} = \overline{nk + n} = \overline{nk} + \overline{n} = \overline{n} \cdot \overline{k} + \overline{n} = \overline{n}(\overline{k} + 1) = \overline{n}(\overline{k} + \overline{1}) = \overline{n}(\overline{k+1}).$$

By the Principle of Mathematical Induction, for all natural numbers $m, \overline{n \cdot m} = \overline{n} \cdot \overline{m}$.

Since $n \in \mathbb{N}$ was arbitrary, we have that for all $n, m \in \mathbb{N}, \overline{n \cdot m} = \overline{n} \cdot \overline{m}$.

We now wish to prove that for all $n, m \in \mathbb{N}, n < m \leftrightarrow \overline{n} < \overline{m}$.

We first note that since $1 > 0$ (because $1 = 1 \cdot 1 > 0$ by Theorem 6.6), we have for any natural number $n, \overline{n+1} = \overline{n} + 1 > \overline{n} + 0 = \overline{n}$ by property (1) of an ordered field.

We now prove by induction on $m \in \mathbb{N}$ that $n < m \rightarrow \overline{n} < \overline{m}$.

The base case $k = 0$ is vacuously true because $n < 0$ never occurs. For the inductive step, let $k \in \mathbb{N}$ and assume that $n < k \rightarrow \overline{n} < \overline{k}$. Now, suppose that $n < k + 1 = k \cup \{k\}$. Then $n < k$ or $n = k$. If $n < k$, then $\overline{n} < \overline{k}$ by the inductive hypothesis. By property (1) of an ordered field and $0 < 1$, we have $\overline{k} = \overline{k} + 0 < \overline{k} + 1 = \overline{k+1}$. By the transitivity of $<, \overline{n} < \overline{k+1}$. If $n = k$, then by the same reasoning in the last sentence, $\overline{k} < \overline{k+1}$. By the Principle of Mathematical Induction, we have for all $m \in \mathbb{N}$, $n < m \rightarrow \overline{n} < \overline{m}$.

Since $n \in \mathbb{N}$ was arbitrary, we have that for all $n, m \in \mathbb{N}, n < m \rightarrow \overline{n} < \overline{m}$.

Finally, we need to prove that $\overline{n} < \overline{m} \rightarrow n < m$. By the law of the contrapositive, this statement is equivalent to $m \leq n \rightarrow \overline{m} \leq \overline{n}$. If $m < n$, then $\overline{m} < \overline{n}$, by the same reasoning used in the last paragraph. Otherwise, $m = n$, in which case $\overline{m} = \overline{n}$. $\quad\square$

Note: The function that sends $n \in \mathbb{N}$ to $\overline{n} \in \overline{\mathbb{N}}$ is called an **isomorphism**. It has the following properties: (i) $\overline{n+m} = \overline{n} + \overline{m}$; (ii) $\overline{n \cdot m} = \overline{n} \cdot \overline{m}$; (iii) $n < m$ if and only if $\overline{n} < \overline{m}$; (iv) the function provides a bijection between the elements of $\mathbb{N}$ and the elements of $\overline{\mathbb{N}}$.

So, when we say that every field contains a "copy" of the natural numbers, we mean that there is a subset $\overline{\mathbb{N}}$ of the field so that $(\overline{\mathbb{N}}, +, \cdot, \leq)$ is isomorphic to $(\mathbb{N}, +, \cdot, \leq)$.

Theorem 6.8: Let $(F, \leq)$ be an ordered field and let $x \in F$ with $x > 0$. Then $\frac{1}{x} > 0$.

Proof: Since $x \neq 0, \frac{1}{x} = x^{-1}$ exists and is nonzero.

Assume toward contradiction that $\frac{1}{x} < 0$. Then $\frac{1}{x} \le 0$, and so, by property (1) of an ordered field, $\frac{1}{x} + \left(-\frac{1}{x}\right) \le -\frac{1}{x}$, or equivalently, $0 \le -\frac{1}{x}$. Using Problem 7 (part (vi)) below, together with commutativity and associativity of multiplication, the multiplicative inverse property, and the multiplicative identity property, $x\left(-\frac{1}{x}\right) = x(-1)x^{-1} = -1xx^{-1} = -1 \cdot 1 = -1$. Since $x > 0$ and $-\frac{1}{x} \ge 0$, by property (2) of an ordered field, $-1 = x\left(-\frac{1}{x}\right) \ge 0$. So, by property (1) of an ordered field, $0 = 1 + (-1) \ge 1$. But by Theorem 6.6, $1 = 1 \cdot 1 > 0$. This is a contradiction. Thus, $\frac{1}{x} > 0$. $\qquad\square$

Why Isn't $\mathbb{Q}$ Enough?

At first glance, it would appear that the ordered field of rational numbers would be sufficient to solve all "real world" problems. However, a long time ago, a group of people called the Pythagoreans showed that this is not the case. The problem was first discovered when applying the now well-known Pythagorean Theorem.

Theorem 6.9 (Pythagorean Theorem): In a right triangle with legs of lengths a and b, and a hypotenuse of length c, $c^2 = a^2 + b^2$.

The picture to the right shows a right triangle. The vertical and horizontal segments (labeled a and b, respectively) are called the **legs** of the right triangle, and the side opposite the right angle (labeled c) is called the **hypotenuse** of the right triangle.

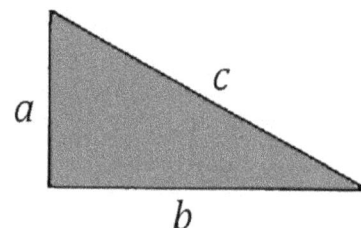

There are many ways to prove the Pythagorean Theorem. Here, we will provide a simple geometric argument. For the proof we will want to recall that the area of a square with side length s is $A = s^2$, and the area of a triangle with base b and height h is $A = \frac{1}{2}bh$. Notice that in our right triangle drawn here, the base is labeled b (how convenient), and the height is labeled a. So, the area of this right triangle is $A = \frac{1}{2}ba = \frac{1}{2}ab$.

Proof of Theorem 6.9: We draw 2 squares, each of side length $a + b$, by rearranging 4 copies of the given triangle in 2 different ways:

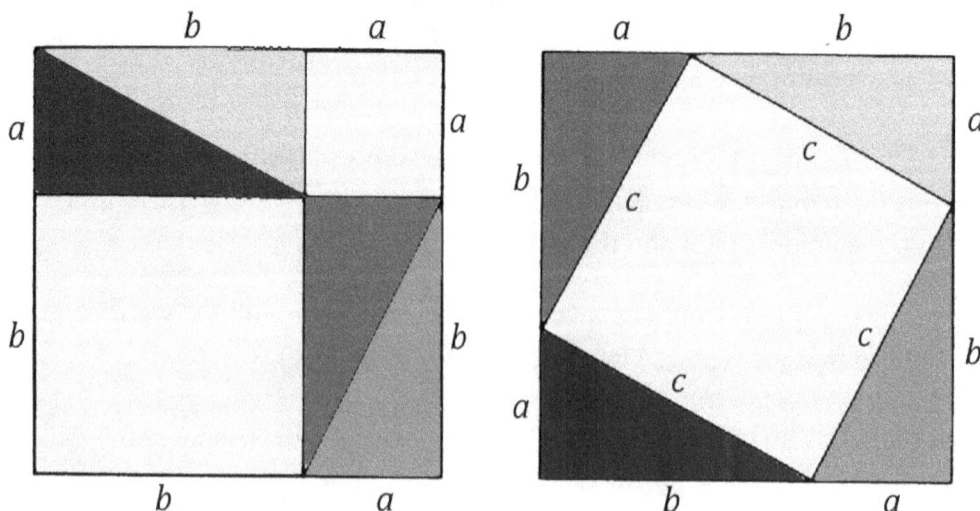

We can get the area of each of these squares by adding the areas of all the figures that comprise each square.

The square on the left consists of 4 copies of the given right triangle, a square of side length a and a square of side length b. It follows that the area of this square is $4 \cdot \frac{1}{2}ab + a^2 + b^2 = 2ab + a^2 + b^2$.

The square on the right consists of 4 copies of the given right triangle, and a square of side length c. It follows that the area of this square is $4 \cdot \frac{1}{2}ab + c^2 = 2ab + c^2$.

Since the areas of both squares of side length $a + b$ are equal (both areas are equal to $(a + b)^2$), $2ab + a^2 + b^2 = 2ab + c^2$. Cancelling $2ab$ from each side of this equation yields $a^2 + b^2 = c^2$. $\quad \square$

Question: In a right triangle where both legs have length 1, what is the length of the hypotenuse?

Let's try to answer this question. If we let c be the length of the hypotenuse of the triangle, then by the Pythagorean Theorem, we have $c^2 = 1^2 + 1^2 = 1 + 1 = 2$. Since $c^2 = c \cdot c$, we need to find a number with the property that when you multiply that number by itself you get 2. The Pythagoreans showed that if we use only numbers in $\mathbb{Q}$, then no such number exists.

Theorem 6.10: There does not exist a rational number a such that $a^2 = 2$.

Analysis: We will prove this theorem by assuming that there is a rational number a such that $a^2 = 2$, and arguing until we reach a contradiction. A first attempt at a proof would be to let $a = \frac{m}{n} \in \mathbb{Q}$ satisfy $\left(\frac{m}{n}\right)^2 = 2$. It follows that $\boldsymbol{m^2 = 2n^2}$ ($\frac{m^2}{n^2} = \frac{m \cdot m}{n \cdot n} = \frac{m}{n} \cdot \frac{m}{n} = \left(\frac{m}{n}\right)^2$ and $2 = \frac{2}{1} \Rightarrow \frac{m^2}{n^2} = \frac{2}{1} \Rightarrow m^2 = 2n^2$), showing that $\boldsymbol{m^2}$ **is even**. We will then use this information to show that both m and n are even (we will need a preliminary Lemma to help us with this).

Now, in our first attempt, the fact that m and n both turned out to be even did not produce a contradiction. However, we can modify the beginning of the argument to make this happen.

Remember that every rational number has infinitely many representations. For example, $\frac{6}{12}$ is the same rational number as $\frac{2}{4}$ (because $6 \cdot 4 = 12 \cdot 2$). Notice that in both representations, the numerator (number on the top) and the denominator (number on the bottom) are even. However, they are both equivalent to $\frac{1}{2}$, which has the property that the numerator is not even.

In Problem 16 below, you will be asked to show that every rational number can be written in the form $\frac{m}{n}$, where at least one of m or n is **not** even. We can now adjust our argument to get the desired contradiction.

Recall that an integer n is **even** if there is another integer b such that $n = 2b$. For example, 6 and -22 are even because $6 = 2 \cdot 3$ and $-22 = 2 \cdot (-11)$. An integer n is **odd** if there is another integer b such that $n = 2b + 1$. For example, 7 and -23 are odd because $7 = 2 \cdot 3 + 1$ and $-23 = 2 \cdot (-12) + 1$.

Lemma 6.11: Every integer is even or odd, but not both.

Proof: We already proved in Theorem 5.15 that every natural number is even or odd. If $n < 0$ is an integer, then $-n > 0$, and so, there is a natural number j such that $-n = 2j$ or $-n = 2j + 1$. If $-n = 2j$, then $n = 2(-j)$ (and since $j \in \mathbb{N}$, $-j \in \mathbb{Z}$). If $-n = 2j + 1$, then

$$n = -(2j + 1) = -2j - 1 = -2j - 1 - 1 + 1 \text{ (SACT)} = -2j - 2 + 1 = 2(-j - 1) + 1.$$

Here we used the fact that $(\mathbb{Z}, +, \cdot)$ is a ring. Since $\mathbb{Z}$ is closed under addition, it follows that $-j - 1 = -j + (-1) \in \mathbb{Z}$.

Now, if $n = 2j$ and $n = 2k + 1$, then $2j = 2k + 1$. So, we have

$$2(j - k) = 2j - 2k = (2k + 1) - 2k = 2k + (1 - 2k) = 2k + (-2k + 1)$$
$$= (2k - 2k) + 1 = 0 + 1 = 1.$$

So, $2(j - k) = 1$. But by part 4 of Example 6.3, 2 does not have a multiplicative inverse in $\mathbb{Z}$, and so, this is a contradiction. $\qquad\square$

Lemma 6.12: The product of two odd integers is odd.

Proof: Let m and n be odd integers. Then there are integers j and k such that $m = 2j + 1$ and $n = 2k + 1$. So,

$$m \cdot n = (2j + 1) \cdot (2k + 1) = (2j + 1)(2k) + (2j + 1)(1) = (2k)(2j + 1) + (2j + 1)$$
$$= \big((2k)(2j) + 2k\big) + (2j + 1) = \big(2(k(2j)) + 2k\big) + (2j + 1) = 2(k(2j) + k) + (2j + 1)$$
$$= (2(k(2j) + k) + 2j) + 1 = 2\big((k(2j) + k) + j\big) + 1.$$

Here we used the fact that $(\mathbb{Z}, +, \cdot)$ is a ring. (Which properties did we use?) Since $\mathbb{Z}$ is closed under addition and multiplication, we have $(k(2j) + k) + j \in \mathbb{Z}$. Therefore, mn is odd. $\qquad\square$

Proof of Theorem 6.10: Assume, toward contradiction, that there is a rational number a such that $a^2 = 2$. Since a is a rational number, there are $m \in \mathbb{Z}$ and $n \in \mathbb{Z}^*$, **not both even**, so that $a = \frac{m}{n}$.

So, we have $\frac{m^2}{n^2} = \frac{m \cdot m}{n \cdot n} = \frac{m}{n} \cdot \frac{m}{n} = a \cdot a = a^2 = 2 = \frac{2}{1}$. Thus, $m^2 \cdot 1 = n^2 \cdot 2$. So, $m^2 = 2n^2$. Therefore, m^2 is even. If m were odd, then by Lemma 6.12, $m^2 = m \cdot m$ would be odd. So, **m is even**.

Since m is even, there is $k \in \mathbb{Z}$ such that $m = 2k$. Replacing m by $2k$ in the equation $m^2 = 2n^2$ gives us $2n^2 = m^2 = (2k)^2 = (2k)(2k) = 2\big(k(2k)\big)$. So, $n^2 = k(2k)$ (we can use part (viii) of Problem 4 from Problem Set 5 to easily prove that if $a, b, c \in \mathbb{Z}$, then $ca = cb \to a = b$). Using associativity and commutativity of multiplication in $\mathbb{Z}$, we have $k(2k) = (k \cdot 2)k = (2k)k = 2(k \cdot k)$. So, $n^2 = 2(k \cdot k)$, and we see that n^2 is even. Again, by Lemma 6.12, **n is even**.

So, we have m even and n even, contrary to our original assumption that m and n are not both even. Therefore, there is no rational number a such that $a^2 = 2$. $\qquad\square$

It turns out that $\mathbb{Q}$ fails to have an element a such that $a^2 = 2$ because $\mathbb{Q}$ is not "complete." Luckily $\mathbb{R}$ doesn't have this deficiency, as we will show shortly.

Completeness

Let F be an ordered field and let S be a nonempty subset of F. We say that S is **bounded above** if there is $M \in F$ such that for all $s \in S$, $s \leq M$. Each such number M is called an **upper bound** of S.

In words, an upper bound of a set S is simply an element from the field that is at least as big as every element in S.

Similarly, we say that S is **bounded below** if there is $K \in F$ such that for all $s \in S$, $K \leq s$. Each such number K is called a **lower bound** of S.

In words, a lower bound of a set S is simply an element from the field that is no bigger than any element in S.

We will say that S is **bounded** if it is both bounded above and bounded below. Otherwise S is **unbounded**.

A **least upper bound** of a set S is an upper bound that is smaller than any other upper bound of S, and a **greatest lower bound** of S is a lower bound that is larger than any other lower bound of S.

Example 6.13: Let F be an ordered field with $\mathbb{Q} \subseteq F$.

Note: The only two examples of F that we are interested in are $\mathbb{Q}$ (the set of rational numbers) and $\mathbb{R}$ (the set of real numbers). As you look at the set in each example below, think about what it looks like as a subset of $\mathbb{Q}$ and as a subset of $\mathbb{R}$.

1. $S = \{1, 2, 3, 4, 5\}$ is bounded.

    5 is an upper bound of S, as is any number larger than 5. The number 5 is special in the sense that there are no upper bounds smaller than it. So, 5 is the **least** upper bound of S.

    Similarly, 1 is a lower bound of S, as is any number smaller than 1. The number 1 is the **greatest** lower bound of S because there are no lower bounds larger than it.

    Notice that the least upper bound and greatest lower bound of S are inside the set S itself. This will always happen when the set S is finite.

2. $T = \{x \in F \mid -2 < x \leq 2\}$ is also bounded. Any number greater than or equal to 2 is an upper bound of T, and any number less than or equal to -2 is a lower bound of T.

    2 is the least upper bound of T and -2 is the greatest lower bound of T.

    Note that the least upper bound of T is in T, whereas the greatest lower bound of T is not in T.

3. $U = \{x \in F \mid x < -3\}$ is bounded above by any number greater than or equal to -3, and -3 is the least upper bound of U. The set U is not bounded below, and therefore, U is unbounded.

4. $V = \{x \in F \mid x^2 < 2\}$ is bounded above by 2. To see this, note that if $x > 2$, then $x^2 > 4 \geq 2$ (the reader should verify that for all $a, b \in F^+$, $a > b \rightarrow a^2 > b^2$), and therefore, $x \notin V$. Any number greater than 2 is also an upper bound. Is 2 the least upper bound of V? It's not! For example, $\frac{3}{2}$ is also an upper bound. Indeed, if $x > \frac{3}{2}$, then $x^2 > \frac{9}{4} \geq 2$.

Does V have a least upper bound? A moment's thought might lead you to suspect that a least upper bound M would satisfy $M^2 = 2$. And it turns out that you are right! (Proving this, however, is quite difficult—see Problem 22 below). Clearly, this least upper bound M is not in the set V. The big question is "Does M exist at all?"

Well, if $F = \mathbb{Q}$, then by Theorem 6.10, M **does not** exist in F. In this case, V is an example of a set which is bounded above in $\mathbb{Q}$, but has no least upper bound in $\mathbb{Q}$.

So, if we want an ordered field F containing $\mathbb{Q}$ where M does exist, we can insist that F has the property that any set which is bounded above in F has a least upper bound in F. It turns out that $\mathbb{R}$ is such an ordered field.

Many authors use the term **supremum** for "least upper bound" and **infimum** for "greatest lower bound," and they may write sup A and inf A for the supremum and infimum of a set A, respectively (if they exist).

In Example 6.13 above, we stated the least upper bound and greatest lower bound of the sets S, T, U, and V without proof. Intuitively, it seems reasonable that those numbers are correct. Let's do one of the examples carefully.

Theorem 6.14: Let $U = \{x \in F \mid x < -3\}$. Then sup $U = -3$.

Analysis: We need to show that -3 is an upper bound of U, and that any number less than -3 is **not** an upper bound of U. That -3 is an upper bound of U follows immediately from the definition of U.

The harder part of the argument is showing that a number less than -3 is not an upper bound of U. However, conceptually it's not hard to see that this is true. If $a < -3$, we simply need to find some number x between a and -3. Here is a picture of the situation.

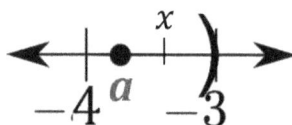

Notice that a can be very close to -3 and we don't know exactly what a is—we know only that it's less than -3. So, we need to be careful how we choose x. The most natural choice for x would be to go midway between a and -3. In other words, we can take the average of a and -3. So, we will let $x = \frac{1}{2}(a + (-3))$. Then we just need to verify that $a < x$ and that $x \in U$ (that is, $x < -3$).

Proof of Theorem 6.14: If $x \in U$, then $x < -3$ by definition, and so, -3 is an upper bound of U.

Suppose that $a < -3$ (or equivalently, $-a - 3 > 0$). We want to show that a is **not** an upper bound of U. To do this, we let $x = \frac{1}{2}(a - 3) = 2^{-1}(a + (-3))$. $x \in F$ because F is closed under addition and multiplication, and the multiplicative inverse property holds in F^*. We will show that $a < x < -3$.

$$x - a = \frac{1}{2}(a - 3) - a = \frac{1}{2}(a - 3) - \frac{1}{2}(2a) = \frac{1}{2}(a - 3 - 2a) = \frac{1}{2}(a - 2a - 3) = \frac{1}{2}(-a - 3).$$

Since $\frac{1}{2} > 0$ (by Theorem 6.8) and $-a - 3 > 0$, it follows that $x - a > 0$, and therefore, $x > a$.

$$-3 - x = -3 - \frac{1}{2}(a - 3) = \frac{1}{2}(-6) - \frac{1}{2}a + \frac{1}{2} \cdot 3 = \frac{1}{2}(-6 - a + 3) = \frac{1}{2}(-a - 3).$$

Again, since $\frac{1}{2} > 0$ and $-a - 3 > 0$, it follows that $-3 - x > 0$, and therefore, $x < -3$. Thus, $x \in U$.

So, we found an element $x \in U$ (because $x < -3$) with $a < x$. This shows that a is **not** an upper bound of U. It follows that $-3 = \sup U$. $\quad\square$

An ordered field F has the **Completeness Property** if every nonempty subset of F that is bounded above in F has a least upper bound in F. In this case, we say that F is a **complete ordered field**.

Our next goal is to provide an outline for a proof that the ordered field $\mathbb{R}$ is complete. In order to do this, we will need a few preliminary results.

Theorem 6.15: Every Cauchy sequence of rational numbers is bounded by a rational number.

Analysis: First note that any finite set A of rational numbers is bounded. If a and b are the least and greatest rational numbers in A, respectively, then for any $q \in A$, we have $a \le q \le b$. In this case, we can also write $|q| \le M$, where $M = \max\{|a|, |b|\}$ (note that $|q| \le M$ is equivalent to $-M \le q \le M$), and we may say that A is bounded by M.

Also, note that if A is bounded by M_1 and B is bounded by M_2, then $A \cup B$ is bounded by $\max\{M_1, M_2\}$.

By the previous remarks, we need only show that some "**tail**" of the Cauchy sequence of rational numbers is bounded (by a tail, we mean we can forget about finitely many terms of the sequence and just consider the infinitely many terms from some point on). By the definition of Cauchy sequence, we can choose a natural number K so that the distance between x_{K+1} and x_m is less than 1 for all $m > K$. That is, we find $K \in \mathbb{N}$ so that $\forall m > K(|x_m - x_{K+1}| < 1)$. Finally, we use the following little trick:

$$|x_m| = |(x_m - x_{K+1}) + x_{K+1}| \le |x_m - x_{K+1}| + |x_{K+1}| < 1 + |x_{K+1}|$$

For the first equality we used the Standard Advanced Calculus Trick (SACT) from Note 7 following the proof of Theorem 5.11. We wanted $x_m - x_{K+1}$ to appear, so we made it appear. We then added x_{K+1} to correct the damage that we did.

We then used the Triangle Inequality, which was described in the Note following Example 5.18.

Let's write out the details of the proof.

Proof: Let (x_n) be a Cauchy sequence. Then for every $k \in \mathbb{N}^+$, there is $K \in \mathbb{N}$ such that $m \ge n > K$ implies $|x_m - x_n| < \frac{1}{k}$. In particular, by letting $k = 1$, we see that there is $K \in \mathbb{N}$ such that $m \ge n > K$ implies $|x_m - x_n| < 1$. By the Triangle Inequality (and SACT), we have

$$|x_m| = |(x_m - x_n) + x_n| \le |x_m - x_n| + |x_n|.$$

Therefore, $m \ge n > K$ implies $|x_m| \le |x_m - x_n| + |x_n| < 1 + |x_n|$. Letting $n = K + 1$, we see that $m > K$ implies $|x_m| < 1 + |x_{K+1}|$.

Let $M = \max\{|x_0|, |x_1|, \dots, |x_K|, 1 + |x_{K+1}|\}$. Then for all $m \in \mathbb{N}$, $|x_m| \le M$. $\quad\square$

By Theorem 6.7, every ordered field F contains an isomorphic copy of $\mathbb{N}$. For example, $\mathbb{Q}$ contains $\left\{\frac{n}{1} \mid n \in \mathbb{N}\right\}$ and $\mathbb{R}$ contains $\left\{\left[\left(\frac{n}{1}\right)\right] \mid n \in \mathbb{N}\right\}$. To avoid using messy notation, we will usually abuse notation just a bit and use the name $\mathbb{N}$ for any isomorphic copy of $\mathbb{N}$. So, if F is an ordered field, we can write $\mathbb{N} \subseteq F$.

We say that an ordered field F has the **Archimedean Property** if $\mathbb{N}$ is unbounded in F. Symbolically, we write $\forall x \in F \, \exists n \in \mathbb{N}(n > x)$.

Theorem 6.16: $\mathbb{Q}$ and $\mathbb{R}$ both have the Archimedean Property.

Proof: We first show that $\mathbb{Q}$ has the Archimedean Property. Let $q \in \mathbb{Q}$. If $q \leq 0$, then since $0 < 1$, by the transitivity of $<$, we have $q < 1$. If $q > 0$, then there are $a, b \in \mathbb{Z}^+$ with $q = \frac{a}{b}$. We have

$$(a+1) - \frac{a}{b} = \frac{(a+1)}{1} + \left(\frac{-a}{b}\right) = \frac{(a+1)b + 1 \cdot (-a)}{1b} = \frac{ab + b - a}{b} = \frac{a(b-1) + b}{b}.$$

Since $b \in \mathbb{Z}^+$, $b - 1 \geq 0$, and so, $a(b-1) \geq 0$. Therefore, $a(b-1) + b \geq 0 + b > 0$. Thus, $(a+1) - \frac{a}{b} = \frac{a(b-1)+b}{b} > 0$. Therefore, we have $a + 1 > \frac{a}{b} = q$. Since $a, 1 \in \mathbb{N}$ and $\mathbb{N}$ is closed under addition, it follows that $a + 1 \in \mathbb{N}$. Since $q \in \mathbb{Q}$ was arbitrary, we have shown that $\mathbb{Q}$ has the Archimedean Property.

Next, we show that $\mathbb{R}$ has the Archimedean Property. Let $[(x_n)] \in \mathbb{R}$. We want to find a natural number $[(t)]$ such that $[(t)] > [(x_n)]$. Suppose toward contradiction, that for all $t \in \mathbb{N}$, $[(t)] \leq [(x_n)]$. Then for each $t \in \mathbb{N}$, there is K_t such that $n > K_t$ implies $t \leq x_n$. Since (x_n) is a Cauchy sequence, by Theorem 6.15, (x_n) is bounded by a rational number M. So, for all $n \in \mathbb{N}$, $x_n \leq M$. Let $t \in \mathbb{N}$ and choose K_t such that $n > K_t$ implies $t \leq x_n$. Since $x_n \leq M$, by the transitivity of $\leq$, we have $t \leq M$. Since $t \in \mathbb{N}$ was arbitrary, we see that $\mathbb{N}$ is bounded by the rational number M, contradicting that $\mathbb{Q}$ has the Archimedean Property. This contradiction proves that $\mathbb{R}$ has the Archimedean Property. $\square$

Theorem 6.17 (The Density Theorem): If $x, y \in \mathbb{R}$ with $x < y$, then there is $q \in \mathbb{Q}$ with $x < q < y$.

In other words, the Density Theorem says that between any two real numbers we can always find a rational number. We say that $\mathbb{Q}$ is **dense** in $\mathbb{R}$.

To help understand the proof, let's first run a simple simulation using a specific example. Let's let $x = \frac{16}{3}$ and $y = \frac{17}{3}$. We begin by subtracting to get $y - x = \frac{1}{3}$. This is the distance between x and y. We wish to find a natural number n such that $\frac{1}{n}$ is smaller than this distance. In other words, we want $\frac{1}{n} < \frac{1}{3}$, or equivalently, $n > 3$. So, we can let n be any natural number greater than 3, say $n = 4$. We now want to "shift" $\frac{1}{n} = \frac{1}{4}$ to the right to get a rational number between x and y. We can do this as follows. We multiply n times x to get $nx = 4 \cdot \frac{16}{3} = \frac{64}{3}$. We then let m be the **least** integer greater than nx. So, $m = \frac{66}{3} = 22$. Finally, we let $q = \frac{m}{n} = \frac{22}{4} = \frac{11}{2}$. And we did it! Indeed, we have $\frac{16}{3} < \frac{11}{2} < \frac{17}{3}$. The reader should confirm that these inequalities hold. Let's write out the details of the proof.

Proof: Let's first consider the case where $0 \leq x < y$. Let $z = y - x = y + (-x)$. Since $\mathbb{R}$ has the additive inverse property and is closed under addition, $z \in \mathbb{R}$. Also, $z > 0$. By the Archimedean Property, there is $n \in \mathbb{N}$ such that $n > \frac{1}{z}$. Using Problem 11 (part (i)) below, we have $\frac{1}{n} < z$.

By the Archimedean Property once again, there is $m \in \mathbb{N}$ such that $m > nx$. Therefore, $\frac{m}{n} > x$ (Check this!). So, $\left\{ m \in \mathbb{N} \mid \frac{m}{n} > x \right\} \neq \emptyset$. By the Well Ordering Principle, $\left\{ m \in \mathbb{N} \mid \frac{m}{n} > x \right\}$ has a least element, let's call it k. Since $k > 0$, (because $x \geq 0$ and $n > 0$) and k is the **least** natural number such that $\frac{k}{n} > x$, it follows that $k - 1 \in \mathbb{N}$ and $\frac{k-1}{n} \leq x$, or equivalently, $\frac{k}{n} - \frac{1}{n} \leq x$. Therefore, we have $\frac{k}{n} \leq x + \frac{1}{n} < x + z = x + (y - x) = y$. Thus, $x < \frac{k}{n} < y$. Since $k, n \in \mathbb{N}$, we have $\frac{k}{n} \in \mathbb{Q}$.

Now, we consider the case where $x < 0$ and $x < y$. By the Archimedean Property, there is $t \in \mathbb{N}$ such that $t > -x$. Then, we have $0 < x + t < y + t$. So, $x + t$ and $y + t$ satisfy the first case above. Thus, there is $q \in \mathbb{Q}$ with $x + t < q < y + t$. It follows that $x < q - t < y$. Since $t \in \mathbb{N}$, $-t \in \mathbb{Z}$. Since $\mathbb{Z} \subseteq \mathbb{Q}$, $-t \in \mathbb{Q}$. So, we have $q, -t \in \mathbb{Q}$. Since $\mathbb{Q}$ is closed under addition, $q - t = q + (-t) \in \mathbb{Q}$. $\quad \square$

Theorem 6.18: $\mathbb{R}$ is a complete ordered field.

The proof of this theorem requires several steps. I will give a brief outline of these steps and leave the details as an exercise for the reader (see Problem 23 below).

Proof outline for Theorem 6.18: We already know that $\mathbb{R}$ is an ordered field. All that's left to show is that $\mathbb{R}$ is complete. So, let $S \subseteq \mathbb{R}$ with $S \neq \emptyset$ be bounded above. Then there is $b_0 \in \mathbb{R}$ such that $\forall s \in S, s \leq b_0$ (b_0 is an upper bound of S). Since $S \neq \emptyset$, there is $a_0 \in S$. Let $c_0 = \frac{1}{2}(a_0 + b_0)$. If c_0 is an upper bound of S, let $a_1 = a_0$ and $b_1 = c_0$. If c_0 is **not** an upper bound of S, let $a_1 = c_0$ and $b_1 = b_0$. We continue in this fashion to inductively create sequences of real numbers (a_n), (b_n), and (c_n), where $c_n = \frac{1}{2}(a_n + b_n)$, if c_n is an upper bound of S, $a_{n+1} = a_n$ and $b_{n+1} = c_n$, and if c_n is **not** an upper bound of S, $a_{n+1} = c_n$ and $b_{n+1} = b_n$. Then let $c_{n+1} = \frac{1}{2}(a_n + b_n)$.

Step 1: Use induction on $\mathbb{N}$ to prove that for all $n \in \mathbb{N}$, $a_{n+1} \geq a_n$ and $b_{n+1} \leq b_n$.

Step 2: Use Step 1 to prove that (a_n) and (b_n) are **real-valued** Cauchy sequences.

Step 3: Use Theorems 6.16 and 6.17 to prove that there is $x \in \mathbb{R}$ such that the following holds:

$$\text{for every } k \in \mathbb{N}^+, \text{ there is } K \in \mathbb{N} \text{ such that } n > K \text{ implies } |b_n - x| < \frac{1}{k}.$$

In this case, we say that (b_n) converges to x and we write $b_n \to x$.

Step 4: Use Theorem 6.16 to prove that $a_n \to x$.

Step 5: Use the sequence (b_n) to prove that x is an upper bound of S.

Step 6: Use the sequence (a_n) to prove that if $y < x$, then y is **not** an upper bound of S.

Completing these steps will prove the theorem. $\quad \square$

Problem Set 6

Full solutions to these problems are available for free download here:
www.SATPrepGet800.com/TFBZLF

LEVEL 1

1. Show that there are exactly two monoids on the set $S = \{e, a\}$, where e is the identity. Which of these monoids are groups? Which of these monoids are commutative?

2. The addition and multiplication tables below are defined on the set $S = \{0, 1\}$. Show that $(S, +, \cdot)$ does **not** define a ring.

+	0	1		$\cdot$	0	1
0	0	1		0	1	0
1	1	0		1	0	1

3. The addition and multiplication tables below are defined on the set $S = \{0, 1, 2\}$. Show that $(S, +, \cdot)$ does **not** define a field.

+	0	1	2		$\cdot$	0	1	2
0	0	1	2		0	0	0	0
1	1	2	0		1	0	1	2
2	2	0	1		2	0	2	2

4. Let $F = \{0, 1\}$, where $0 \neq 1$. Show that there is exactly one field $(F, +, \cdot)$, where 0 is the additive identity and 1 is the multiplicative identity.

LEVEL 2

5. Let $G = \{e, a, b\}$ and let $(G, \star)$ be a group with identity element e. Draw a multiplication table for $(G, \star)$.

6. Prove that in any monoid $(M, \star)$, the identity element is unique.

7. Let $(F, +, \cdot)$ be a field. Prove each of the following:

   (i)   If $a, b \in F$ with $a + b = b$, then $a = 0$.

   (ii) If $a \in F$, $b \in F^*$, and $ab = b$, then $a = 1$.

   (iii) If $a \in F$, then $a \cdot 0 = 0$.

   (iv) If $a \in F^*$, $b \in F$, and $ab = 1$, then $b = \frac{1}{a}$.

   (v)   If $a, b \in F$ and $ab = 0$, then $a = 0$ or $b = 0$.

   (vi) If $a \in F$, then $-a = -1a$.

   (vii) $(-1)(-1) = 1$.

8. Let $(F, +, \cdot)$ be a field with $\mathbb{N} \subseteq F$. Prove that $\mathbb{Q} \subseteq F$.

LEVEL 3

9. Assume that a group $(G, \star)$ of order 4 exists with $G = \{e, a, b, c\}$, where e is the identity, $a^2 = b$ and $b^2 = e$. Construct the multiplication table for the operation of such a group.

10. Prove that in any group $(G, \star)$, each element has a unique inverse.

11. Let $(F, +, \cdot, \leq)$ be an ordered field. Prove each of the following:

    (i)   If $a, b \in F^+$ and $a > b$, then $\frac{1}{a} < \frac{1}{b}$.

    (ii) If $a, b \in F$, then $a \geq b$ if and only if $-a \leq -b$.

12. Let $(F, +, \cdot)$ be a field. Show that $(F, \cdot)$ is a commutative monoid.

LEVEL 4

13. Let $(G, \star)$ be a group with $a, b \in G$, and let a^{-1} and b^{-1} be the inverses of a and b, respectively. Prove

    (i) $(a \star b)^{-1} = b^{-1} \star a^{-1}$.

    (ii) the inverse of a^{-1} is a.

14. Prove that there is no smallest positive real number.

15. Let a be a nonnegative real number. Prove that $a = 0$ if and only if a is less than every positive real number. (Note: a nonnegative means $a \geq 0$.)

16. Prove that every rational number can be written in the form $\frac{m}{n}$, where $m \in \mathbb{Z}$, $n \in \mathbb{Z}^*$, and at least one of m or n is **not** even.

LEVEL 5

17. Prove that $(\mathbb{Q}, +, \cdot, \leq)$ and $(\mathbb{R}, +, \cdot, \leq)$ are ordered fields. Also prove that $(\mathbb{C}, +, \cdot)$ is field that cannot be ordered.

18. Prove that every nonempty set of real numbers that is bounded below has a greatest lower bound in $\mathbb{R}$.

19. Show that between any two real numbers there is a real number that is **not** rational.

20. Let $T = \{x \in F \mid -2 < x \leq 2\}$. Prove $\sup T = 2$ and $\inf T = -2$.

21. Prove that every interval of real numbers is of one of the nine types described after Example 3.14.

22. Let $V = \{x \in F \mid x^2 < 2\}$ and let $a = \sup V$. Prove that $a^2 = 2$.

23. Prove that $\mathbb{R}$ is a complete ordered field.

24. Let $D = \{A, B\}$ be a partition of $\mathbb{Q}$ such that $A \neq \emptyset$, $A \neq \mathbb{Q}$, A has no greatest element, and every element of A is less than every element of B. D is called a **Dedekind cut**. Let $X = \{D \mid D \text{ is a Dedekind cut}\}$. Prove that $+$, $\cdot$, and $\leq$ can be defined on X so that $(X, +, \cdot, \leq)$ is a complete ordered field that is isomorphic to $\mathbb{R}$.

25. Prove that any two complete ordered fields are isomorphic.

LESSON 7
BASIC TOPOLOGY OF $\mathbb{R}$ AND $\mathbb{C}$

Absolute Value and Distance

If x and y are real or complex numbers such that $y = x^2$, then we call x a **square root** of y. If x is a positive real number, then we say that x is the **positive square root** of y and we write $x = \sqrt{y}$.

Note: In this book, we will use the square root symbol only for the positive square root of a positive real number and occasionally for 0 ($0 = \sqrt{0}$ because $0 = 0^2$).

Example 7.1:

1. Since $2^2 = 4$, $2 \in \mathbb{R}$, and $2 > 0$, we see that 2 is the positive square root of 4 and we write $2 = \sqrt{4}$.

2. We have $(-2)^2 = 4$, but $-2 < 0$, and so we **do not** write $-2 = \sqrt{4}$. However, -2 is still a square root of 4, and we can write $-2 = -\sqrt{4}$.

3. Since $i^2 = -1$, we see that i is a square root of -1.

4. Since $(-i)^2 = (-i)(-i) = (-1)(-1)i^2 = 1(-1) = -1$, we see that $-i$ is also a square root of -1.

5. $(1+i)^2 = (1+i)(1+i) = (1-1) + (1+1)i = 0 + 2i = 2i$. So, $1 + i$ is a square root of $2i$.

The **absolute value** or **modulus** of the complex number $z = a + bi$ is the nonnegative real number

$$|z| = \sqrt{a^2 + b^2} = \sqrt{(\text{Re } z)^2 + (\text{Im } z)^2}.$$

Note: If $z = a + 0i = a$ is a real number, then $|a| = \sqrt{a^2}$. This is equal to a if $a \geq 0$ and $-a$ if $a < 0$.

For example, $|4| = \sqrt{4^2} = \sqrt{16} = 4$ and $|-4| = \sqrt{(-4)^2} = \sqrt{16} = 4 = -(-4)$.

The statement "$|a| = -a$ for $a < 0$" often confuses students. This confusion is understandable, as a minus sign is usually used to indicate that an expression is negative, whereas here we are negating a negative number to make it positive. Unfortunately, this is the simplest way to say, "delete the minus sign in front of the number" using basic notation.

Geometrically, the absolute value of a complex number z is the distance between the point z and the origin.

Example 7.2: Which of the following complex numbers is closest to the origin? $1 + 2i$, $-3 + i$, or $-2 + 3i$?

$$|1 + 2i| = \sqrt{1^2 + 2^2} = \sqrt{1 + 4} = \sqrt{5}$$

$$|-3 + i| = \sqrt{(-3)^2 + 1^2} = \sqrt{9 + 1} = \sqrt{10}$$

$$|-2 + 3i| = \sqrt{(-2)^2 + 3^2} = \sqrt{4 + 9} = \sqrt{13}$$

Since $\sqrt{5} < \sqrt{10} < \sqrt{13}$, we see that $1 + 2i$ is closest to the origin.

Notes: (1) Here we have used the following theorem: If $a, b \in \mathbb{R}^+$, then $a < b$ if and only if $a^2 < b^2$. To see this, observe that $a^2 < b^2$ if and only if $b^2 - a^2 > 0$ if and only if $(b + a)(b - a) > 0$. Since $a > 0$ and $b > 0$, by property (1) of an ordered field, $b + a > 0$. It follows that $a^2 < b^2$ if and only if $b - a > 0$ if and only if $b > a$ if and only if $a < b$.

Applying this theorem to $5 < 10 < 13$, we get $\sqrt{5} < \sqrt{10} < \sqrt{13}$.

(2) Below is a picture displaying $1 + 2i$, $-3 + i$, and $-2 + 3i$ in the Complex Plane.

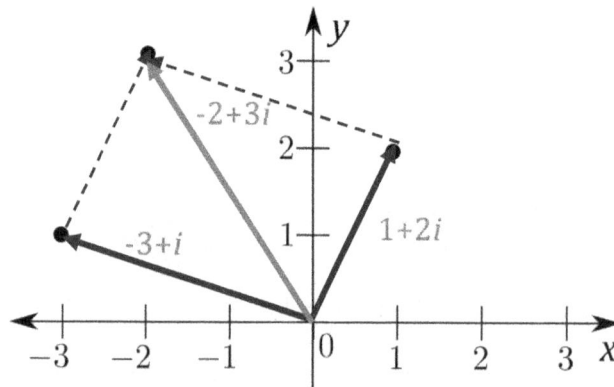

(3) The definition of the absolute value of a complex number is motivated by the Pythagorean Theorem.

As an example, look at $-3 + i$ in the figure below. Observe that to get from the origin to the point $(-3, 1)$, we move to the left 3 units and then up 1 unit. This gives us a right triangle with legs of lengths 3 and 1. By the Pythagorean Theorem, the hypotenuse has length $\sqrt{3^2 + 1^2} = \sqrt{9 + 1} = \sqrt{10}$.

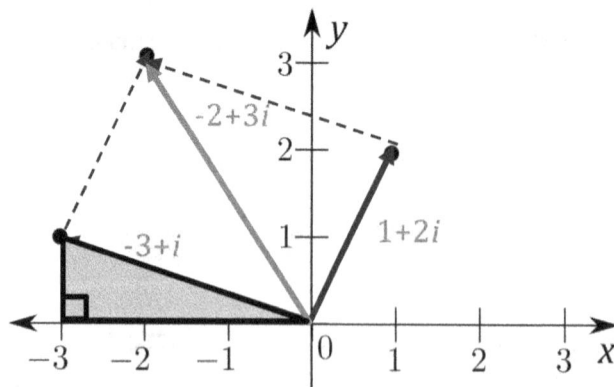

Recall that if a and b are in a field F, then we define subtraction as follows: $a - b = a + (-b)$. The expression $a - b$ is called a **difference**.

In particular, if $z, w \in \mathbb{C}$, with $z = a + bi$ and $w = c + di$, then the difference $z - w$ is defined by

$$z - w = z + (-w) = (a + bi) + (-c - di) = (a - c) + (b - d)i.$$

As a point, this difference is $(a - c, b - d)$. Here is an example illustrating how subtraction works using the computation $(1 + 2i) - (2 - i) = -1 + 3i$.

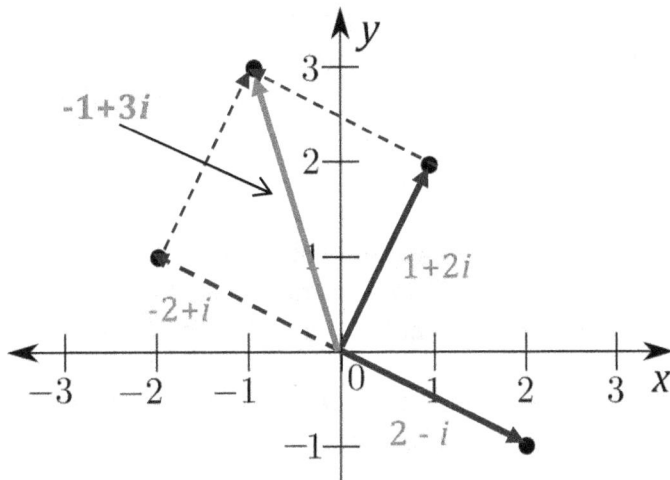

Observe how we first replaced $2 - i$ by $-2 + i$ so that we could change the subtraction problem to the addition problem: $(1 + 2i) + (-2 + i)$. We then formed a parallelogram using $1 + 2i$ and $-2 + i$ as edges, and finally, we drew the diagonal of that parallelogram to see the result.

Note: Geometrically, we can translate the vector $-1 + 3i$ so that its initial point coincides with the terminal point of the vector $2 - i$. When we do this, the terminal point of $-1 + 3i$ coincides with the terminal point of the vector $1 + 2i$. See the figure below.

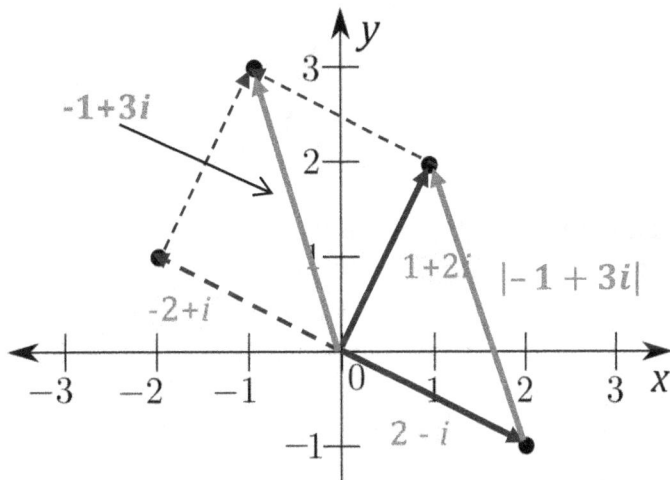

The **distance** between the complex numbers $z = a + bi$ and $w = c + di$ is

$$d(z, w) = |z - w| = \sqrt{(a - c)^2 + (b - d)^2}.$$

Note: The distance between the real numbers x and y is simply $d(x, y) = |x - y| = \sqrt{(x - y)^2}$.

Example 7.3:

1. The distance between 2 and 5 is $|2 - 5| = |-3| = \mathbf{3}$.

   Observe that the distance between 5 and 2 is the same. Indeed, we have $|5 - 2| = |3| = 3$.

2. The distance between $1 + 2i$ and $2 - i$ is

$$|(1 + 2i) - (2 - i)| = \sqrt{(1 - 2)^2 + \left(2 - (-1)\right)^2} = \sqrt{(-1)^2 + 3^2} = \sqrt{1 + 9} = \sqrt{\mathbf{10}}.$$

Notes: (1) In general, for real numbers x and y, $|x - y| = |y - x|$. This follows immediately from the fact that $y - x = -(x - y)$. The same will be true if x and y are complex numbers.

(2) Take a look one more time at the figure we drew above for $(1 + 2i) - (2 - i) = -1 + 3i$ with the solution vector $z - w = -1 + 3i$ translated so that the directed line segment begins at the terminal point of $w = 2 - i$ and ends at the terminal point of $z = 1 + 2i$.

Notice that the expression for the distance between two complex numbers follows from a simple application of the Pythagorean Theorem. Let's continue to use the same example to help us see this.

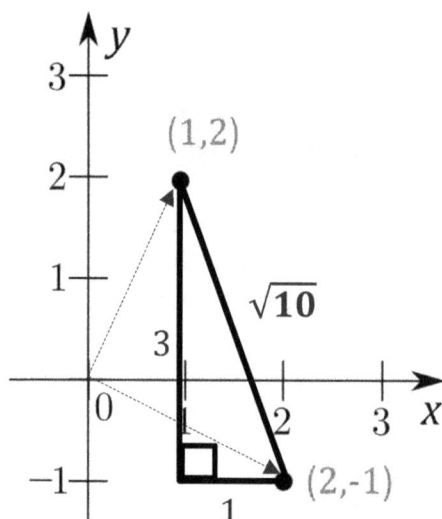

In the figure above, we can get the lengths of the legs of the triangle either by simply counting the units, or by subtracting the appropriate coordinates. For example, the length of the horizontal leg is $2 - 1 = 1$ and the length of the vertical leg is $2 - (-1) = 2 + 1 = 3$. We can then use the Pythagorean Theorem to get the length of the hypotenuse of the triangle: $c = \sqrt{1^2 + 3^2} = \sqrt{1 + 9} = \sqrt{10}$.

Compare this geometric procedure to the formula for distance given above.

While we're on the subject of triangles, the next theorem involving arbitrary triangles is very useful.

Theorem 7.4 (The Triangle Inequality): For all $z, w \in \mathbb{C}$, $|z + w| \leq |z| + |w|$.

Geometrically, the Triangle Inequality says that the length of the third side of a triangle is less than or equal to the sum of the lengths of the other two sides of the triangle. We leave the proof as an exercise (see Problem 6 below).

As an example, let's look at the sum $(1 + 2i) + (-3 + i) = -2 + 3i$. In Example 7.2, we computed

$$|1 + 2i| = \sqrt{5}, |-3 + i| = \sqrt{10}, \text{ and } |-2 + 3i| = \sqrt{13}.$$

Note that $\sqrt{5} + \sqrt{10} > \sqrt{4} + \sqrt{9} = 2 + 3 = 5$, whereas $\sqrt{13} < \sqrt{16} = 4$. So, we see that

$$|(1 + 2i) + (-3 + i)| = |-2 + 3i| = \sqrt{13} < 4 < 5 < \sqrt{5} + \sqrt{10} = |1 + 2i| + |-3 + i|.$$

In the following picture, there are two triangles. We've put dark bold lines around the leftmost triangle and labeled the sides with their lengths.

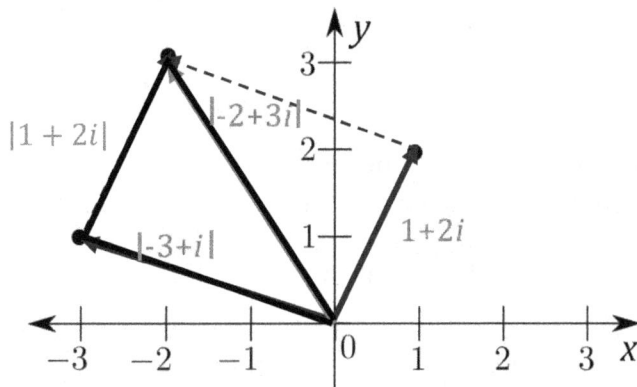

Neighborhoods

Recall from Lesson 5 that a **circle** in the Complex Plane is the set of all points that are at a fixed distance from a fixed point. The fixed distance is called the **radius** of the circle and the fixed point is called the **center** of the circle.

If a circle has radius $r > 0$ and center $c = a + bi$, then any point $z = x + yi$ on the circle must satisfy $|z - c| = r$, or equivalently, $(x - a)^2 + (y - b)^2 = r^2$.

Note: The equation $|z - c| = r$ says "The distance between z and c is equal to r." In other words, the distance between any point on the circle and the center of the circle is equal to the radius of the circle.

Example 7.5: The circle with equation $|z + 2 - i| = 2$ has center $c = -(2 - i) = -2 + i$ and radius $r = 2$.

Note: $|z + 2 - i| = |z - (-2 + i)|$. So, if we rewrite the equation as $|z - (-2 + i)| = 2$, it is easy to pick out the center and radius of the circle.

A picture of the circle is shown to the right. The center is labeled, and a typical radius is drawn.

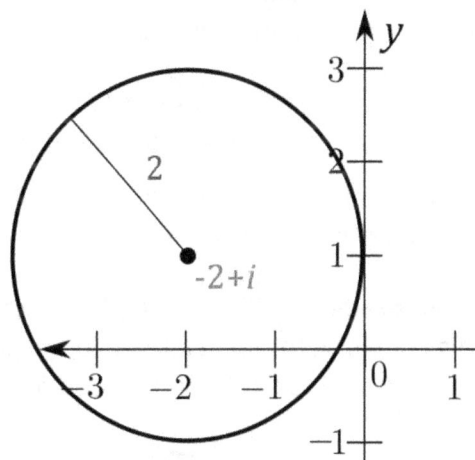

If x and a are real numbers, then the equation $|x - a| = r$ is equivalent to the statement "$x - a = r$ or $x - a = -r$," or equivalently, "$x = a + r$ or $x = a - r$." So, the real-valued version of a "circle" is two points that are equidistant from the "center" of the "circle."

Example 7.6: The equation $|x - 1| = 2$ is equvalent to "$x = 1 + 2 = 3$ or $x = 1 - 2 = -1$". In other words, there are exactly two real numbers that are at a distance of 2 from 1. These two numbers are 3 and -1.

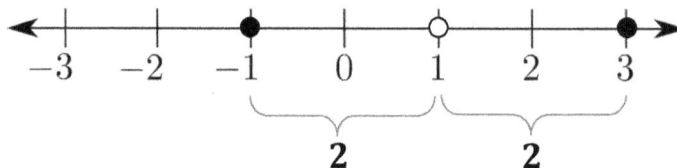

An **open disk** in $\mathbb{C}$ consists of all the points in the interior of a circle. If c is the center of the open disk and r is the radius of the open disk, then any point z inside the disk satisfies $|z - c| < r$.

$N_r(c) = \{z \in \mathbb{C} \mid |z - c| < r\}$ is also called the **r-neighborhood of c**.

Example 7.7: $N_2(-2 + i) = \{z \in \mathbb{C} \mid |z + 2 - i| < 2\}$ is the 2 neighborhood of $-2 + i$. It consists of all points inside the circle $|z + 2 - i| = 2$.

Notes: (1) A picture of the 2-neighborhood of $-2 + i$ is shown to the right. The center is labeled and a typical radius is drawn. We drew the boundary of the disk with dashes to indicate that points on the circle are **not** in the neighborhood and we shaded the interior of the disk to indicate that every point inside the circle is in the neighborhood.

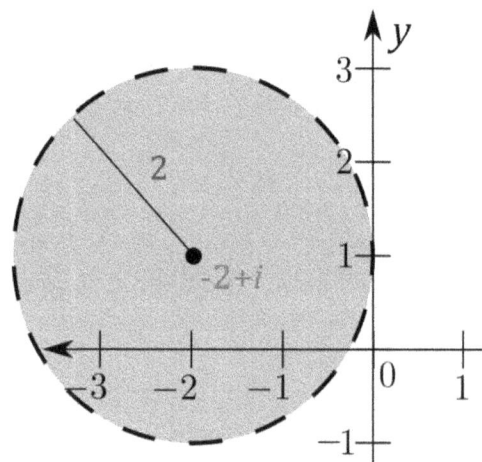

(2) The definitions of open disk and r-neighborhood of a also make sense in $\mathbb{R}$, but the geometry looks quite different. An open disk in $\mathbb{R}$ is simply an open interval. If x and a are real numbers, then we have

$$x \in N_r(a) \Leftrightarrow |x - a| < r \Leftrightarrow \sqrt{(x - a)^2} < r \Leftrightarrow 0 \leq (x - a)^2 < r^2$$
$$\Leftrightarrow -r < x - a < r \Leftrightarrow a - r < x < a + r \Leftrightarrow x \in (a - r, a + r).$$

So, in $\mathbb{R}$, an r-neighborhood of a is the open interval $N_r(a) = (a - r, a + r)$. Notice that the length (or **diameter**) of this interval is $2r$.

As an example, let's draw a picture of $N_2(1) = (1 - 2, 1 + 2) = (-1, 3)$. Observe that the center of this open disk (or open interval or neighborhood) in $\mathbb{R}$ is the real number 1, the radius of the open disk is 2, and the diameter of the open disk (or length of the interval) is 4.

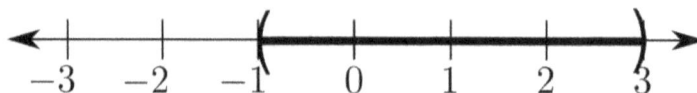

A **closed disk** in $\mathbb{C}$ is the interior of a circle together with the circle itself (the **boundary** is included). If c is the center of the closed disk and r is the radius of the closed disk, then any point z inside the closed disk satisfies $|z - c| \leq r$.

Notes: (1) In this case, the circle itself would be drawn solid to indicate that all points on the circle are included.

(2) Just like an open disk in $\mathbb{R}$ is an open interval, a closed disk in $\mathbb{R}$ is a closed interval.

(3) The reader is encouraged to draw a few open and closed disks in both $\mathbb{C}$ and $\mathbb{R}$, and to write down the corresponding sets of points using set builder notation and, in the case of $\mathbb{R}$, interval notation.

A **punctured open disk** in $\mathbb{C}$ consists of all the points in the interior of a circle **except** for the center of the circle. If c is the center of the punctured open disk and r is the radius of the open disk, then any point z inside the punctured disk satisfies $|z - c| < r$ and $z \neq c$.

Note that $z \neq c$ is equivalent to $z - c \neq 0$. In turn, this is equivalent to $|z - c| \neq 0$. Since $|z - c|$ must be nonnegative, $|z - c| \neq 0$ is equivalent to $|z - c| > 0$ or $0 < |z - c|$.

Therefore, a punctured open disk with center c and radius r consists of all points z that satisfy

$$0 < |z - c| < r.$$

$N_r^{\odot}(c) = \{z \mid 0 < |z - c| < r\}$ is also called a **deleted r-neighborhood** of c.

Example 7.8: $N_2^{\odot}(-2 + i) = \{z \in \mathbb{C} \mid 0 < |z + 2 - i| < 2\}$ is the deleted 2 neighborhood of $-2 + i$. It consists of all points inside the circle $|z + 2 - i| = 2$, **except for** $-2 + i$.

Notes: (1) A picture of the deleted 2-neighborhood of $-2 + i$ is shown to the right. Notice that this time we excluded the center of the disk $-2 + i$, as this point is not included in the set.

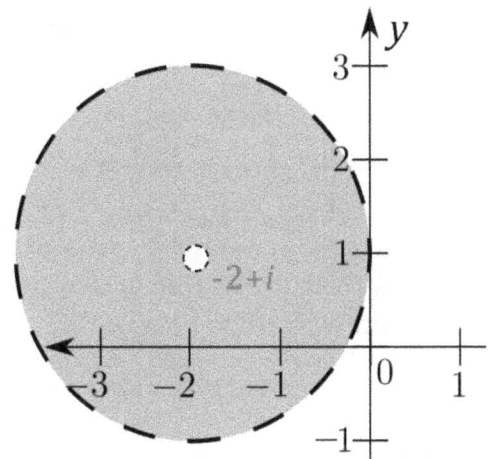

(2) In $\mathbb{R}$, we have

$$N_r^{\odot}(a) = (a - r, a + r) \setminus \{a\} = (a - r, a) \cup (a, a + r).$$

This is the open interval centered at a of length (or diameter) $2r$ with a removed.

Let's draw a picture of $N_2^{\odot}(1) = (-1, 3) \setminus \{1\} = (-1, 1) \cup (1, 3)$.

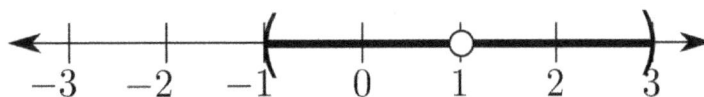

(3) Notice how all the topological definitions we are presenting make sense in both $\mathbb{R}$ and $\mathbb{C}$, but the geometry in each case looks different. You will continue to see this happen. In fact, these definitions make sense for many, many sets and structures, all with their own "look." In general, topology allows us to make definitions and prove theorems that can be applied very broadly and can be used in many (if not all) branches of mathematics.

Open and Closed Sets in $\mathbb{R}$

A subset X of $\mathbb{R}$ is said to be **open** in $\mathbb{R}$ if for every real number $x \in X$, there is an open interval (a, b) with $x \in (a, b)$ and $(a, b) \subseteq X$.

In words, a set is open in $\mathbb{R}$ if every number in the set has "some space" on both sides of that number inside the set. If you think of each point in the set as an animal, then each animal in the set should be able to move a little to the left and a little to the right without ever leaving the set. Another way to think of this is that no number is on "the edge" or "the boundary" of the set, about to fall out of it.

Example 7.9:

1. Every bounded open interval is open in $\mathbb{R}$. To see this, let $X = (a, b)$ and let $x \in X$. Then $X = (a, b)$ itself is an open interval with $x \in (a, b)$ and $(a, b) \subseteq X$. For example, $(0, 1)$ and $\left(-\sqrt{2}, \frac{3}{5}\right)$ are open in $\mathbb{R}$.

2. We will prove in the theorems below that **all** open intervals are open in $\mathbb{R}$. For example, $(-2, \infty)$, $(-\infty, 5)$, and $(-\infty, \infty)$ are all open in $\mathbb{R}$.

3. $(0, 1]$ is **not** open in $\mathbb{R}$ because the "boundary point" 1 is included in the set. If (a, b) is any open interval containing 1, then $(a, b) \not\subseteq (0, 1]$ because there are numbers greater than 1 inside (a, b). For example, let $x = \frac{1}{2}(1 + b)$ (the average of 1 and b). Since $b > 1$, we have that $x > \frac{1}{2}(1 + 1) = \frac{1}{2} \cdot 2 = 1$. So, $x > 1$. Also, since $1 > a$, $x > a$. Now, since $1 < b$, we have that $x < \frac{1}{2}(b + b) = \frac{1}{2}(2b) = \left(\frac{1}{2} \cdot 2\right)b = 1b = b$. So, $x \in (a, b)$.

4. We can use reasoning similar to that used in 3 to see that all half-open intervals and closed intervals are **not** open in $\mathbb{R}$.

Theorem 7.10: Let $a \in \mathbb{R}$. The infinite interval (a, ∞) is open in $\mathbb{R}$.

The idea behind the proof is quite simple. If $x \in (a, \infty)$, then $(a, x + 1)$ is an open interval with x inside of it and with $(a, x + 1) \subseteq (a, \infty)$.

Proof of Theorem 7.10: Let $x \in (a, \infty)$ and let $b = x + 1$.

Since $x \in (a, \infty)$, $x > a$. Since $(x + 1) - x = 1 > 0$, we have $b = x + 1 > x$.

So, we have $a < x < b$. That is, $x \in (a, b)$. Also, $(a, b) \subseteq (a, \infty)$. Since $x \in (a, \infty)$ was arbitrary, (a, ∞) is open in $\mathbb{R}$. $\qquad\square$

In Problem 5 below (part (i)), you will be asked to show that an interval of the form $(-\infty, b)$ is also open in $\mathbb{R}$.

Theorem 7.11: $\emptyset$ and $\mathbb{R}$ are both open in $\mathbb{R}$.

Proof: The statement that $\emptyset$ is open in $\mathbb{R}$ is vacuously true (since $\emptyset$ has no elements, there is nothing to check).

If $x \in \mathbb{R}$, then $x \in (x - 1, x + 1)$ and $(x - 1, x + 1) \subseteq \mathbb{R}$. Since x was an arbitrary element of $\mathbb{R}$, we have shown that for every $x \in \mathbb{R}$, there is an open interval (a, b) with $x \in (a, b)$ and $(a, b) \subseteq \mathbb{R}$. So, $\mathbb{R}$ is open in $\mathbb{R}$. $\qquad\square$

Many authors define "open" in a slightly different way from the definition we've been using. This next theorem will show that the definition we have been using is equivalent to theirs.

Theorem 7.12: A subset X of $\mathbb{R}$ is open in $\mathbb{R}$ if and only if for every real number $x \in X$, there is a positive real number c such that $(x - c, x + c) \subseteq X$.

Analysis: The harder direction of the proof is showing that if X is open in $\mathbb{R}$, then for every real number $x \in X$, there is a positive real number c such that $(x - c, x + c) \subseteq X$.

To see this, suppose that X is open in $\mathbb{R}$ and let $x \in X$. Then there is an open interval (a, b) with $x \in (a, b)$ and $(a, b) \subseteq X$. We want to replace the interval (a, b) by an interval that has x right in the center.

The following picture should help us to come up with an argument.

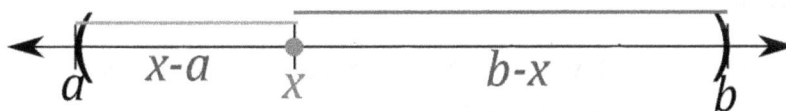

In the picture, we have an open interval (a, b), containing x. In this particular picture, x is a bit closer to a than it is to b. However, we should remember to be careful that our argument doesn't assume this (as we have no control over where x "sits" inside of (a, b)).

In the picture, we see that $x - a$ is the distance from a to x, and $b - x$ is the distance from x to b. Since the distance from a to x is smaller, let's let c be that smaller distance. In other words, we let $c = x - a$. From the picture, it looks like the interval $(x - c, x + c)$ will be inside the interval (a, b).

In general, if x is closer to a, we would let $c = x - a$, and if x is closer to b, we would let $c = b - x$. We can simply define c to be the smaller of $x - a$ and $b - x$. That is, $c = \min\{x - a, b - x\}$. From the picture, it seems like with this choice of c, the interval $(x - c, x + c)$ should give us what we want.

Proof of Theorem 7.12: Let X be an open subset of $\mathbb{R}$ and let $x \in X$. Then there is an open interval (a, b) with $x \in (a, b)$ and $(a, b) \subseteq X$. Let $c = \min\{x - a, b - x\}$. We claim that $(x - c, x + c)$ is an open interval containing x and contained in (a, b). We need to show $a \leq x - c < x < x + c \leq b$.

Since $c = \min\{x - a, b - x\}$, $c \leq x - a$. So, $-c \geq -(x - a)$. It follows that

$$(x - c) - a \geq \big(x - (x - a)\big) - a = (x - x + a) - a = a - a = 0.$$

So, $x - c \geq a$.

Since $c = \min\{x - a, b - x\}$, $c \leq b - x$. So, $-c \geq -(b - x)$. It follows that

$$b - (x + c) = b - x - c \geq b - x - (b - x) = 0.$$

So, $b \geq x + c$, or equivalently, $x + c \leq b$.

Note that $x > a$, so that $x - a > 0$, and $x < b$, so that $b - x > 0$. It follows that $c > 0$.

We have $x - (x - c) = c > 0$, so that $x > x - c$. We also have $(x + c) - x = c > 0$, so that $x + c > x$.

We have shown $a \leq x - c < x < x + c \leq b$, as desired.

Since $(x - c, x + c) \subseteq (a, b)$ and $(a, b) \subseteq X$, by the transitivity of $\subseteq$ (Theorem 1.14), we have $(x - c, x + c) \subseteq X$.

The converse is immediate since for $x \in X$, $(x - c, x + c)$ is an open interval containing x. □

Theorem 7.13: The union of two open sets in $\mathbb{R}$ is open in $\mathbb{R}$.

Proof: Let A and B be open sets in $\mathbb{R}$, and let $x \in A \cup B$. Then $x \in A$ or $x \in B$. **Without loss of generality**, we may assume that $x \in A$. Since A is open in $\mathbb{R}$, there is an interval (a, b) with $x \in (a, b)$ and $(a, b) \subseteq A$. By Theorem 2.5, $A \subseteq A \cup B$. Since $\subseteq$ is transitive (Theorem 1.14), $(a, b) \subseteq A \cup B$. Therefore, $A \cup B$ is open. □

Note: In the proof of Theorem 7.13, we used the expression "Without loss of generality." Recall from Note 1 following Example 4.4 that this expression can be used when an argument can be split up into 2 or more cases, and the proof of each of the cases is nearly identical.

For Theorem 7.13, the two cases are (i) $x \in A$ and (ii) $x \in B$. The argument for case (ii) is the same as the argument for case (i), essentially word for word—only the roles of A and B are interchanged.

Example 7.14: $(-5, 2)$ is open in $\mathbb{R}$ by part 1 of Example 7.9 and $(7, \infty)$ is open in $\mathbb{R}$ by Theorem 7.10. Therefore, by Theorem 7.13, $(-5, 2) \cup (7, \infty)$ is also open in $\mathbb{R}$.

If you look at the proof of Theorem 7.13 closely, you should notice that the proof would still work if we were taking a union of more than 2 sets. In fact, **any** union of open sets is open, as we now prove.

Theorem 7.15: Let X be a set of open subsets of $\mathbb{R}$. Then $\bigcup X$ is open in $\mathbb{R}$.

Proof: Let X be a set of open subsets of $\mathbb{R}$ and let $x \in \bigcup X$. Then $x \in A$ for some $A \in X$. Since A is open in $\mathbb{R}$, there is an interval (a, b) with $x \in (a, b)$ and $(a, b) \subseteq A$. By part 1 of Problem 9 from Problem Set 2, $A \subseteq \bigcup X$. Since $\subseteq$ is transitive (Theorem 1.14), $(a, b) \subseteq \bigcup X$. Therefore, $\bigcup X$ is open in $\mathbb{R}$. □

Example 7.16:

1. $(1, 2) \cup (2, 3) \cup (3, 4) \cup (4, \infty)$ is open in $\mathbb{R}$.

2. $\mathbb{R} \setminus \mathbb{Z}$ is open because it is a union of open intervals. It looks like this:

$$\cdots (-2, -1) \cup (-1, 0) \cup (0, 1) \cup (1, 2) \cup \cdots$$

$\mathbb{R} \setminus \mathbb{Z}$ can also be written as

$$\bigcup \{(n, n+1) \mid n \in \mathbb{Z}\} \qquad \text{or} \qquad \bigcup_{n \in \mathbb{Z}} (n, n+1)$$

3. If we take the union of all intervals of the form $\left(\frac{1}{n+1}, \frac{1}{n}\right)$ for positive integers n, we get an open set. We can visualize this open set as follows:

$$\bigcup \left\{ \left(\frac{1}{n+1}, \frac{1}{n}\right) \,\Big|\, n \in \mathbb{Z}^+ \right\} = \cdots \cup \left(\frac{1}{5}, \frac{1}{4}\right) \cup \left(\frac{1}{4}, \frac{1}{3}\right) \cup \left(\frac{1}{3}, \frac{1}{2}\right) \cup \left(\frac{1}{2}, 1\right)$$

Theorem 7.17: Every nonempty open set in $\mathbb{R}$ can be expressed as a union of bounded open intervals.

The main idea of the argument will be the following. Every real number that is in an open set is inside an open interval that is a subset of the set. Just take the union of all these open intervals (one interval for each real number in the set).

Proof of Theorem 7.17: Let X be a nonempty open set in $\mathbb{R}$. Since X is open, for each $x \in X$, there is an interval (a_x, b_x) with $x \in (a_x, b_x)$ and $(a_x, b_x) \subseteq X$. We Let $Y = \{(a_x, b_x) \mid x \in X\}$. We will show that $X = \bigcup Y$.

First, let $x \in X$. Then $x \in (a_x, b_x)$. Since $(a_x, b_x) \in Y$, $x \in \bigcup Y$. Since x was arbitrary, $X \subseteq \bigcup Y$.

Now, let $x \in \bigcup Y$. Then there is $z \in X$ with $x \in (a_z, b_z)$. Since $(a_z, b_z) \subseteq X$, $x \in X$. Since $x \in X$ was arbitrary, $\bigcup Y \subseteq X$.

Since $X \subseteq \bigcup Y$ and $\bigcup Y \subseteq X$, it follows that $X = \bigcup Y$. $\qquad \square$

We can actually get even more specific than Theorem 7.17.

Theorem 7.18: Every nonempty open set in $\mathbb{R}$ can be expressed as a countable union of pairwise disjoint open intervals.

The proof of Theorem 7.18 is more difficult than the proof of Theorem 7.17. We leave it as an exercise for the reader. See Problem 21 below.

Theorem 7.19: The intersection of two open sets in $\mathbb{R}$ is open in $\mathbb{R}$.

Proof: Let A and B be open sets in $\mathbb{R}$ and let $x \in A \cap B$. Then $x \in A$ and $x \in B$. Since A is open, there is an open interval (a, b) with $x \in (a, b)$ and $(a, b) \subseteq A$. Since B is open, there is an open interval (c, d) with $x \in (c, d)$ and $(c, d) \subseteq B$. Let $C = (a, b) \cap (c, d)$. Since $x \in (a, b)$ and $x \in (c, d)$, $x \in C$. By Problem 5 below (part (ii)), C is an open interval. By Problem 6 and part (ii) of Problem 4, both from Problem Set 2, $C \subseteq A$ and $C \subseteq B$. It follows that $C \subseteq A \cap B$ (Prove this!). Since $x \in A \cap B$ was arbitrary, $A \cap B$ is open in $\mathbb{R}$. $\qquad \square$

In Problem 5 below (part (iii)), you will be asked to show that the intersection of **finitely** many open sets in $\mathbb{R}$ is open in $\mathbb{R}$. In problem 11 below, you will be asked to show that an **arbitrary** intersection of open sets in $\mathbb{R}$ does **not** need to be open in $\mathbb{R}$.

A subset X of $\mathbb{R}$ is said to be **closed** in $\mathbb{R}$ if $\mathbb{R} \setminus X$ is open in $\mathbb{R}$.

$\mathbb{R} \setminus X$ is called the **complement** of X in $\mathbb{R}$, or simply the complement of X. It consists of all real numbers **not** in X.

Example 7.20:

1. Every closed interval is closed in $\mathbb{R}$. For example, $[0, 1]$ is closed in $\mathbb{R}$ because its complement in $\mathbb{R}$ is $\mathbb{R} \setminus [0, 1] = (-\infty, 0) \cup (1, \infty)$. This is a union of open intervals, which is open in $\mathbb{R}$.

   Similarly, $[3, \infty)$ is closed in $\mathbb{R}$ because $\mathbb{R} \setminus [3, \infty) = (-\infty, 3)$, which is open in $\mathbb{R}$.

2. Half-open intervals are neither open nor closed in $\mathbb{R}$. For example, we saw in part 3 of Example 7.9 that $(0, 1]$ is **not** open in $\mathbb{R}$. We see that $(0, 1]$ is not closed in $\mathbb{R}$ by observing that $\mathbb{R} \setminus (0,1] = (-\infty, 0] \cup (1, \infty)$, which is not open in $\mathbb{R}$.

3. $\emptyset$ is closed in $\mathbb{R}$ because $\mathbb{R} \setminus \emptyset = \mathbb{R}$ is open in $\mathbb{R}$. Also, $\mathbb{R}$ is closed in $\mathbb{R}$ because $\mathbb{R} \setminus \mathbb{R} = \emptyset$ is open in $\mathbb{R}$. $\emptyset$ and $\mathbb{R}$ are the only two sets of real numbers that are both open and closed.

4. If $a \in \mathbb{R}$, then the set $\{a\}$ consisting of just one real number is closed in $\mathbb{R}$ because it is the complement of the open set $(-\infty, a) \cup (a, \infty)$.

5. By Theorem 7.18, every closed set in $\mathbb{R}$ can be expressed as $\mathbb{R} \setminus U$, where U is a countable union of pairwise disjoint open intervals. In other words, we can form a closed set by removing countably many pairwise disjoint open intervals from the real line. This procedure sounds innocent enough, but we can create some strange looking sets this way. See part 2 of Example 7.22 below.

Theorem 7.21: The intersection of two closed sets in $\mathbb{R}$ is closed in $\mathbb{R}$.

Proof: Let A and B be closed in $\mathbb{R}$. Then $\mathbb{R} \setminus A$ and $\mathbb{R} \setminus B$ are open in $\mathbb{R}$. By Theorem 7.13 (or 7.15), $(\mathbb{R} \setminus A) \cup (\mathbb{R} \setminus B)$ is open in $\mathbb{R}$. Therefore, $\mathbb{R} \setminus [(\mathbb{R} \setminus A) \cup (\mathbb{R} \setminus B)]$ is closed in $\mathbb{R}$. So, it suffices to show that $A \cap B = \mathbb{R} \setminus [(\mathbb{R} \setminus A) \cup (\mathbb{R} \setminus B)]$. Well, $x \in A \cap B$ if and only if $x \in A$ and $x \in B$ if and only if $x \notin \mathbb{R} \setminus A$ and $x \notin \mathbb{R} \setminus B$ if and only if $x \notin (\mathbb{R} \setminus A) \cup (\mathbb{R} \setminus B)$ if and only if $x \in \mathbb{R} \setminus [(\mathbb{R} \setminus A) \cup (\mathbb{R} \setminus B)]$. So, $A \cap B = \mathbb{R} \setminus [(\mathbb{R} \setminus A) \cup (\mathbb{R} \setminus B)]$, completing the proof. $\square$

A similar argument can be used to show that the union of two closed sets in $\mathbb{R}$ is a closed set in $\mathbb{R}$. This result can be extended to the union of finitely many closed sets in $\mathbb{R}$ with the help of Problem 5 below (part (iii)). The dedicated reader should prove this. In Problem 10 below, you will be asked to show that an arbitrary intersection of closed sets in $\mathbb{R}$ is closed in $\mathbb{R}$. In Problem 11 you will be asked to show that an arbitrary union of closed sets in $\mathbb{R}$ does **not** need to be closed.

Example 7.22:

1. Since the union of finitely many closed sets is closed and a set containing a single point is closed (by part 4 of Example 7.20), it follows that every finite set is closed.

2. The **Cantor set**, C, can be defined as follows: Let C_0 be the closed unit interval $[0, 1]$. Remove the open interval $\left(\frac{1}{3}, \frac{2}{3}\right)$ from C_0 to form the set $C_1 = \left[0, \frac{1}{3}\right] \cup \left[\frac{2}{3}, 1\right]$. Note that we have removed the middle third from C_0. We now delete the middle thirds from each of $\left[0, \frac{1}{3}\right]$ and $\left[\frac{2}{3}, 1\right]$ to form the set $C_2 = \left[0, \frac{1}{9}\right] \cup \left[\frac{2}{9}, \frac{1}{3}\right] \cup \left[\frac{2}{3}, \frac{7}{9}\right] \cup \left[\frac{8}{9}, 1\right]$. We continue in this fashion, creating a sequence (C_n) of sets. Each C_n is a finite union of closed sets in $\mathbb{R}$, and therefore, each C_n is closed in $\mathbb{R}$. The Cantor set is defined to be $C = \cap\{C_n | n \in \mathbb{N}\}$. Since an arbitrary intersection of closed sets in $\mathbb{R}$ is closed in $\mathbb{R}$, the Cantor set is closed in $\mathbb{R}$.

Which real numbers are in C? Well, C certainly contains the endpoints of each interval that make up each C_n. For example, $0, \frac{1}{3}, \frac{2}{3}, 1 \in C$ because these real numbers are endpoints of the intervals that make up C_0. Similarly, $\frac{1}{9}, \frac{2}{9}, \frac{7}{9}, \frac{8}{9} \in C$.

Are there any other points that are in C besides these endpoints? Well, there are only countably many such endpoints, and it turns out that C is uncountable. So, there are "many" other points in C. You will be asked to prove that C is uncountable in Problem 19 below.

We would now like to give an alternate definition of a closed set in $\mathbb{R}$ in terms of "accumulation points."

If $S \subseteq \mathbb{R}$, then $x \in \mathbb{R}$ is called an **accumulation point** of S if every open interval containing x contains at least one point of S different from x. Symbolically, we have that x is an accumulation point of S if and only if

$$\forall a, b \in \mathbb{R}(a < x < b \rightarrow \exists y \in S(a < y < b \wedge y \neq x)).$$

In words, an accumulation point of S is really, really close to the set S. It's so close to the set S that given any positive distance, we can find an element of S that is within that distance of the element. For example, suppose that 3 is an accumulation point of S and let's choose a distance of $r = 0.001$. Since 3 is an accumulation point, we can find a point $y \in S \cap (2.999, 3.001)$ that is different from 3. Notice that the distance between y and 3 is less than 0.001. That is, $|y - 3| < 0.001$.

Example 7.23:

1. Let $S = \left\{1, \frac{1}{2}, \frac{1}{3}, \dots\right\}$. Then 0 is an accumulation point of S. To see this, let (a, b) be an open interval containing 0. By the Archimedean Property of the real numbers (Theorem 6.16), there is $n \in \mathbb{N}$ such that $n > \frac{1}{b}$, or equivalently, $\frac{1}{n} < b$. So, $\frac{1}{n} \in S$, $\frac{1}{n} \in (a, b)$, and $\frac{1}{n} \neq 0$.

   It is not too hard to check that 0 is the **only** accumulation point of S. I leave it to the reader to show that S has no other accumulation points.

2. The half-open interval $(0, 1]$ has many accumulation points. In fact, the set of accumulation points of $(0,1]$ is the closed interval $[0, 1]$. Notice that 0 is an accumulation point of $(0, 1]$ that is **not** in the set, whereas 1 is an accumulation point of $(0, 1]$ that is in the set.

3. The set of rational numbers, $\mathbb{Q}$, has every real number as an accumulation point. To see this, let $x \in \mathbb{R}$ and let (a, b) be an open interval containing x. By the Density Theorem (Theorem 6.17), there is a rational number q with $a < q < x$. So, $q \in \mathbb{Q}$, $q \in (a, b)$ and $q \neq x$.

4. The set of integers, $\mathbb{Z}$, has **no** accumulation points. Indeed, if $x \in \mathbb{R}$, let n be the least integer such that $x < n$. Then $n - 1 \leq x < n$. If $x = n - 1$, then $(x - 1, x + 1)$ is an interval containing x that contains no integers other than x. Otherwise, if we let $c = \min\{x - (n - 1), n - x\}$, then $c > 0$ and the interval $(x - c, x + c)$ is an interval containing x that contains no integers. See the analysis after Theorem 7.12 as well as the proof of Theorem 7.12 for details.

Theorem 7.24: A subset C of $\mathbb{R}$ is closed in $\mathbb{R}$ if and only if C contains each of its accumulation points.

Proof: First assume that C is closed in $\mathbb{R}$. Then $\mathbb{R} \setminus C$ is open in $\mathbb{R}$. Let x be any point **not** in C. Then $x \in \mathbb{R} \setminus C$. Since $\mathbb{R} \setminus C$ is open, there is an open interval (a, b) with $x \in (a, b)$ and $(a, b) \subseteq \mathbb{R} \setminus C$. So, if $y \in (a, b)$, then $y \notin C$. It follows that x is **not** an accumulation point of C. So, we have shown that $x \notin C$ implies that x is not an accumulation point of C. The contrapositive of this statement is "if x is an accumulation point of C, then $x \in C$." So, C contains each of its accumulation points.

Conversely, assume that C contains each of its accumulation points. We need to show that $\mathbb{R} \setminus C$ is open in $\mathbb{R}$. Since $\emptyset$ is open in $\mathbb{R}$, we can assume that $\mathbb{R} \setminus C \neq \emptyset$. Let $x \in \mathbb{R} \setminus C$. Then x is **not** an accumulation point of C. So, there is an open interval (a, b) containing x and no points of C. So, $(a, b) \subseteq \mathbb{R} \setminus C$. Since $x \in \mathbb{R} \setminus C$ was arbitrary, we have shown that $\mathbb{R} \setminus C$ is open in $\mathbb{R}$. $\square$

If $S \subseteq \mathbb{R}$, then the **closure** of S in $\mathbb{R}$, written $\overline{S}$, is the intersection of all closed sets containing S. Symbolically, we have

$$\overline{S} = \cap\{C \mid S \subseteq C \wedge C \text{ is closed in } \mathbb{R}\}.$$

Since $\overline{S}$ is an intersection of closed sets in $\mathbb{R}$, $\overline{S}$ is closed in $\mathbb{R}$. We now prove a few basic facts about the closure of a set S.

Theorem 7.25: Let $S \subseteq \mathbb{R}$.

1. $S \subseteq \overline{S}$.

2. If C is closed in $\mathbb{R}$ with $S \subseteq C$, then $\overline{S} \subseteq C$.

3. $\overline{S} = S \cup \{x \in \mathbb{R} \mid x \text{ is an accumulation point of } S\}$.

4. S is closed in $\mathbb{R}$ if and only if $S = \overline{S}$.

5. $x \in \overline{S}$ if and only if every open interval containing x contains at least one point of S.

Proof:

1. Let $x \in S$. If $S \subseteq C$ with C closed, then $x \in C$. So, $x \in \cap\{C \mid S \subseteq C \wedge C \text{ is closed}\} = \overline{S}$. Since $x \in S$ was arbitrary, $\forall x (x \in S \rightarrow x \in \overline{S})$. Therefore, $S \subseteq \overline{S}$.

2. Let C be closed in $\mathbb{R}$ with $S \subseteq C$ and let $x \in \overline{S}$. By the definition of $\overline{S}$, $x \in C$. Since $x \in \overline{S}$ was arbitrary, $\forall x (x \in \overline{S} \rightarrow x \in C)$. Therefore, $\overline{S} \subseteq C$.

3. Let $C = S \cup \{x \in \mathbb{R} \mid x \text{ is an accumulation point of } S\}$. We first prove that C is closed in $\mathbb{R}$. Let x be an accumulation point of C and let (a, b) be an open interval containing x. Let $y \in C$ with $y \in (a, b)$ and $y \neq x$. Without loss of generality, assume that $x < y$. Since $y \in C$, $y \in S$ or y is an accumulation point of S. If y is an accumulation point of S, then there is $z \in S$ with $z \in (x, b)$. In either case, we have found an element of S in (a, b) different from x. So, x is an accumulation point of S, and therefore, $x \in C$. Since x was an arbitrary accumulation point of C, it follows that C contains each of its accumulation points. By Theorem 7.24, C is closed in $\mathbb{R}$.

Since $S \subseteq C$ and C is closed in $\mathbb{R}$, it follows from 2 that $\overline{S} \subseteq C$.

Now suppose $x \notin \overline{S}$. Since $S \subseteq \overline{S}$ (from 1), $x \notin S$. Since $x \notin \overline{S}$, x is in the open set $\mathbb{R} \setminus \overline{S}$. Therefore, there is an interval (a, b) with $x \in (a, b)$ and $(a, b) \subseteq \mathbb{R} \setminus \overline{S}$. Since $S \subseteq \overline{S}$, we have $\mathbb{R} \setminus \overline{S} \subseteq \mathbb{R} \setminus S$. By Theorem 1.14, $(a, b) \subseteq \mathbb{R} \setminus S$. So, x is **not** an accumulation point of S. Since $x \notin S$ and x is not an accumulation point of S, $x \notin C$. Since $x \notin \overline{S}$ was arbitrary, we have $\forall x (x \notin \overline{S} \rightarrow x \notin C)$. By the law of the contrapositive, $\forall x (x \in C \rightarrow x \in \overline{S})$. So, $C \subseteq \overline{S}$.

Since $\overline{S} \subseteq C$ and $C \subseteq \overline{S}$, we have $\overline{S} = C$

4. First assume that S is closed in $\mathbb{R}$. By 1 above, $S \subseteq \overline{S}$. Since S is closed and $S \subseteq S$, by 2 above, we have $\overline{S} \subseteq S$. So, $S = \overline{S}$.

Conversely, assume that $S = \overline{S}$. Since $\overline{S}$ is closed in $\mathbb{R}$ and $S = \overline{S}$, S is closed in $\mathbb{R}$.

5. First let $x \in \overline{S}$ and let (a, b) be an open interval containing x. By 3, $x \in S$ or x is an accumulation point of S. If $x \in S$, then we have found a point in (a, b) that is in S, namely x. If x is an accumulation point of S, then there is $y \in (a, b)$ with $y \in S$ and $y \neq x$.

Conversely, assume that every open interval containing x contains at least one point of S. If $x \in S$, then by 1, $x \in \overline{S}$. So, assume that $x \notin S$. Then every open interval containing x contains a point of S different from x. Thus, x is an accumulation point of S. By 3, $x \in \overline{S}$. $\qquad\square$

Example 7.26:

1. Let $A = (a, b)$. Then $\overline{A} = [a, b]$.

2. Let C be the Cantor set from part 2 of Example 7.22. Since C is closed in $\mathbb{R}$, by part 4 of Theorem 7.23, we have $\overline{C} = C$.

3. Let $S = \left\{ 1, \frac{1}{2}, \frac{1}{3}, \ldots \right\}$. In part 1 of Example 7.23, we saw that the only accumulation point of S is 0. It follows from part 3 of Theorem 7.25 that $\overline{S} = S \cup \{0\}$.

4. $\overline{\mathbb{Q}} = \mathbb{R}$. This follows immediately from part 3 of Example 7.23 and part 3 of Theorem 7.25.

5. $\mathbb{Z}$ has no accumulation points by part 4 of Example 7.23. It follows from part 3 of Theorem 7.25 that $\overline{\mathbb{Z}} = \mathbb{Z} \cup \emptyset = \mathbb{Z}$.

<u>Open and Closed Sets in $\mathbb{C}$</u>

A subset X of $\mathbb{C}$ is said to be **open** in $\mathbb{C}$ if for every complex number $z \in X$, there is an open disk D with $z \in D$ and $D \subseteq X$.

In words, a set is open in $\mathbb{C}$ if every point in the set has "space" all around it inside the set. If you think of each point in the set as an animal, then each animal in the set should be able to move a little in any direction it chooses without leaving the set. Another way to think of this is that no number is right on "the edge" or "the boundary" of the set, about to fall out of it.

Example 7.27:

1. Every open disk D is open in $\mathbb{C}$. To see this, simply observe that if $z \in D$, then D itself is an open disk with $z \in D$ and $D \subseteq D$.

2. A closed disk is **not** open in $\mathbb{C}$ because it contains its "boundary." As an example, let's look at the closed unit disk $D = \{z \in \mathbb{C} \mid |z| \leq 1\}$. Let's focus on the point i. First note that $i \in D$ because $|i| = \sqrt{0^2 + 1^2} = \sqrt{1} = 1$ and $1 \leq 1$. Now, any open disk N containing i will contain points above i. Let's say $(1 + \epsilon)i \in N$ for some positive real number ϵ. Now, we have $|(1 + \epsilon)i| = \sqrt{0^2 + (1 + \epsilon)^2} = 1 + \epsilon$, which is greater than 1. Therefore, $(1 + \epsilon)i \notin D$. It follows that $N \nsubseteq D$, and so, D is not open.

3. We can use reasoning similar to that used in 2 to see that if we take any subset of a disk that contains any points on the bounding circle, then that set will **not** be open in $\mathbb{C}$.

4. $\emptyset$ and $\mathbb{C}$ are both open in $\mathbb{C}$. You will be asked to prove this in Problem 12 below (parts (i) and (ii)).

As we mentioned right before Theorem 7.12, many authors define "open" in a slightly different way from the definition we've been using. Once again, let's show that the definition we have been using is equivalent to theirs.

Theorem 7.28: A subset X of $\mathbb{C}$ is open in $\mathbb{C}$ if and only if for every complex number $w \in X$, there is a positive real number d such that $N_d(w) \subseteq X$.

Analysis: The harder direction of the proof is showing that if X is open in $\mathbb{C}$, then for every complex number $w \in X$, there is a positive real number d such that $N_d(w) \subseteq X$.

To see this, suppose that X is open and let $w \in X$. Then there is an open disk $D = \{z \in \mathbb{C} \mid |z - a| < r\}$ with $w \in D$ and $D \subseteq X$. We want to replace the disk D with a disk that has w right in the center.

To accomplish this, we let c be the distance from w to a. Then $r - c$ is the distance from w to the boundary of D. We will show that the disk with center w and radius $r - c$ is a subset of D.

The picture to the right illustrates this idea. Notice that $c + (r - c) = r$, the radius of disk D.

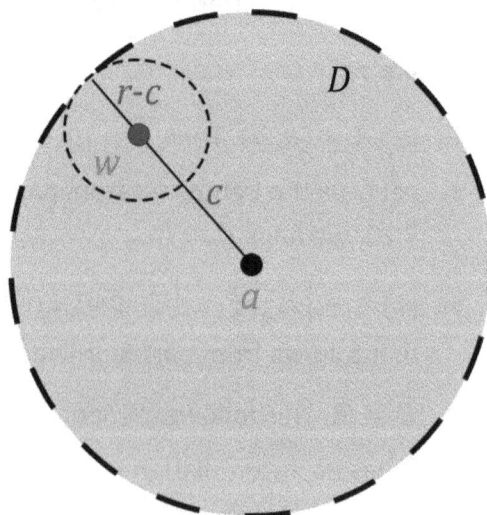

Proof of Theorem 7.28: Let X be an open subset of $\mathbb{C}$ and let $w \in X$. Then there is an open disk D with $w \in D$ and $D \subseteq X$.

Suppose that D has center a and radius r. So, $D = \{z \in \mathbb{C} \mid |z - a| < r\}$.

Let $c = |w - a|$ and let $d = r - c$. We will show that $N_d(w) \subseteq D$.

Let $z \in N_d(w)$. Then $|z - w| < d = r - c$.

By the Triangle Inequality (and SACT—see Note 2 below),
$$|z - a| = |(z - w) + (w - a)| \le |z - w| + |w - a| < (r - c) + c = r.$$

So, $z \in D$. Since z was an arbitrary element of $N_d(w)$, we showed that $N_d(w) \subseteq D$.

So, we have $N_d(w) \subseteq D$ and $D \subseteq X$. By the transitivity of $\subseteq$ (Theorem 1.14), we have $N_d(w) \subseteq X$.

The converse is immediate since for $w \in X$, $N_d(w)$ is an open disk containing w. $\quad\square$

Notes: (1) The picture to the right shows how we used the Triangle Inequality. The three sides of the triangle have lengths $|z - w|$, $|w - a|$, and $|z - a|$.

(2) Notice how we used SACT (the Standard Advanced Calculus Trick) here. Starting with $z - a$, we wanted to make $z - w$ and $w - a$ "appear." We were able to do this simply by subtracting and then adding w between z and a. We often use this trick when applying the Triangle Inequality. SACT was introduced in Lesson 5 (Note 7 following the proof of Theorem 5.11).

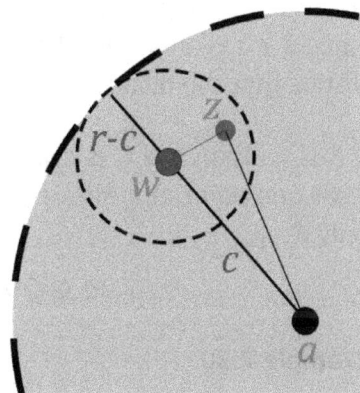

(3) The same proof used here can be used to prove Theorem 7.12. The geometry looks different (disks and neighborhoods are open intervals instead of the interiors of circles, and points appear on the real line instead of in the Complex Plane), but the argument is identical. Compare this proof to the proof we used in Theorem 7.12.

In Problem 13 below (parts (i), (ii), and (v)), you will be asked to show that an arbitrary union of open sets in $\mathbb{C}$ is open in $\mathbb{C}$, a finite intersection of open sets in $\mathbb{C}$ is open in $\mathbb{C}$, and every open set in $\mathbb{C}$ can be expressed as a union of open disks.

A subset X of $\mathbb{C}$ is said to be **closed** in $\mathbb{C}$ if $\mathbb{C} \setminus X$ is open in $\mathbb{C}$.

$\mathbb{C} \setminus X$ is called the **complement** of X in $\mathbb{C}$, or simply the complement of X. It consists of all complex numbers **not** in X.

Example 7.29:

1. Every closed disk is closed in $\mathbb{C}$. For example, $D = \{z \in \mathbb{C} \mid |z| \le 1\}$ is closed because its complement in $\mathbb{C}$ is $\mathbb{C} \setminus D = \{z \in \mathbb{C} \mid |z| > 1\}$. You will be asked to prove that this set D is open in $\mathbb{C}$ in Problem 12 below (part (iii)).

2. If we take any subset of a closed disk that includes the interior of the disk but is missing at least one point on the bounding circle, then that set will **not** be closed in $\mathbb{C}$. You will be asked to prove this for the closed unit disk $\{z \in \mathbb{C} \mid |z| \leq 1\}$ in Problem 16 below.

3. $\emptyset$ is closed in $\mathbb{C}$ because $\mathbb{C} \setminus \emptyset = \mathbb{C}$ is open in $\mathbb{C}$. Also, $\mathbb{C}$ is closed in $\mathbb{C}$ because $\mathbb{C} \setminus \mathbb{C} = \emptyset$ is open in $\mathbb{C}$. $\emptyset$ and $\mathbb{C}$ are the only two sets of complex numbers that are both open and closed in $\mathbb{C}$.

In Problem 13 below (parts (iii) and (iv)), you will be asked to show that an arbitrary intersection of closed sets in $\mathbb{C}$ is closed in $\mathbb{C}$, and a finite union of closed sets in $\mathbb{C}$ is closed in $\mathbb{C}$.

Just like we did for $\mathbb{R}$, we will give an alternate definition of a closed set in $\mathbb{C}$ in terms of accumulation points.

As before, an accumulation point of a set S is really, really close to the set S. It's so close to the set S that given any positive distance, we can find an element of S that is within that distance of the element. For example, suppose that $2 + i$ is an accumulation point of S and let's choose a distance of $r = 0.05$. Since $2 + i$ is an accumulation point, we can find a point $w \in N_{0.05}(2 + i)$ that is different from $2 + i$. Notice that the distance between w and $2 + i$ is less than 0.05. That is, $|w - (2 + i)| < 0.05$.

If $S \subseteq \mathbb{C}$, then $z \in \mathbb{C}$ is called an **accumulation point** of S if every open disk containing z contains at least one point of S different from z. Symbolically, we have that z is an accumulation point of S if and only if

$$\forall a \in \mathbb{C} \, \forall r \in \mathbb{R}^+ (z \in N_r(a) \rightarrow \exists w \in S(w \in N_r(a) \wedge w \neq z)).$$

Example 7.30:

1. Let $S = \left\{ \frac{i^n}{n} \mid n \in \mathbb{Z}^+ \right\}$. Then 0 is an accumulation point of S. To see this, let D be an open disk containing 0. Since D is open, by Theorem 7.28, there is a positive real number d such that $N_d(0) \subseteq D$. By the Archimedean Property of the real numbers (Theorem 6.16), there is $n \in \mathbb{N}$ such that $n > \frac{1}{d}$, or equivalently, $\frac{1}{n} < d$. Then we have $\left| \frac{i^n}{n} - 0 \right| = \frac{1}{n} < d$, and so, $\frac{i^n}{n} \in N_d(0)$. Since $N_d(0) \subseteq D, \frac{i^n}{n} \in D$. By the definition of $S, \frac{i^n}{n} \in S$. Finally, it is clear that $\frac{i^n}{n} \neq 0$.

   I leave it to the reader to show that S does not have any other accumulation points.

2. Let $a \in \mathbb{C}$, let r be a positive real number, and let $S = N_r(a) = \{z \in \mathbb{C} \mid |z - a| < r\}$ (S is the r-neighborhood of a). The set of accumulation points of S is $A = \{z \in \mathbb{C} \mid |z - a| \leq r\}$ (A is the closed disk with center a and radius r). I leave the verification of this to the reader.

3. Let $S = \{a + bi \mid a, b \in \mathbb{Q}\}$ and let $T = \{a + bi \mid a, b \in \mathbb{Z}\}$. I leave it to the reader to verify that every complex number is an accumulation point of S and that T has no accumulation points.

The proofs of the next two theorems are similar to the proofs of Theorems 7.24 and 7.25. Therefore, they are left to the reader.

Theorem 7.31: A subset C of $\mathbb{C}$ is closed in $\mathbb{C}$ if and only if C contains each of its accumulation points.

Theorem 7.32: Let $S \subseteq \mathbb{C}$ and let $\overline{S} = \cap\{C \mid S \subseteq C \wedge C \text{ is closed}\}$ ($\overline{S}$ is called the **closure** of S in $\mathbb{C}$).

1. $S \subseteq \overline{S}$.

2. If C is closed in $\mathbb{C}$ with $S \subseteq C$, then $\overline{S} \subseteq C$.

3. $\overline{S} = S \cup \{z \in \mathbb{C} \mid z \text{ is an accumulation point of } S\}$.

4. S is closed in $\mathbb{C}$ if and only if $S = \overline{S}$.

5. $z \in \overline{S}$ if and only if every open disk containing z contains at least one point of S.

Example 7.33:

1. Let $S = N_r(a) = \{z \in \mathbb{C} \mid |z - a| < r\}$. By part 2 of Example 7.30 and part 3 of Theorem 7.32, $\overline{S} = \{z \in \mathbb{C} \mid |z - a| \leq r\}$.

2. Let $S = \left\{\frac{i^n}{n} \mid n \in \mathbb{Z}^+\right\}$. In part 1 of Example 7.30, we saw that the only accumulation point of S is 0. It follows from part 3 of Theorem 7.32 that $\overline{S} = S \cup \{0\}$.

3. Let $S = \{a + bi \mid a, b \in \mathbb{Q}\}$ and let $T = \{a + bi \mid a, b \in \mathbb{Z}\}$. Then $\overline{S} = \mathbb{C}$ and $\overline{T} = T$. These results follow immediately from part 3 of Example 7.30 and part 3 of Theorem 7.32.

Euclidean and Unitary Spaces

Recall from part 3 of Example 1.6 that we can identify a complex number $x + yi$ with the ordered pair (x, y) of real numbers. It follows that the function $f : \mathbb{C} \to \mathbb{R}^2$ defined by $f(x + yi) = (x, y)$ is a bijection. Therefore, it is natural to identify the set $\mathbb{R}^2 = \{(x, y) \mid x, y \in \mathbb{R}\}$ of ordered pairs of real numbers with the set $\mathbb{C} = \{x + yi \mid x, y \in \mathbb{R}\}$ of complex numbers.

We will sometimes describe an element of $\mathbb{R}^2$ as a vector, say $\mathbf{x}$. When we use a boldface variable like this, we mean that $\mathbf{x}$ is an n-tuple of real numbers, for some n. For the moment, sticking with $n = 2$, we have $\mathbf{x} \in \mathbb{R}^2$ means that $\mathbf{x} = (x, y)$ for some $x, y \in \mathbb{R}$. From now on, we will use (x_1, x_2) instead of (x, y) because the subscript notation generalizes more nicely to $\mathbb{R}^n$ for larger values of n, as we will see shortly.

If $\mathbf{x} \in \mathbb{R}^2$ with $\mathbf{x} = (x_1, x_2)$, we define the **norm** (or **length** or **magnitude**) of $\mathbf{x}$ to be the nonnegative real number

$$|\mathbf{x}| = \sqrt{x_1^2 + x_2^2}.$$

Notice that this definition is pretty much identical to the definition of the absolute value of a complex number.

We can now define addition, subtraction, and other operations on vectors in $\mathbb{R}^2$ in the obvious way (essentially the same as the corresponding operations on elements of $\mathbb{C}$). For example, the **distance** between the vectors $\mathbf{x} = (x_1, x_2)$ and $\mathbf{y} = (y_1, y_2)$ is

$$d(\mathbf{x}, \mathbf{y}) = |\mathbf{x} - \mathbf{y}| = \sqrt{(x_1 - y_1)^2 + (x_2 - y_2)^2}.$$

Note: As is the case for complex numbers, it is easy to see that $|\mathbf{x} - \mathbf{y}| = |\mathbf{y} - \mathbf{x}|$.

Everything we developed in this lesson for $\mathbb{C}$ works equally well for $\mathbb{R}^2$. Once again, other than the slight change in notation (renaming of the elements), the two sets are indistinguishable. As one example, if r is a positive real number and $\mathbf{x} \in \mathbb{R}^2$, then the r-neighborhood of $\mathbf{x}$ is

$$N_r(\mathbf{x}) = \{\mathbf{y} \in \mathbb{R}^2 \mid |\mathbf{y} - \mathbf{x}| < r\}.$$

Example 7.34: Let $\mathbf{x} = (-2, 1)$ and $r = 2$. Then the 2-neighborhood of $\mathbf{x}$ is

$$N_2(\mathbf{x}) = N_2\big((-2, 1)\big) = \{\mathbf{y} \in \mathbb{R}^2 \mid |\mathbf{y} - \mathbf{x}| < 2\}$$

$$= \left\{(y_1, y_2) \in \mathbb{R}^2 \,\middle|\, \sqrt{\left(y_1 - (-2)\right)^2 + (y_2 - 1)^2} < 2\right\}$$

$$= \{(y_1, y_2) \in \mathbb{R}^2 \mid (y_1 + 2)^2 + (y_2 - 1)^2 < 4\}.$$

A picture of the deleted 2-neighborhood of $\mathbf{x} = (-2, 1)$ is shown to the right.

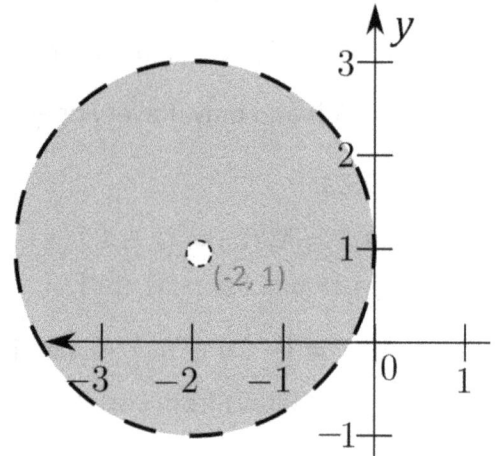

We now generalize the previous theory to $\mathbb{R}^n = \{(x_1, x_2, \ldots, x_n) \mid x_1, x_2, \ldots, x_n \in \mathbb{R}\}$. Once again, when we say that the vector $\mathbf{x} \in \mathbb{R}^n$, we mean that there are $x_1, x_2, \ldots, x_n \in \mathbb{R}$ such that $\mathbf{x} = (x_1, x_2, \ldots, x_n)$. We will always make the italicized variables with subscripts (the components of the vector) agree with the boldface variable (the vector itself). For example, if $\mathbf{y} \in \mathbb{R}^n$, we will assume that $\mathbf{y}$ can be written as $(y_1, y_2, \ldots, y_n)$.

The **sum** of two vectors $\mathbf{x}, \mathbf{y} \in \mathbb{R}^n$ is $\mathbf{x} + \mathbf{y} = (x_1 + y_1, x_2 + y_2, \ldots, x_n + y_n)$.

If $k \in \mathbb{R}$ and $\mathbf{x} \in \mathbb{R}^n$, then the **scalar multiple** of $\mathbf{x}$ by k is $k\mathbf{x} = (kx_1, kx_2, \ldots, kx_n)$.

The **negative** of a vector $\mathbf{x} \in \mathbb{R}^n$ is the vector $-\mathbf{x} = (-x_1, -x_2, \ldots, -x_n)$.

We can then define the **difference** $\mathbf{x} - \mathbf{y}$ to be the vector $\mathbf{x} + (-\mathbf{y}) = (x_1 - y_1, x_2 - y_2, \ldots, x_n - y_n)$.

The **norm** (or **length** or **magnitude**) of $\mathbf{x} \in \mathbb{R}^n$ is the nonnegative real number

$$|\mathbf{x}| = \sqrt{x_1^2 + x_2^2 + \cdots + x_n^2}.$$

$\mathbb{R}^n$ with the definitions of sum, scalar multiple, and norm given above is called **n-dimensional Euclidean space**. $\mathbb{R}^2$ is sometimes called the **Euclidean plane**.

Similarly, $\mathbb{C}^n$ with the analogous definitions of sum, scalar multiple (using complex numbers as scalars), and norm is called **n-dimensional unitary space**.

Notes: (1) We will usually write the norm of $\mathbf{x}$, $|\mathbf{x}| = \sqrt{x_1^2 + x_2^2 + \cdots + x_n^2}$ using **summation notation**:

$$|\mathbf{x}| = \sqrt{x_1^2 + x_2^2 + \cdots + x_n^2} = \left(\sum_{k=1}^{n} x_k^2\right)^{\frac{1}{2}}$$

The symbol Σ is the Greek letter Sigma. In mathematics, this symbol is often used to denote a sum. Σ is generally used to abbreviate a very large sum or a sum of unknown length by specifying what a typical term of the sum looks like.

(2) Let's look at a simpler example of a sum first before we analyze the more complicated sum in Note 1 above:

$$\sum_{k=1}^{5} k^2 = 1^2 + 2^2 + 3^2 + 4^2 + 5^2 = 1 + 4 + 9 + 16 + 25 = 55.$$

The expression "$k = 1$" written underneath the symbol indicates that we get the first term of the sum by replacing k by 1 in the given expression. When we replace k by 1 in the expression k^2, we get 1^2.

For the second term, we simply increase k by 1 to get $k = 2$. So, we replace k by 2 to get $k^2 = 2^2$.

We continue in this fashion, increasing k by 1 each time until we reach the number written above the symbol. In this case, that is $k = 5$.

(3) Let's now get back to the expression that we're interested in:

$$|\mathbf{x}| = \sqrt{x_1^2 + x_2^2 + \cdots + x_n^2} = \left(\sum_{k=1}^{n} x_k^2\right)^{\frac{1}{2}}$$

Let's first remove the square root symbol and $\frac{1}{2}$ power. In this way we can analyze just the sum.

$$x_1^2 + x_2^2 + \cdots + x_n^2 = \sum_{k=1}^{n} x_k^2$$

Once again, the expression "$k = 1$" written underneath the symbol indicates that we get the first term of the sum by replacing k by 1 in the given expression. When we replace k by 1 in the expression x_k^2, we get x_1^2. Notice that this is the first term.

For the second term, we simply increase k by 1 to get $k = 2$. So, we replace k by 2 to get x_2^2.

We continue in this fashion, increasing k by 1 each time until we reach the number written above the symbol. In this case, that is $k = n$. So, the last term is x_n^2.

(4) We get the expression for |x| by taking the square root of each side of the equation:

$$\sqrt{x_1^2 + x_2^2 + \cdots x_n^2} = \sqrt{\sum_{k=1}^{n} x_k^2}$$

(5) In general, for any nonnegative real number x, $\sqrt{x} = x^{\frac{1}{2}}$ (by definition). In this case, using an exponent of $\frac{1}{2}$ with the summation symbol looks a bit less messy than using the square root symbol. So, we get the following:

$$\sqrt{x_1^2 + x_2^2 + \cdots x_n^2} = \left(\sum_{k=1}^n x_k^2\right)^{\frac{1}{2}}$$

Example 7.35: Let $\mathbf{x}, \mathbf{y} \in \mathbb{R}^4$ with $\mathbf{x} = (1, 0, -3, 5)$ and $\mathbf{y} = (3, -2, 1, 3)$. We have the following:

$$\mathbf{x} + \mathbf{y} = (4, -2, -2, 8)$$
$$2\mathbf{x} = (2, 0, -6, 10)$$
$$|\mathbf{x}| = \sqrt{1^2 + 0^2 + (-3)^2 + 5^2} = \sqrt{1 + 0 + 9 + 25} = \sqrt{35}$$
$$|2\mathbf{x} - \mathbf{y}| = \sqrt{(2-3)^2 + \left(0 - (-2)\right)^2 + (-6-1)^2 + (10-3)^2} = \sqrt{1 + 4 + 49 + 49} = \sqrt{103}$$

To prove many of the theorems about $\mathbb{R}^n$ and $\mathbb{C}^n$ that are analogous to the results for $\mathbb{R}$ and $\mathbb{C}$, we will need a higher dimensional version of the Triangle Inequality. This is much harder to prove, but luckily it is true. You will be asked to prove this version of the Triangle Inequality in Problem 23 below. For now, we state the theorem.

Theorem 7.36 (Generalized Triangle Inequality): Let $\mathbf{x}, \mathbf{y} \in \mathbb{R}^n$ or $\mathbb{C}^n$. Then $|\mathbf{x} + \mathbf{y}| \leq |\mathbf{x}| + |\mathbf{y}|$.

To avoid saying "$\mathbb{R}^n$ or $\mathbb{C}^n$" over and over again, we will restrict our attention to $\mathbb{R}^n$. However, everything we say for the rest of this lesson about $\mathbb{R}^n$ will be true for $\mathbb{C}^n$ as well.

We can now define the r-neighborhood of $\mathbf{x} \in \mathbb{R}^n$ by $N_r(\mathbf{x}) = \{\mathbf{y} \in \mathbb{R}^n \mid |\mathbf{y} - \mathbf{x}| < r\}$. In higher dimensions, we will usually refer to an r-neighborhood as an **open ball** (some authors use the expression **open sphere**) and we will may the notation $B_r(\mathbf{x})$ in place of $N_r(\mathbf{x})$.

Note: When $n = 1$, an open ball is an open interval, when $n = 2$, an open ball is an open disk, and when $n = 3$, an open ball actually looks like a ball (or sphere) in 3-dimensional space. For $n > 3$, we can no longer expect to have a precise visualization of what an "open ball" looks like. However, even in higher dimensions, it is still useful to draw pictures. When proving theorems about $\mathbb{R}^n$, it's usually most useful to visualize open balls as open disks in $\mathbb{C}$. This does **not** mean that the space looks like $\mathbb{C}$. The visualization should be used as evidence that a theorem might be true. Of course, a detailed proof still needs to be written.

For each definition for $\mathbb{R}$ and $\mathbb{C}$, we have analogous definitions for $\mathbb{R}^n$ as follows:

A subset X of $\mathbb{R}^n$ is said to be **open** in $\mathbb{R}^n$ if for every $\mathbf{x} \in X$, there is an open ball B with $\mathbf{x} \in B$ and $B \subseteq X$.

A subset X of $\mathbb{R}^n$ is **closed** in $\mathbb{R}^n$ if $\mathbb{R}^n \setminus X$ is open in $\mathbb{R}^n$.

If $S \subseteq \mathbb{R}^n$, then $\mathbf{x} \in \mathbb{R}^n$ is called an **accumulation point** of S if every open ball containing $\mathbf{x}$ contains at least one point of S different from $\mathbf{x}$.

If $S \subseteq \mathbb{R}^n$, then the **closure** of S in $\mathbb{R}^n$, written $\overline{S}$, is equal to $\cap\{C \mid S \subseteq C \wedge C \text{ is closed in } \mathbb{R}^n\}$.

Example 7.37:

1. Every open ball $B_r(\mathbf{x}) = \{\mathbf{y} \in \mathbb{R}^n \mid |\mathbf{y} - \mathbf{x}| < r\}$ is open in $\mathbb{R}^n$. Similarly, every closed ball $\overline{B_r(\mathbf{x})} = \{\mathbf{y} \in \mathbb{R}^n \mid |\mathbf{y} - \mathbf{x}| \leq r\}$ is closed in $\mathbb{R}^n$. Note that the set of accumulation points of $B_r(\mathbf{x})$ is $\overline{B_r(\mathbf{x})}$.

2. $\emptyset$ and $\mathbb{R}^n$ are the only two subsets of $\mathbb{R}^n$ that are both open and closed in $\mathbb{R}^n$.

3. If $S = \{\mathbf{x} \in \mathbb{R}^n \mid x_k \in \mathbb{Q} \text{ for each } k = 1, \ldots, n\}$, then every $\mathbf{y} \in \mathbb{R}^n$ is an accumulation point of S (Prove this!). It follows that $\overline{S} = \mathbb{R}^n$.

4. If $T = \{\mathbf{x} \in \mathbb{R}^n \mid x_k \in \mathbb{Z} \text{ for each } k = 1, 2, \ldots, n\}$, then T has no accumulation points (Prove this!). It follows that $\overline{T} = T$.

For most of the theorems we proved about $\mathbb{R}$ and $\mathbb{C}$, we have analogous theorems for $\mathbb{R}^n$ as well. We state some of these results in the following theorem. The proofs are nearly identical to the proofs for $\mathbb{R}$ and $\mathbb{C}$, and so, I leave them to the reader.

Theorem 7.38:

1. A subset X of $\mathbb{R}^n$ is open in $\mathbb{R}^n$ if and only if for every $\mathbf{x} \in \mathbb{R}^n$, there is a positive real number d such that $B_d(\mathbf{x}) \subseteq X$.

2. $\emptyset$ and $\mathbb{R}^n$ are both open in $\mathbb{R}^n$. It follows that $\emptyset$ and $\mathbb{R}^n$ are both closed in $\mathbb{R}^n$.

3. An arbitrary union of open sets in $\mathbb{R}^n$ is open in $\mathbb{R}^n$.

4. A finite intersection of open sets in $\mathbb{R}^n$ is open in $\mathbb{R}^n$.

5. An arbitrary intersection of closed sets in $\mathbb{R}^n$ is closed in $\mathbb{R}^n$.

6. A finite union of closed sets in $\mathbb{R}^n$ is closed in $\mathbb{R}^n$.

7. A subset C of $\mathbb{R}^n$ is closed in $\mathbb{R}^n$ if and only if C contains each of its accumulation points.

8. Let $S \subseteq \mathbb{R}^n$. Then $S \subseteq \overline{S}$.

9. If C is closed in $\mathbb{R}^n$ with $S \subseteq C$, then $\overline{S} \subseteq C$.

10. Let $S \subseteq \mathbb{R}^n$. Then $\overline{S} = S \cup \{x \in \mathbb{R}^n \mid x \text{ is an accumulation point of } S\}$.

11. S is closed in $\mathbb{R}^n$ if and only if $S = \overline{S}$.

12. $\mathbf{x} \in \overline{S}$ if and only if every open ball containing $\mathbf{x}$ contains at least one point of S.

Once again, note that if we replace $\mathbb{R}^n$ by $\mathbb{C}^n$ everywhere in Theorem 7.38, each of the given statements is still true.

Problem Set 7

Full solutions to these problems are available for free download here:

www.SATPrepGet800.com/TFBZLF

LEVEL 1

1. Let $z = -4 - i$ and $w = 3 - 5i$. Compute the distance between z and w.

2. Define a set of real numbers with exactly two accumulation points.

LEVEL 2

3. Let S be a set of real numbers and let S' be the set of accumulation points of S. Prove that S' is closed in $\mathbb{R}$.

4. Determine the accumulation points of each of the following subsets of $\mathbb{C}$:

   (i) $\left\{ \frac{1}{n} \,\middle|\, n \in \mathbb{Z}^+ \right\}$

   (ii) $\left\{ \frac{i}{n} \,\middle|\, n \in \mathbb{Z}^+ \right\}$

   (iii) $\{ i^n \mid n \in \mathbb{Z}^+ \}$

   (iv) $\{ z \mid |z| < 1 \}$

   (v) $\{ z \mid 0 < |z - 2| \le 3 \}$

LEVEL 3

5. Prove the following:

   (i)   For all $b \in \mathbb{R}$, the infinite interval $(-\infty, b)$ is open in $\mathbb{R}$.

   (ii)   The intersection of two open intervals in $\mathbb{R}$ is either empty or an open interval in $\mathbb{R}$.

   (iii) The intersection of finitely many open sets in $\mathbb{R}$ is an open set in $\mathbb{R}$.

6. Prove the Triangle Inequality (Theorem 7.4).

7. Let z and w be complex numbers. Prove $\big| |z| - |w| \big| \le |z \pm w| \le |z| + |w|$.

8. Let $x \in \mathbb{R}$ be an accumulation point of a set S. Prove that every open interval containing x contains infinitely many points of S.

9. Let $X, Y,$ and Z be sets of real numbers with $Z = X \cup Y$. Prove that $\overline{Z} = \overline{X} \cup \overline{Y}$.

10. Prove that if $\boldsymbol{X}$ is a nonempty set of closed subsets of $\mathbb{R}$, then $\cap \boldsymbol{X}$ is closed.

11. Give an example of an infinite collection of open sets in $\mathbb{R}$ whose intersection is not open in $\mathbb{R}$. Also, give an example of an infinite collection of closed sets in $\mathbb{R}$ whose union is not closed in $\mathbb{R}$. Provide a proof for each example.

12. Determine if each of the following subsets of $\mathbb{C}$ is open, closed, both, or neither. Give a proof in each case.

    (i) $\emptyset$

    (ii) $\mathbb{C}$

    (iii) $\{z \in \mathbb{C} \mid |z| > 1\}$

    (iv) $\{z \in \mathbb{C} \mid \operatorname{Im} z \le -2\}$

    (v) $\{i^n \mid n \in \mathbb{Z}^+\}$

    (vi) $\{z \in \mathbb{C} \mid 2 < |z - 2| < 4\}$

13. Prove the following:

    (i)   An arbitrary union of open sets in $\mathbb{C}$ is an open set in $\mathbb{C}$.

    (ii)   A finite intersection of open sets in $\mathbb{C}$ is an open set in $\mathbb{C}$.

    (iii) An arbitrary intersection of closed sets in $\mathbb{C}$ is a closed set in $\mathbb{C}$.

    (iv) A finite union of closed sets in $\mathbb{C}$ is a closed set in $\mathbb{C}$.

    (v)   Every open set in $\mathbb{C}$ can be expressed as a union of open disks.

14. Prove that every closed set in $\mathbb{R}$ can be written as an intersection $\bigcap X$, where each element of X is a union of at most 2 closed intervals.

15. A complex number z is an **interior point** of a set S of complex numbers if there is a neighborhood of z that contains only points in S, whereas w is a **boundary point** of S if each neighborhood of w contains at least one point in S and one point not in S. Prove the following:

    (i)   A set of complex numbers is open in $\mathbb{C}$ if and only if each point in S is an interior point of S.

    (ii)   A set of complex numbers is open in $\mathbb{C}$ if and only if it contains none of its boundary points.

    (iii) A set of complex numbers is closed in $\mathbb{C}$ if and only if it contains all its boundary points.

16. Let $D = \{z \in \mathbb{C} \mid |z| \le 1\}$ be the closed unit disk and let S be a subset of D that includes the interior of the disk but is missing at least one point on the bounding circle of the disk. Show that S is not a closed set.

17. Prove that a subset C of $\mathbb{C}$ is closed in $\mathbb{C}$ if and only if C contains each of its accumulation points (this is Theorem 7.31).

18. Prove that a set consisting of finitely many complex numbers is a closed set in $\mathbb{C}$. (Hint: Show that a finite set has no accumulation points.)

19. Let C be the Cantor set. Prove the following:

    (i) $\frac{1}{4} \in C$.

    (ii) C is uncountable.

20. If $X \subseteq Y$, then X is said to be **dense** in Y if $\overline{X} = Y$. Prove that $\mathbb{C}$ has a countable dense subset. Does $\mathbb{R}^n$ have a countable dense subset for each $n \in \mathbb{N}$?

CHALLENGE PROBLEMS

21. Prove that every nonempty open set of real numbers can be expressed as a union of pairwise disjoint open intervals.

22. Let X and Y be closed subsets of $\mathbb{R}^n$ or $\mathbb{C}^n$ with $X \cap Y = \emptyset$. Prove that there are disjoint open sets of complex numbers U and V such that $X \subseteq U$ and $Y \subseteq V$.

23. Prove the **Generalized Triangle Inequality** (Theorem 7.36). Hint: You should first prove the **Cauchy-Schwarz Inequality**. This inequality says that for all $\mathbf{x}, \mathbf{y} \in \mathbb{R}^n$ or $\mathbb{C}^n$,

$$\sum_{k=1}^{n} |x_i y_i| \leq |\mathbf{x}| + |\mathbf{y}|.$$

LESSON 8
CONTINUITY IN $\mathbb{R}$ AND $\mathbb{C}$

Strips and Rectangles in $\mathbb{R}$

A **horizontal strip** in $\mathbb{R} \times \mathbb{R}$ is a set of the form $\mathbb{R} \times (c, d) = \{(x, y) \mid c < y < d\}$.

Example 8.1: The horizontal strips $\mathbb{R} \times (-2, 1)$ and $\mathbb{R} \times (2.25, 2.75)$ can be visualized in the xy-**plane** as follows:

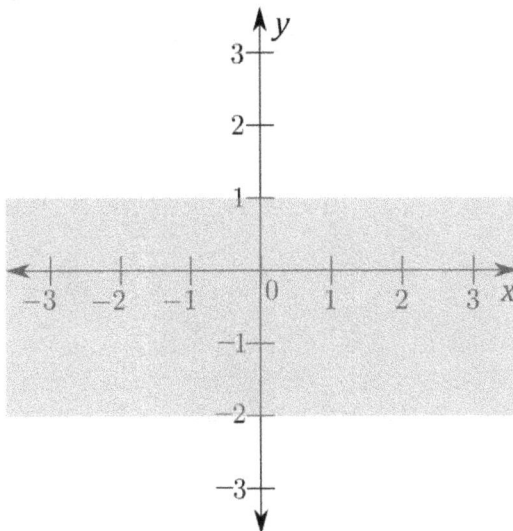

$\mathbb{R} \times (-2, 1)$

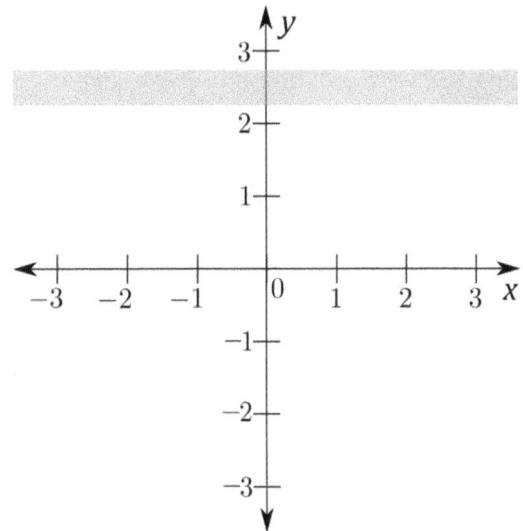

$\mathbb{R} \times (2.25, 2.75)$

Similarly, a **vertical strip** in $\mathbb{R} \times \mathbb{R}$ is a set of the form $(a, b) \times \mathbb{R} = \{(x, y) \mid a < x < b\}$.

Example 8.2: The vertical strips $(-3, 0) \times \mathbb{R}$ and $(0.8, 1) \times \mathbb{R}$ can be visualized in the xy-**plane** as follows:

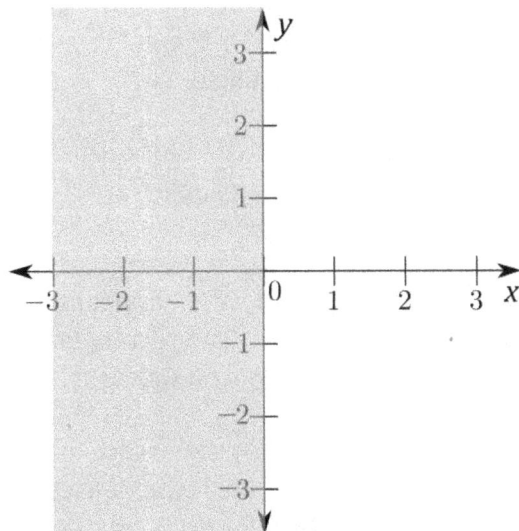

$(-3, 0) \times \mathbb{R}$

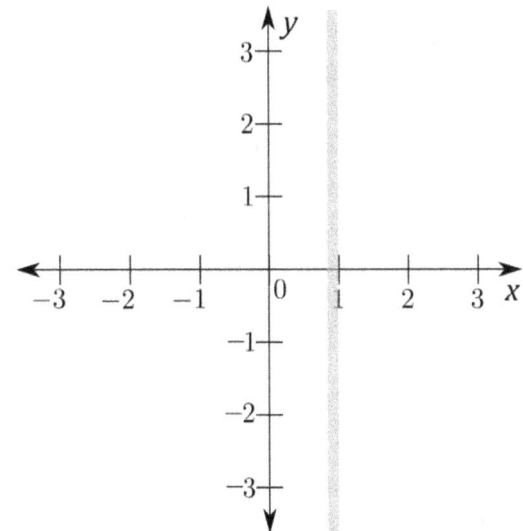

$(0.8, 1) \times \mathbb{R}$

135

We will say that the horizontal strip $\mathbb{R} \times (c, d)$ **contains** y if $y \in \mathbb{R}$ and $c < y < d$. Otherwise, we will say that the horizontal strip **excludes** y.

Similarly, we will say that the vertical strip $(a, b) \times \mathbb{R}$ **contains** x if $x \in \mathbb{R}$ and $a < x < b$. Otherwise, we will say that the vertical strip **excludes** x.

Example 8.3: The horizontal strip $\mathbb{R} \times (2.25, 2.75)$ contains 2.5 and excludes 3. One way to visualize this is to draw the horizontal lines $y = 2.5$ and $y = 3$. Below in the figure on the left, we used a solid line for the line $y = 2.5$ because it is contained in the horizontal strip and we used a dashed line for the line $y = 3$ because it is not contained in the horizontal strip.

Similarly, the vertical strip $(0.8, 1) \times \mathbb{R}$ contains 0.9 and excludes 2. Again, we can visualize this by drawing the vertical lines $x = 0.9$ and $x = 2$. These vertical lines are shown below in the figure on the right.

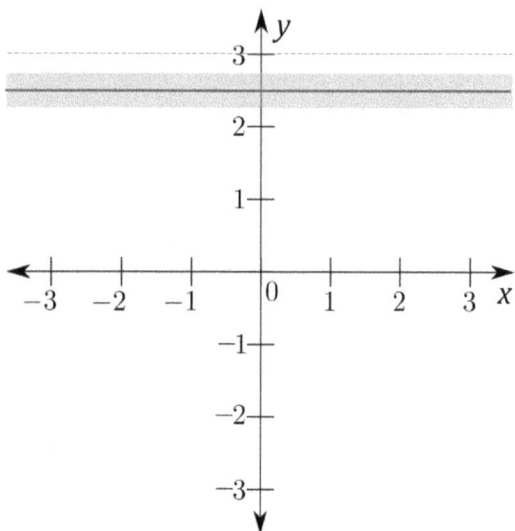

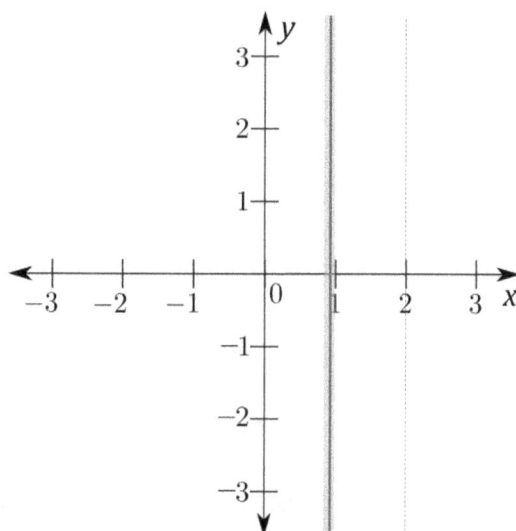

An **open rectangle** is a set of the form $(a, b) \times (c, d) = \{(x, y) \mid a < x < b \wedge c < y < d\}$. Note that the open rectangle $(a, b) \times (c, d)$ is the intersection of the horizontal strip $\mathbb{R} \times (c, d)$ and the vertical strip $(a, b) \times \mathbb{R}$. We will say that an open rectangle **traps** the point (x, y) if $x, y \in \mathbb{R}$ and (x, y) is in the open rectangle. Otherwise, we will say that (x, y) **escapes** from the open rectangle.

Example 8.4: The open rectangle $R = (-3, 0) \times (-2, 1)$ is the intersection of the horizontal strip $H = \mathbb{R} \times (-2, 1)$ and the vertical strip $V = (-3, 0) \times \mathbb{R}$. So, $R = H \cap V$. The rectangle R traps $(-1, 0)$, whereas $(-2, 3)$ escapes from R. This can be seen in the figure below on the left.

The open rectangle $R = (0.8, 1) \times (2.25, 2.75)$ is the intersection of the horizontal strip $H = \mathbb{R} \times (2.25, 2.75)$ and the vertical strip $V = (0.8, 1) \times \mathbb{R}$. So, $R = H \cap V$. The rectangle R traps $(0.9, 2.5)$, whereas $(0.9, 2)$ escapes from R. This can be seen in the figure below on the right.

Observe that in this example, I chose points that escape the given rectangles in the vertical direction. They fall outside the rectangle because they're too high or too low. This is the only type of escape that we will be interested in here. We do not care about points that escape to the left or right of a rectangle.

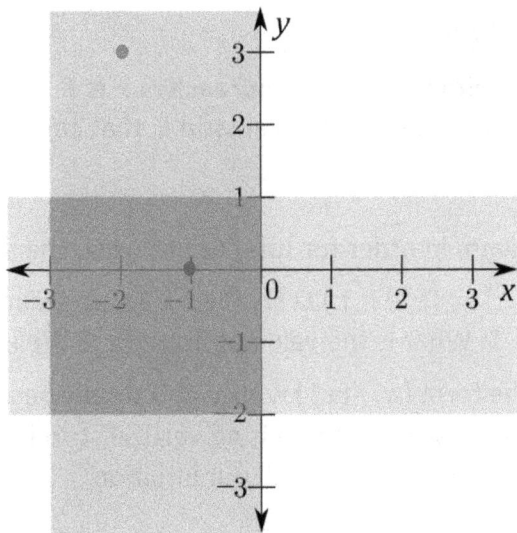

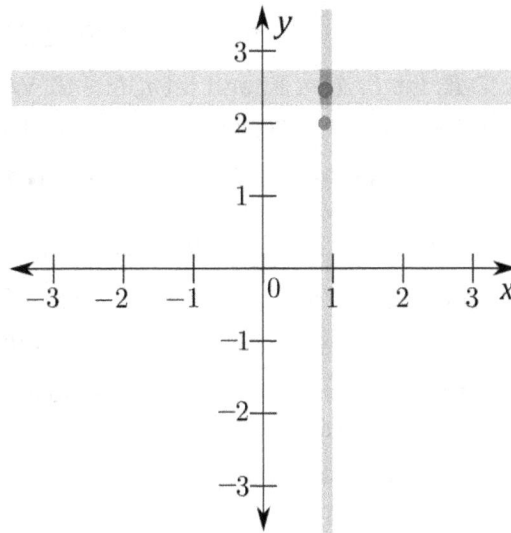

Let $A \subseteq \mathbb{R}$, let $f: A \to \mathbb{R}$, and let $R = (a, b) \times (c, d)$ be an open rectangle. We say that R **traps** f if for all $x \in (a, b)$, R traps $(x, f(x))$. Otherwise we say that f **escapes** from R.

Example 8.5: Let $f: \mathbb{R} \to \mathbb{R}$ be defined by $f(x) = x + 1$. Consider the open rectangles $R = (0, 2) \times (1, 3)$ and $S = (0, 2) \times (0, 2)$. Then R traps f, as can be seen in the figure below on the left, whereas f escapes from S, as can be seen in the figure below on the right. I put a box around the points of the form $(x, f(x))$ that escape from S. For example, the point $(1.2, f(1.2)) = (1.2, 2.2)$ escapes from S because $0 < 1.2 < 2$, but $f(1.2) = 2.2 \geq 2$.

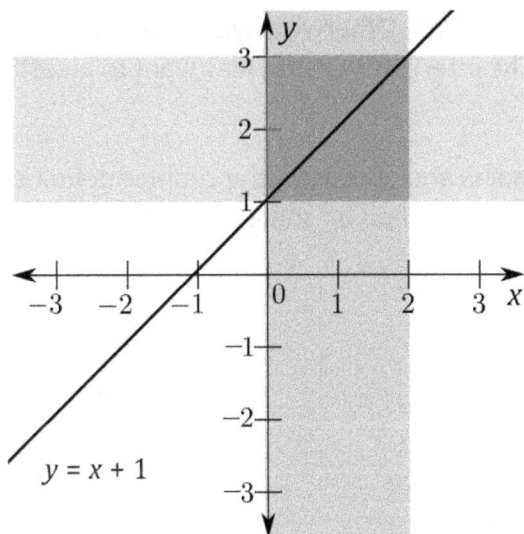

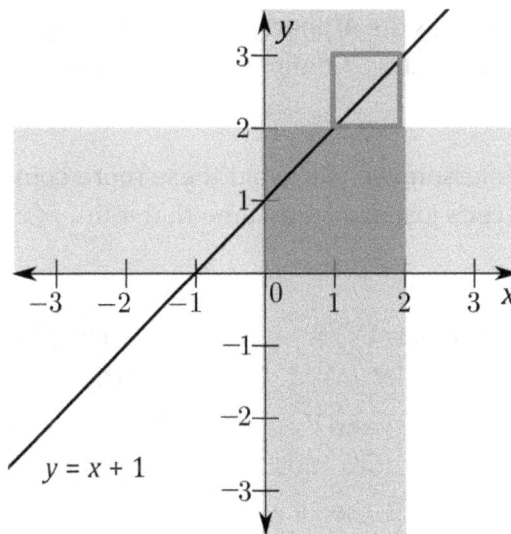

When we are checking the limiting behavior near a real number r, we don't care if the point $(r, f(r))$ escapes. Therefore, before we define a limit, we need to modify our definitions of "traps" and "escapes" slightly to account for this.

Let $A \subseteq \mathbb{R}$, let $f: A \to \mathbb{R}$, and let $R = (a, b) \times (c, d)$ be an open rectangle. We say that R **traps** f **around** r if for all $x \in (a, b) \setminus \{r\}$, R traps $(x, f(x))$. Otherwise, we say f **escapes from** R **around** r.

Limits and Continuity in $\mathbb{R}$

Let $A \subseteq \mathbb{R}$, let $f: A \to \mathbb{R}$, and let $r, L \in \mathbb{R}$. We say that the **limit of f as x approaches r is L**, written $\lim_{x \to r} f(x) = L$, if for every horizontal strip H that contains L there is a vertical strip V that contains r such that the rectangle $H \cap V$ traps f around r.

Technical note: According to the definition of limit just given, in order for $\lim_{x \to r} f(x)$ to exist, the set A needs to contain a deleted neighborhood of r, say $N_\epsilon^\odot(r) = (r - \epsilon, r) \cup (r, r + \epsilon)$. As an example, suppose that $A = \{0\}$ and $f: A \to \mathbb{R}$ is defined by $f(0) = 1$. What is the value of $\lim_{x \to 0} f(x)$? Well, any rectangle of the form $H \cap V$ does not trap any points of the form $(x, f(x))$ with $x \neq 0$ simply because $f(x)$ is not defined when $x \neq 0$. Therefore, given a horizontal strip H, there is no vertical strip V such that $H \cap V$ traps f around r, and so, $\lim_{x \to r} f(x)$ does not exist. This agrees with our intuition.

As a less extreme example, suppose that $A = \mathbb{Q}$ and $g: \mathbb{Q} \to \mathbb{R}$ is the constant function where $g(x) = 1$ for all $x \in \mathbb{Q}$. Then for any $r \in \mathbb{Q}$, we should probably have $\lim_{x \to r} g(x) = 1$. But if we use our current definition of limit, then $\lim_{x \to r} g(x)$ does not exist. A more general definition of limit would yield finite values for limits defined on certain sets (like $\mathbb{Q}$) that do not contain a neighborhood of r.

Specifically, we really should insist only that for each $j \in \mathbb{R}^+$, $A \cap \left((r - j, r + j) \setminus \{r\} \right) \neq \emptyset$. The definition of limit given above could be modified slightly to accommodate this more general situation. For example, we could change "R traps f around r" to "for all $x \in A \cap \left((a, b) \setminus \{r\} \right)$, R traps $(x, f(x))$." If we were to use this more general definition, it is very important that we also insist that the set A has the property given at the beginning of this paragraph. Otherwise, we would have an issue with the function f defined at the beginning of this note. The interested reader may want to investigate this.

In this lesson, we will avoid these more complicated domains and stick with the simpler definition of limit. Let's just always assume that if $\lim_{x \to r} f(x)$ exists, then f is defined on some deleted neighborhood of r.

Example 8.6: Let $f: \mathbb{R} \to \mathbb{R}$ be defined by $f(x) = x + 1$, let $r = 1.5$, and let $L = 1$. Let's show that $\lim_{x \to 1.5} f(x) \neq 1$. If $H = \mathbb{R} \times (0, 2)$ and V is any vertical strip that contains 1.5, then $H \cap V$ does **not** trap f around 1.5. Indeed, if $V = (a, b) \times \mathbb{R}$, then if we let $x = \frac{1}{2}(1.5 + b)$, we will show that $x \in (a, b)$ and $f(x) = \frac{1}{2}(1.5 + b) + 1 > 2$ (see the figure to the right).

To see that $x \in (a, b)$, note that since $b > 1.5$, we have $x = \frac{1}{2}(1.5 + b) > \frac{1}{2}(1.5 + 1.5) = \frac{1}{2} \cdot 3 = 1.5 > a$, and we have $x = \frac{1}{2}(1.5 + b) < \frac{1}{2}(b + b) = \frac{1}{2} \cdot 2b = b$.

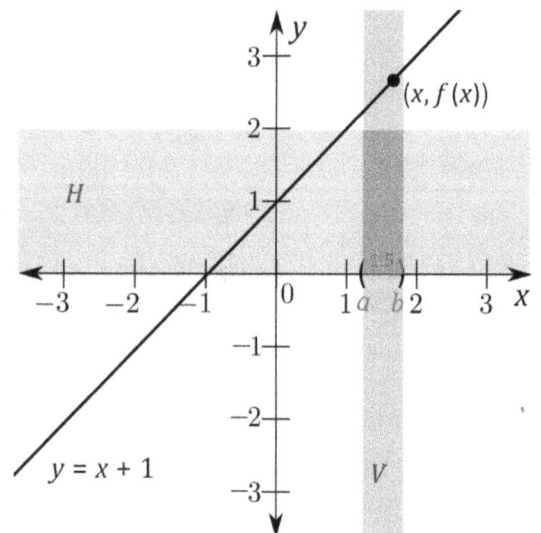

To see that $f(x) > 2$, note that

$$f(x) = \tfrac{1}{2}(1.5 + b) + 1 > \tfrac{1}{2}(1.5 + 1.5) + 1 = \tfrac{1}{2} \cdot 3 + 1 = 1.5 + 1 = 2.5 > 2.$$

So, what is $\lim_{x \to 1.5} f(x)$ equal to? From the picture above, a good guess would be 2.5. To verify that this is true, let $H = \mathbb{R} \times (c, d)$ be a horizontal strip that contains 2.5. Next, let $V = (c - 1, d - 1) \times \mathbb{R}$. We will show that $H \cap V = (c - 1, d - 1) \times (c, d)$ traps f around 1.5. Let $x \in (c - 1, d - 1) \setminus \{1.5\}$, so that $c - 1 < x < d - 1$ and $x \neq 1.5$. Adding 1 to each part of this sequence of inequalities gives $c < x + 1 < d$, so that $c < f(x) < d$, or equivalently, $f(x) \in (c, d)$. Since $x \in (c - 1, d - 1) \setminus \{1.5\}$ and $f(x) \in (c, d)$, it follows that $(x, f(x)) \in (c - 1, d - 1) \times (c, d) = H \cap V$. Therefore, $H \cap V$ traps f around 1.5.

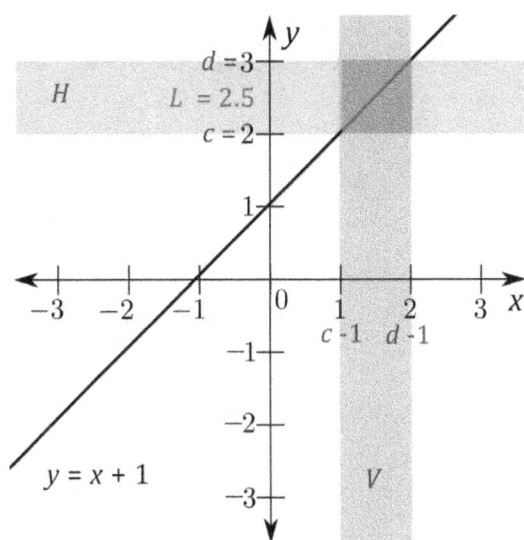

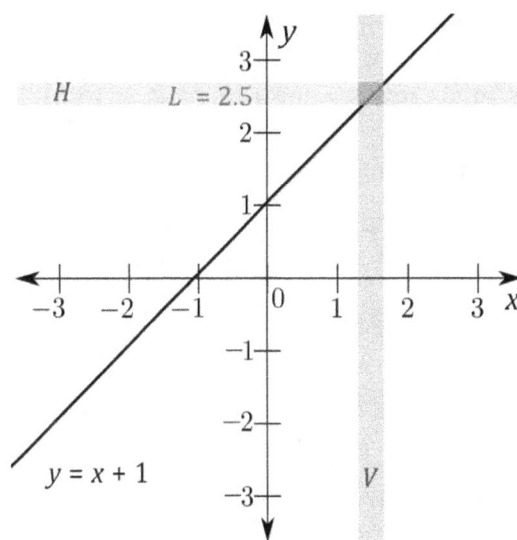

Notes: (1) The figures above give a visual representation of the argument just presented. In the figure on the left, we let $c = 2$ and $d = 3$, so that $H = \mathbb{R} \times (2, 3)$. Our choice of V is then $(1, 2) \times \mathbb{R}$, and therefore, $H \cap V = (1, 2) \times (2, 3)$. Now, if $1 < x < 2$, then $2 < x + 1 < 3$. So, $(x, f(x)) \in H \cap V$.

In the figure on the right, we started with a thinner horizontal strip without being specific about its exact definition. Notice that we then need to use a thinner vertical strip to prevent f from escaping. If the vertical strip were just a little wider on the right, then some points of the form $(x, f(x))$ would escape the rectangle because they would be too high. If the vertical strip were just a little wider on the left, then some points of the form $(x, f(x))$ would escape the rectangle because they would be too low.

(2) Notice that in this example, the point $(1.5, f(1.5))$ itself always stays in the rectangle. In the argument given, we excluded this point from consideration. Even if $(1.5, f(1.5))$ were to escape the rectangle, it would not change the result here. We would still have $\lim_{x \to 1.5} f(x) = 2.5$. I indicated the parts of the argument where $(1.5, f(1.5))$ was being excluded from consideration in Example 8.6 above by placing rectangles around that part of the text. If we delete all the parts of the argument inside those rectangles, the resulting argument would still be correct. We will examine this situation more carefully in the next example.

If we modify the definition of limit by getting rid of "around r," insisting that $r \in A$, and replacing L by $f(r)$, we get the definition of continuity. Specifically, we have the following definition.

Let $A \subseteq \mathbb{R}$, let $f: A \to \mathbb{R}$, and let $r \in A$. We say that the function f is **continuous** at r if for every horizontal strip H that contains $f(r)$, there is a vertical strip V that contains r such that the rectangle $H \cap V$ traps f.

Example 8.7:

1. If we delete all the text that I placed in rectangles in Example 8.6 above, then the resulting argument shows that the function f defined by $f(x) = x + 1$ is continuous at $x = 1.5$.

   To summarize, given a horizontal strip H containing $f(1.5) = 2.5$, we found a vertical strip V containing 1.5 such that $H \cap V$ traps f. Notice once again that in this example we do not exclude $x = 1.5$ from consideration, and when we mention trapping f, we do not say "around 1.5." We need to trap $(1.5, f(1.5)) = (1.5, 2.5)$ as well.

2. Let's consider the function $g: \mathbb{R} \to \mathbb{R}$ defined by $g(x) = \begin{cases} x + 1 & \text{if } x \neq 1.5 \\ -2 & \text{if } x = 1.5 \end{cases}$. This function is nearly identical to the function f we have been discussing. It differs from the previous function only at $x = 1.5$. It should follow that $\lim_{x \to 1.5} g(x) = \lim_{x \to 1.5} f(x)$. And, in fact it does. The same exact argument that we gave in Example 8.6 shows that $\lim_{x \to 1.5} g(x) = 2.5$. The figures below illustrate the situation.

   This time however, we cannot delete the text inside the rectangles in Example 8.6. $x = 1.5$ needs to be excluded from consideration for the argument to go through. In the leftmost figure below, we see that if H is the horizontal strip $H = \mathbb{R} \times (2, 3)$, then for any vertical strip $V = (a, b) \times \mathbb{R}$ that contains 1.5, the point $(1.5, -2)$ will escape the rectangle $H \cap V$. Indeed, $H \cap V = (a, b) \times (2, 3)$, and $(1.5, g(1.5)) = (1.5, -2) \notin (a, b) \times (2, 3)$ because $-2 < 2$. This shows that g is **not** continuous at $x = 1.5$.

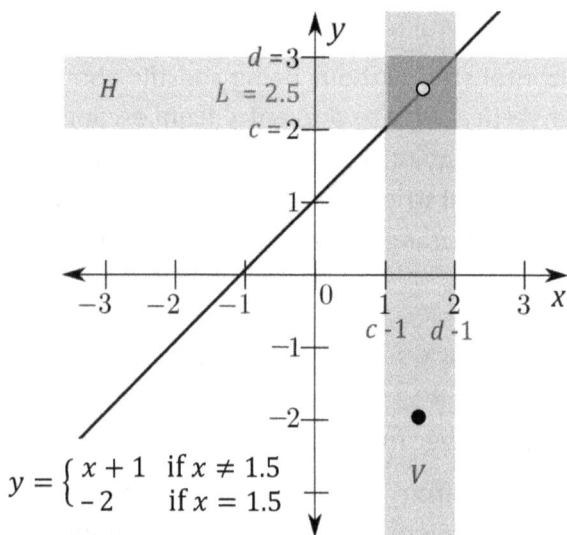

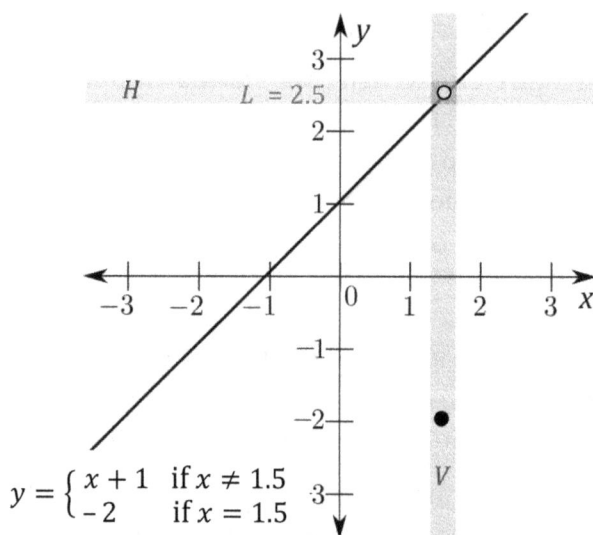

The strip game: Suppose we want to determine if $\lim_{x \to r} f(x) = L$. Consider the following game between two players: Player 1 "attacks" by choosing a horizontal strip H_0 containing L. Player 2 then tries to "defend" by choosing a vertical strip V_0 containing r such that $H_0 \cap V_0$ traps f around r. If Player 2 cannot find such a vertical strip, then Player 1 wins and $\lim_{x \to r} f(x) \neq L$. If Player 2 defends successfully, then Player 1 chooses a new horizontal strip H_1 containing L. If Player 1 is smart, then he/she will choose a "much thinner" horizontal strip that is contained in H_0 (compare the two figures above). The thinner the strip, the harder it will be for Player 2 to defend. Player 2 once again tries to choose a vertical strip V_1 such that $H_1 \cap V_1$ traps f around r. This process continues indefinitely. Player 1 wins the strip game if at some stage, Player 2 cannot defend successfully. Player 2 wins the strip game if he or she defends successfully at every stage.

Player 1 has a winning strategy for the strip game if and only if $\lim_{x \to r} f(x) \neq L$, while Player 2 has a winning strategy for the strip game if and only if $\lim_{x \to r} f(x) = L$.

Note that if it's possible for Player 1 to win the strip game, then Player 1 can win with a single move — just choose the horizontal strip immediately that Player 2 cannot defend against.

For example, if $f(x) = x + 1$, then $\lim_{x \to 1.5} f(x) \neq 1$. Player 1 can win the appropriate strip game immediately by choosing the horizontal strip $H = \mathbb{R} \times (0, 2)$. Indeed, if Player 2 chooses any vertical strip $V = (a, b) \times \mathbb{R}$ that contains 1.5, let $x \in (a, b)$ with $x > 1.5$. Then we have

$$f(x) = x + 1 > 1.5 + 1 = 2.5 > 2.$$

So, $(x, f(x))$ escapes $H \cap V$. In the figure to the right, we see that Player 1 has chosen $H = \mathbb{R} \times (0, 2)$ and Player 2 chose $V = (a, b) \times \mathbb{R}$ for some $a, b \in \mathbb{R}$ with $a < 1.5 < b$. The part of the line inside the square is an illustration of where f escapes $H \cap V$ between a and b. Observe that no matter how much thinner we try to make that vertical strip, if it contains 1.5, then it will contain a portion of the line that is inside the square.

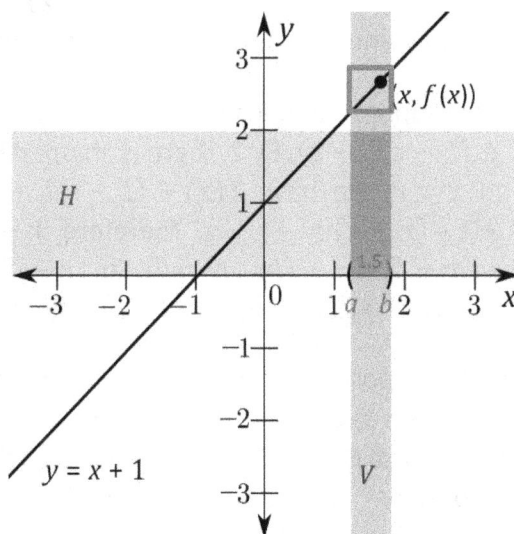

Now, if it's possible for Player 2 to win the game, then we need to describe how Player 2 defends against an arbitrary attack from Player 1. Suppose again that $f(x) = x + 1$ and we are trying to show that $\lim_{x \to 1.5} f(x) = 2.5$. We have already seen how Player 2 can defend against an arbitrary attack from Player 1 in Example 8.6. If at stage n, Player 1 attacks with the horizontal strip $H_n = \mathbb{R} \times (a, b)$, then Player 2 can successfully defend with the vertical strip $V_n = (a - 1, b - 1) \times \mathbb{R}$.

Equivalent Definitions of Limits and Continuity in $\mathbb{R}$

The definitions of limit and continuity can be written using open intervals instead of strips. Specifically, we have the following:

141

Theorem 8.8: Let $A \subseteq \mathbb{R}$, let $f: A \to \mathbb{R}$, and let $r, L \in \mathbb{R}$. The following are equivalent:

1. $\lim_{x \to r} f(x) = L$.

2. For every open interval (c, d) with $L \in (c, d)$, there is an open interval (a, b) with $r \in (a, b)$ such that whenever $x \in (a, b)$ and $x \neq r$, $f(x) \in (c, d)$.

3. For every positive real number ϵ, there is a positive real number δ such that whenever $x \in (r - \delta, r + \delta)$ and $x \neq r$, $f(x) \in (L - \epsilon, L + \epsilon)$.

This is the first Theorem where we want to prove more than two statements equivalent. We will do this with the following chain: $1 \to 2 \to 3 \to 1$. In other words, we will assume statement 1 and use it to prove statement 2. We will then assume statement 2 and use it to prove statement 3. Finally, we will assume statement 3 and use it to prove statement 1.

Proof of Theorem 8.8: ($1 \to 2$) Suppose that $\lim_{x \to r} f(x) = L$ and let $L \in (c, d)$. Then the horizontal strip $\mathbb{R} \times (c, d)$ contains L. Since $\lim_{x \to r} f(x) = L$, there is a vertical strip $(a, b) \times \mathbb{R}$ that contains r such that the rectangle $R = (a, b) \times (c, d)$ traps f around r. Since the vertical strip $(a, b) \times \mathbb{R}$ contains r, $r \in (a, b)$. Since the rectangle R traps f around r, for all $x \in (a, b) \setminus \{r\}$, R traps $(x, f(x))$. In other words, whenever $x \in (a, b)$ and $x \neq r$, we have $(x, f(x)) \in (a, b) \times (c, d)$, and thus, $f(x) \in (c, d)$.

($2 \to 3$) Suppose 2 holds and let ϵ be a positive real number. Then $L - \epsilon < L < L + \epsilon$, or equivalently, $L \in (L - \epsilon, L + \epsilon)$. By 2, there is an open interval (a, b) with $r \in (a, b)$ such that whenever $x \in (a, b)$ and $x \neq r$, we have $f(x) \in (L - \epsilon, L + \epsilon)$. Let $\delta = \min\{r - a, b - r\}$. Since $\delta \leq r - a$, we have $-\delta \geq -(r - a) = -r + a$. Therefore, $r - \delta \geq r + (-r + a) = a$. Furthermore, since $\delta \leq b - r$, we have $r + \delta \leq r + (b - r) = b$. So, $(r - \delta, r + \delta) \subseteq (a, b)$. If $x \in (r - \delta, r + \delta)$ and $x \neq r$, then since $(r - \delta, r + \delta) \subseteq (a, b)$, $x \in (a, b)$. Therefore, $f(x) \in (L - \epsilon, L + \epsilon)$.

($3 \to 1$) Suppose 3 holds and $H = \mathbb{R} \times (c, d)$ is a horizontal strip that contains L. Since $c < L < d$, we have $L - c > 0$ and $d - L > 0$. Therefore, $\epsilon = \min\{L - c, d - L\} > 0$. So, there is $\delta > 0$ such that whenever $x \in (r - \delta, r + \delta)$ and $x \neq r$, then $f(x) \in (L - \epsilon, L + \epsilon)$. Let $V = (r - \delta, r + \delta) \times \mathbb{R}$. Then V contains r. We now show that $H \cap V = (r - \delta, r + \delta) \times (c, d)$ traps f around r. Let $x \in (r - \delta, r + \delta)$ with $x \neq r$. Then $f(x) \in (L - \epsilon, L + \epsilon)$. So, $f(x) > L - \epsilon \geq L - (L - c) = c$ and $f(x) < L + \epsilon \leq L + (d - L) = d$. Therefore, $f(x) \in (c, d)$, and so, $H \cap V$ traps f around r. $\square$

Notes: (1) ϵ and δ are Greek letters pronounced "epsilon" and "delta," respectively. Mathematicians tend to use these two symbols to represent arbitrarily small numbers.

(2) Recall from lesson 7 that if $a \in \mathbb{R}$ and $\epsilon > 0$, then the ϵ-neighborhood of a is the interval $N_\epsilon(a) = (a - \epsilon, a + \epsilon)$ and the deleted ϵ-neighborhood of a is the "punctured" interval $N_\epsilon^\odot(a) = (a - \epsilon, a) \cup (a, a + \epsilon)$. We can visualize the deleted ϵ-neighborhood $N_\epsilon^\odot(a)$ as follows:

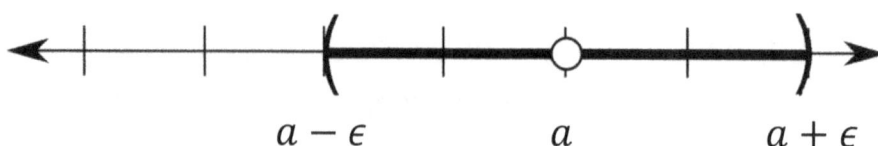

$$a - \epsilon \qquad\qquad a \qquad\qquad a + \epsilon$$

For a specific example, let's look at $N_2^{\odot}(1) = (1-2,1) \cup (1,1+2) = (-1,1) \cup (1,3)$.

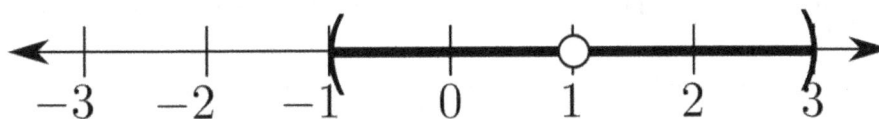

(3) The third part of Theorem 8.8 can be written in terms of neighborhoods as follows:

"For every positive real number ϵ, there is a positive real number δ such that whenever $x \in N_\delta^{\odot}(r)$, $f(x) \in N_\epsilon(L)$."

(4) $x \in (a-\epsilon, a+\epsilon)$ is equivalent to $a-\epsilon < x < a+\epsilon$. If we subtract a from each part of this inequality, we get $-\epsilon < x - a < \epsilon$. This last expression is equivalent to $|x-a| < \epsilon$. So, we have the following sequence of equivalences:

$$x \in N_\epsilon(a) \Leftrightarrow x \in (a-\epsilon, a+\epsilon) \Leftrightarrow a-\epsilon < x < a+\epsilon \Leftrightarrow |x-a| < \epsilon.$$

(5) $x \neq a$ is equivalent to $x - a \neq 0$. Since the absolute value of a real number can never be negative, $x - a \neq 0$ is equivalent to $|x-a| > 0$. This can also be written $0 < |x-a|$. So, we have the following sequence of equivalences:

$$x \in N_\epsilon^{\odot}(a) \Leftrightarrow x \in (a-\epsilon, a) \cup (a, a+\epsilon) \Leftrightarrow 0 < |x-a| < \epsilon.$$

(6) The third part of Theorem 8.8 can be written using absolute values as follows:

"For every positive real number ϵ, there is a positive real number δ such that whenever $0 < |x-r| < \delta$, $|f(x) - L| < \epsilon$."

(7) We can abbreviate the expression from Note 6 using quantifiers as follows:

$$\forall \epsilon > 0 \, \exists \delta > 0 \, (0 < |x-r| < \delta \rightarrow |f(x) - L| < \epsilon)$$

We will refer to this expression as the $\epsilon - \delta$ **definition of a limit**.

For each equivalent formulation of a limit, we have a corresponding formulation for the definition of continuity.

Theorem 8.9: Let $A \subseteq \mathbb{R}$, let $f: A \to \mathbb{R}$, and let $r \in A$. The following are equivalent:

1. f is continuous at r.
2. For every open interval (c, d) with $f(r) \in (c, d)$, there is an open interval (a, b) with $r \in (a, b)$ such that whenever $x \in (a, b)$, $f(x) \in (c, d)$.
3. For every positive real number ϵ, there is a positive real number δ such that whenever $x \in (r - \delta, r + \delta)$, $f(x) \in (f(r) - \epsilon, f(r) + \epsilon)$.
4. $\forall \epsilon > 0 \, \exists \delta > 0 \, (|x-r| < \delta \rightarrow |f(x) - f(r)| < \epsilon)$.

The proof of Theorem 8.9 is left to the reader. It is very similar to the proof of Theorem 8.8.

Basic Examples in $\mathbb{R}$

Example 8.10: Let's use the $\epsilon - \delta$ definition of a limit to prove that $\lim\limits_{x \to 1}(2x + 1) = 3$.

Analysis: Given $\epsilon > 0$, we need to find $\delta > 0$ so that $0 < |x - 1| < \delta$ implies $|(2x + 1) - 3| < \epsilon$. First note that $|(2x + 1) - 3| = |2x - 2| = |2(x - 1)| = |2||x - 1| = 2|x - 1|$. So, $|(2x + 1) - 3| < \epsilon$ is equivalent to $|x - 1| < \frac{\epsilon}{2}$. Therefore, $\delta = \frac{\epsilon}{2}$ should work.

Proof: Let $\epsilon > 0$ and let $\delta = \frac{\epsilon}{2}$. Suppose that $0 < |x - 1| < \delta$. Then we have

$$|(2x + 1) - 3| = |2x - 2| = |2(x - 1)| = |2||x - 1| = 2|x - 1| < 2\delta = 2 \cdot \frac{\epsilon}{2} = \epsilon.$$

Since $\epsilon > 0$ was arbitrary, we have $\forall \epsilon > 0 \, \exists \delta > 0 \, (0 < |x - 1| < \delta \to |(2x + 1) - 3| < \epsilon)$.

Therefore, $\lim\limits_{x \to 1}(2x + 1) = 3$. $\qquad\qquad\qquad\qquad\qquad\qquad\qquad\qquad\qquad\qquad\qquad\qquad$ $\square$

Notes: (1) Even though we're using the "$\epsilon - \delta$ definition" instead of the "strip definition," we can still visualize the situation in terms of the strip game. When we say "Let $\epsilon > 0$," we can think of this as Player 1 "attacking" with the horizontal strip $H = \mathbb{R} \times (3 - \epsilon, 3 + \epsilon)$. In the proof above, Player 2 is then "defending" with the vertical strip $V = \left(1 - \frac{\epsilon}{2}, 1 + \frac{\epsilon}{2}\right) \times \mathbb{R}$. This defense is successful because when $1 - \frac{\epsilon}{2} < x < 1 + \frac{\epsilon}{2}$, we have $2 - \epsilon < 2x < 2 + \epsilon$, and so, $3 - \epsilon < 2x + 1 < 3 + \epsilon$, or equivalently, $2x + 1 \in (3 - \epsilon, 3 + \epsilon)$. In other words, for $x \in \left(1 - \frac{\epsilon}{2}, 1 + \frac{\epsilon}{2}\right)$, $H \cap V$ traps f.

(2) Instead of playing the strip game, we can play the $\epsilon - \delta$ game instead. The idea is the same. Suppose we are trying to figure out if $\lim\limits_{x \to r} f(x) = L$. Player 1 "attacks" by choosing a positive number ϵ. This is equivalent to Player 1 choosing the horizontal strip $H = \mathbb{R} \times (L - \epsilon, L + \epsilon)$. Player 2 then tries to "defend" by finding a positive number δ. This is equivalent to Player 2 choosing the vertical strip $V = (r - \delta, r + \delta) \times \mathbb{R}$. The defense is successful if whenever $x \in (r - \delta, r + \delta)$, $x \neq r$, we have $f(x) \in (L - \epsilon, L + \epsilon)$. This is equivalent to $H \cap V$ trapping f around r.

The figure to the right shows what happens during one round of the $\epsilon - \delta$ game corresponding to checking if $\lim\limits_{x \to 1}(2x + 1) = 3$. In the figure, Player 1 chooses $\epsilon = 0.5$, so that $L - \epsilon = 3 - 0.5 = 2.5$ and $L + \epsilon = 3 + 0.5 = 3.5$. Notice how we drew the corresponding horizontal strip $H = \mathbb{R} \times (2.5, 3.5)$. According to our proof, Player 1 chooses $\delta = \frac{\epsilon}{2} = \frac{0.5}{2} = 0.25$. So $r - \delta = 1 - 0.25 = 0.75$ and $r + \delta = 1 + 0.25 = 1.25$. Notice how we drew the corresponding vertical strip $V = (0.75, 1.25) \times \mathbb{R}$. Also notice how the rectangle $H \cap V$ traps f.

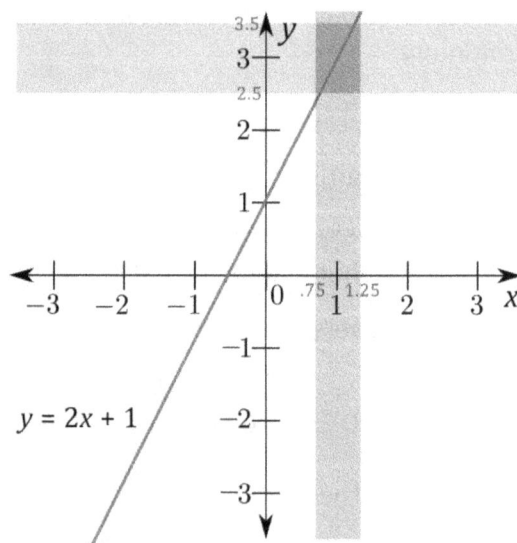

(3) Observe that the value for δ that Player 2 chose here is the largest value of δ that would result in a successful defense. If we widen the vertical strip at all on either side, then f would escape from the resulting rectangle. However, any smaller value of δ will still work. If we shrink the vertical strip, then f is still trapped. After all, we have less that we need to trap.

(4) In the next round, Player 1 will want to choose a smaller value for ϵ. If Player 1 chooses a larger value for ϵ, then the same δ that was already played will work to defend against that larger ϵ. But for this problem, it doesn't matter how small a value for ϵ Player 1 chooses—Player 1 simply cannot win. All Player 2 needs to do is defend with $\delta = \frac{\epsilon}{2}$ (or any smaller positive number).

(5) Essentially the same argument can be used to show that the function f defined by $f(x) = 2x + 1$ is continuous at $x = 1$. Simply replace the expression $0 < |x - 1| < \delta$ by the expression $|x - 1| < \delta$ everywhere it appears in the proof. The point is that $f(1) = 2 \cdot 1 + 1 = 3$. Since this value is equal to $\lim_{x \to 1}(2x + 1)$, we don't need to exclude $x = 1$ from consideration when trying to trap f.

Example 8.11: Let's use the $\epsilon - \delta$ definition of a limit to prove that $\lim_{x \to 3}(x^2 - 2x + 1) = 4$.

Analysis: This is quite a bit more difficult than Example 8.10.

Given $\epsilon > 0$, we need to find $\delta > 0$ so that $0 < |x - 3| < \delta$ implies $|(x^2 - 2x + 1) - 4| < \epsilon$. First note that $|(x^2 - 2x + 1) - 4| = |x^2 - 2x - 3| = |(x - 3)(x + 1)| = |x - 3||x + 1|$. Therefore, $|(x^2 - 2x + 1) - 4| < \epsilon$ is equivalent to $|x - 3||x + 1| < \epsilon$.

There is a small complication here. The $|x - 3|$ is not an issue because we're going to be choosing δ so that this expression is small enough. But to make the argument work we need to make $|x + 1|$ small too. Remember from Note 3 after Example 8.10 that if we find a value for δ that works, then any smaller positive number will work too. This allows us to start by assuming that δ is smaller than any positive number we choose. So, let's just assume that $\delta \leq 1$ and see what effect that has on $|x + 1|$.

Well, if $\delta \leq 1$ and $0 < |x - 3| < \delta$, then $|x - 3| < 1$. Therefore, $-1 < x - 3 < 1$. We now add 4 to each part of this inequality to get $3 < x + 1 < 5$. Since $-5 < 3$, this implies that $-5 < x + 1 < 5$, which is equivalent to $|x + 1| < 5$.

So, if we assume that $\delta \leq 1$, then $|(x^2 - 2x + 1) - 4| = |x - 3||x + 1| < \delta \cdot 5 = 5\delta$. Therefore, if we want to make sure that $|(x^2 - 2x + 1) - 4| < \epsilon$, then is suffices to choose δ so that $5\delta \leq \epsilon$, as long as we also have $\delta \leq 1$. So, we will let $\delta = \min\left\{1, \frac{\epsilon}{5}\right\}$.

Proof: Let $\epsilon > 0$ and let $\delta = \min\left\{1, \frac{\epsilon}{5}\right\}$. Suppose that $0 < |x - 3| < \delta$. Then since $\delta \leq 1$, we have $|x - 3| < 1$, and so, $|x + 1| < 5$ (see the algebra in the analysis above). Also, since $\delta \leq \frac{\epsilon}{5}$, we have $|x - 3| < \frac{\epsilon}{5}$. It follows that $|(x^2 - 2x + 1) - 4| = |x^2 - 2x - 3| = |x - 3||x + 1| < \frac{\epsilon}{5} \cdot 5 = \epsilon$.

Since $\epsilon > 0$ was arbitrary, we have $\forall \epsilon > 0 \,\exists \delta > 0 \,(0 < |x - 3| < \delta \to |(x^2 - 2x + 1) - 4| < \epsilon)$. Therefore, $\lim_{x \to 1}(x^2 - 2x + 1) = 4$. $\qquad\square$

Example 8.12: Let $m, b \in \mathbb{R}$ with $m \neq 0$. Let's use the $\epsilon - \delta$ definition of continuity to prove that the function $f: \mathbb{R} \to \mathbb{R}$ defined by $f(x) = mx + b$ is continuous everywhere.

A function of the form $f(x) = mx + b$, where $m, b \in \mathbb{R}$ and $m \neq 0$ is called a **linear function**. So, we will now show that every linear function is continuous everywhere.

Analysis: Given $a \in \mathbb{R}$ and $\epsilon > 0$, we will find $\delta > 0$ so that $|x - a| < \delta$ implies $|f(x) - f(a)| < \epsilon$. First note that $|f(x) - f(a)| = |(mx + b) - (ma + b)| = |mx - ma| = |m||x - a|$. Therefore, $|f(x) - f(a)| < \epsilon$ is equivalent to $|x - a| < \frac{\epsilon}{|m|}$. So, $\delta = \frac{\epsilon}{|m|}$ should work.

Proof: Let $a \in \mathbb{R}$, let $\epsilon > 0$, and let $\delta = \frac{\epsilon}{|m|}$. Suppose that $|x - a| < \delta$. Then we have

$$|f(x) - f(a)| = |(mx + b) - (ma + b)| = |mx - ma| = |m||x - a| < |m|\delta = |m| \cdot \frac{\epsilon}{|m|} = \epsilon.$$

Since $\epsilon > 0$ was arbitrary, we have $\forall \epsilon > 0 \, \exists \delta > 0 \, (|x - a| < \delta \to |f(x) - f(a)| < \epsilon)$. Therefore, f is continuous at $x = a$. Since $a \in \mathbb{R}$ was arbitrary, f is continuous everywhere. $\square$

Notes: (1) We proved $\forall a \in \mathbb{R} \, \forall \epsilon > 0 \, \exists \delta > 0 \, \forall x \in \mathbb{R}(|x - a| < \delta \to |f(x) - f(a)| < \epsilon)$. In words, we proved that for every real number a, given a positive real number ϵ, we can find a positive real number δ such that whenever the distance between x and a is less than δ, the distance between $f(x)$ and $f(a)$ is less than ϵ. And of course, a simpler way to say this is "for every real number a, f is continuous at a," or $\forall a \in \mathbb{R} \, (f$ is continuous at $a)$."

(2) If we move the expression $\forall a \in \mathbb{R}$ next to $\forall x \in \mathbb{R}$, we get a concept that is stronger than continuity. We say that a function $f: A \to \mathbb{R}$ is **uniformly continuous** on A if

$$\forall \epsilon > 0 \, \exists \delta > 0 \, \forall a, x \in A \, (|x - a| < \delta \to |f(x) - f(a)| < \epsilon).$$

(3) As a quick example of uniform continuity, every linear function is uniformly continuous on $\mathbb{R}$. We can see this by modifying the proof above just slightly:

New proof: Let $\epsilon > 0$ and let $\delta = \frac{\epsilon}{|m|}$. Let $a, x \in \mathbb{R}$ and suppose that $|x - a| < \delta$. Then we have

$$|f(x) - f(a)| = |(mx + b) - (ma + b)| = |mx - ma| = |m||x - a| < |m|\delta = |m| \cdot \frac{\epsilon}{|m|} = \epsilon.$$

Since $\epsilon > 0$ was arbitrary, we have $\forall \epsilon > 0 \, \exists \delta > 0 \, \forall a, x \in \mathbb{R} \, (|x - a| < \delta \to |f(x) - f(a)| < \epsilon)$. Therefore, f is uniformly continuous on $\mathbb{R}$.

(4) The difference between continuity and uniform continuity on a set A can be described as follows: In both cases, an ϵ is given and then a δ is chosen. For continuity, for each value of x, we are allowed to choose a different δ. For uniform continuity, once we choose a δ for some value of x, we need to be able to use the **same** δ for every other value of x in A.

In terms of strips, once a horizontal strip is given, we need to be more careful how we choose a vertical strip. As we check different x-values, we can move the vertical strip left and right. However, we are not allowed to decrease the width of the vertical strip.

Try to come up with a function that is continuous on a set A, but not uniformly continuous on A. This will be explored a little more in the Problem Set below.

(5) The function $f: \mathbb{R} \to \mathbb{R}$ defined by $f(x) = mx + b$ with $m \neq 0$ is a bijection. To see that f is injective, note that if $x \neq y$, then since $m \neq 0$, $mx \neq my$. So, $f(x) = mx + b \neq my + b = f(y)$. To see that f is surjective, observe that if $y \in \mathbb{R}$, then since $m \neq 0$, $\frac{y-b}{m} \in \mathbb{R}$ and $f\left(\frac{y-b}{m}\right) = y$. In fact, the inverse of f is the function $f^{-1}: \mathbb{R} \to \mathbb{R}$ defined by $f(x) = \frac{x-b}{m} = \frac{1}{m}x - \frac{b}{m}$. Notice that this function has the same form as the original function f and therefore, f^{-1} is also continuous everywhere. In this case, we say that f is a **homeomorphism**. In general, a homeomorphism is a bijective continuous function with a continuous inverse.

Limit and Continuity Theorems in $\mathbb{R}$

Theorem 8.13: Let $A, B \subseteq \mathbb{R}$, let $f: A \to \mathbb{R}$, $g: B \to \mathbb{R}$, let $r \in \mathbb{R}$, and suppose that $\lim_{x \to r}[f(x)]$ and $\lim_{x \to r}[g(x)]$ are both finite real numbers. Then $\lim_{x \to r}[f(x) + g(x)] = \lim_{x \to r}[f(x)] + \lim_{x \to r}[g(x)]$.

Analysis: If $\lim_{x \to r}[f(x)] = L$, then given $\epsilon > 0$, there is $\delta > 0$ such that $0 < |x - r| < \delta$ implies $|f(x) - L| < \epsilon$. If $\lim_{x \to r}[g(x)] = K$, then given $\epsilon > 0$, there is $\delta > 0$ such that $0 < |x - r| < \delta$ implies $|g(x) - K| < \epsilon$. We should acknowledge something here. If we are given a single positive real value for ϵ, there is no reason that we would necessarily choose the same δ for both f and g. However, using the fact that once we find a δ that works, any smaller δ will also work, it is easy to see that we **could** choose a single value for δ that would work for both f and g. This should be acknowledged in some way in the proof. There are several ways to work this into the argument. The way we will handle this is to use δ_1 for f and δ_2 for g, and then let δ be the smaller of δ_1 and δ_2.

Recall from Theorem 7.4 that the Triangle Inequality says that for all $x, y \in \mathbb{R}$, $|x + y| \leq |x| + |y|$. (The theorem is stated to be true for all complex numbers, but since $\mathbb{R} \subseteq \mathbb{C}$, it is equally true for all real numbers.) After assuming $0 < |x - r| < \delta$, we will use the Triangle Inequality to write

$$|f(x) + g(x) - (L + K)| = |(f(x) - L) + (g(x) - K)| \leq |f(x) - L| + |g(x) - K| < \epsilon + \epsilon = 2\epsilon.$$

It seems that we wound up with 2ϵ on the right-hand side instead of ϵ. Now, if ϵ is an arbitrarily small positive real number, then so is 2ϵ, and vice versa. So, getting 2ϵ on the right-hand side instead of ϵ really isn't too big of a deal. However, to be rigorous, we should prove that it is okay. There are at least two ways we can handle this. One possibility is to prove a theorem that says 2ϵ works just as well as ϵ. A second possibility (and the way I usually teach it in basic analysis courses) is to edit the original ϵ's, so it all works out to ϵ in the end. The idea is simple. If ϵ is a positive real number, then so is $\frac{\epsilon}{2}$. So, after we are given ϵ, we can pretend that Player 1 (in the $\epsilon - \delta$ game) is "attacking" with $\frac{\epsilon}{2}$ instead. Let's see how this all plays out in the proof.

Proof: Suppose that $\lim_{x \to r}[f(x)] = L$ and $\lim_{x \to r}[g(x)] = K$, and let $\epsilon > 0$. Since $\lim_{x \to r}[f(x)] = L$, there is $\delta_1 > 0$ such that $0 < |x - r| < \delta_1$ implies $|f(x) - L| < \frac{\epsilon}{2}$. Since $\lim_{x \to r}[g(x)] = K$, there is $\delta_2 > 0$ such that $0 < |x - r| < \delta_2$ implies $|g(x) - L| < \frac{\epsilon}{2}$. Let $\delta = \min\{\delta_1, \delta_2\}$ and suppose that $0 < |x - r| < \delta$. Then since $\delta \leq \delta_1$, $|f(x) - L| < \frac{\epsilon}{2}$. Since $\delta \leq \delta_2$, $|g(x) - K| < \frac{\epsilon}{2}$. By the Triangle Inequality, we have

$$|f(x) + g(x) - (L + K)| = |(f(x) - L) + (g(x) - K)| \leq |f(x) - L| + |g(x) - K| < \frac{\epsilon}{2} + \frac{\epsilon}{2} = \epsilon.$$

So, $\lim_{x \to r}[f(x) + g(x)] = L + K = \lim_{x \to r}[f(x)] + \lim_{x \to r}[g(x)]$. □

Theorem 8.14: Let $A, B \subseteq \mathbb{R}$, let $f: A \to \mathbb{R}$, $g: B \to \mathbb{R}$, let $r \in \mathbb{R}$, and suppose that $\lim_{x \to r}[f(x)]$ and $\lim_{x \to r}[g(x)]$ are both finite real numbers. Then $\lim_{x \to r}[f(x)g(x)] = \lim_{x \to r}[f(x)] \cdot \lim_{x \to r}[g(x)]$.

Analysis: As in Theorem 8.13, we let $\lim_{x \to r}[f(x)] = L$ and $\lim_{x \to r}[g(x)] = K$. If $\epsilon > 0$ is given, we will find a single $\delta > 0$ such that $0 < |x - r| < \delta$ implies $|f(x) - L| < \epsilon$ and $|g(x) - K| < \epsilon$ (like we did for Theorem 8.13). Now, we want to show that whenever $0 < |x - r| < \delta$, $|f(x)g(x) - LK| < \epsilon$. This is quite a bit more challenging than anything we had to do in Theorem 8.13.

To show that $|f(x)g(x) - LK| < \epsilon$ we will apply the Standard Advanced Calculus Trick (SACT – see Note 7 following the proof of Theorem 5.11). We would like for $|f(x) - L|$ and $|g(x) - K|$ to appear as factors in our expression. To make this happen, we subtract $Lg(x)$ from $f(x)g(x)$ to get $f(x)g(x) - Lg(x) = (f(x) - L)g(x)$. To "undo the damage," we then add back $Lg(x)$. The application of SACT together with the Triangle Inequality looks like this:

$$|f(x)g(x) - LK| = |(f(x)g(x) - Lg(x)) + (Lg(x) - LK)|$$
$$\leq |f(x)g(x) - Lg(x)| + |Lg(x) - LK| = |f(x) - L||g(x)| + |L||g(x) - K|$$
$$< \epsilon|g(x)| + |L|\epsilon = \epsilon(|g(x)| + |L|).$$

Uh oh! How can we possibly get rid of $|g(x)| + |L|$? We have seen how to handle a constant multiple of ϵ in the proof of Theorem 8.13. But this time we are multiplying ϵ by a function of x. We will resolve this issue by making sure we choose δ small enough so that $g(x)$ is sufficiently **bounded**.

We do this by taking a specific value for ϵ, and then using the fact that $\lim_{x \to r}[g(x)] = K$ to come up with a $\delta > 0$ and a bound M for g on the deleted δ-neighborhood of r. For simplicity, let's choose $\epsilon = 1$. Then since $\lim_{x \to r}[g(x)] = K$, we can find $\delta > 0$ such that $0 < |x - r| < \delta$ implies $|g(x) - K| < 1$. Now, $|g(x) - K| < 1 \Leftrightarrow -1 < g(x) - K < 1 \Leftrightarrow K - 1 < g(x) < K + 1$. For example, if $K = 5$, we would have $4 < g(x) < 6$. Since this implies $-6 < g(x) < 6$, or equivalently, $|g(x)| < 6$, we could choose $M = 6$. If, on the other hand, $K = -3$, we would have $-4 < g(x) < -2$. Since this implies $-4 < g(x) < 4$, or equivalently, $|g(x)| < 4$, we could choose $M = 4$. In general, we will let $M = \max\{|K - 1|, |K + 1|\}$.

We will now be able to get $|f(x)g(x) - (LK)| < \epsilon(|g(x)| + |L|) < \epsilon(M + |L|)$. Great! Now it looks just like the situation we had in Theorem 8.13. The number $M + |L|$ looks messier, but it is just a number, and so we can finish cleaning up the argument by replacing Player 1's ϵ-attacks by $\frac{\epsilon}{M + |L|}$.

Proof: Suppose that $\lim_{x \to r}[f(x)] = L$ and $\lim_{x \to r}[g(x)] = K$, and let $\epsilon > 0$. Since $\lim_{x \to r}[g(x)] = K$, there is $\delta_1 > 0$ such that $0 < |x - r| < \delta_1$ implies $|g(x) - K| < 1$. Now, $|g(x) - K| < 1$ is equivalent to $-1 < g(x) - K < 1$, or by adding K, $K - 1 < g(x) < K + 1$. Let $M = \max\{|K - 1|, |K + 1|\}$. Then, $0 < |x - r| < \delta_1$ implies $-M < g(x) < M$, or equivalently, $|g(x)| < M$. Note also that $M > 0$. Therefore, $M + |L| > 0$.

Now, since $\lim_{x \to r}[f(x)] = L$, there is $\delta_2 > 0$ such that $0 < |x - r| < \delta_2$ implies $|f(x) - L| < \frac{\epsilon}{M + |L|}$. Since $\lim_{x \to r}[g(x)] = K$, there is $\delta_3 > 0$ such that $0 < |x - r| < \delta_3$ implies $|g(x) - L| < \frac{\epsilon}{M + |L|}$. Let $\delta = \min\{\delta_1, \delta_2, \delta_3\}$ and suppose that $0 < |x - r| < \delta$. Then since $\delta \leq \delta_1$, $|g(x)| < M$. Since $\delta \leq \delta_2$,

$|f(x) - L| < \frac{\epsilon}{M + |L|}$. Since $\delta \leq \delta_3$, $|g(x) - K| < \frac{\epsilon}{M + |L|}$. By the Triangle Inequality (and SACT), we have

$$|f(x)g(x) - LK| = |(f(x)g(x) - Lg(x)) + (Lg(x) - LK)|$$
$$\leq |f(x)g(x) - Lg(x)| + |Lg(x) - LK| = |f(x) - L||g(x)| + |L||g(x) - K|$$
$$< \frac{\epsilon}{M + |L|} \cdot M + |L| \frac{\epsilon}{M + |L|} = \frac{\epsilon}{M + |L|}(M + |L|) = \epsilon.$$

So, $\lim_{x \to r}[f(x)g(x)] = LK = \lim_{x \to r}[f(x)] \cdot \lim_{x \to r}[g(x)]$. $\square$

Theorem 8.15: If $\lim_{x \to r} f(x)$ exists, then it is unique.

Proof: Suppose that $\lim_{x \to r} f(x) = L$ and $\lim_{x \to r} f(x) = K$. Let $\epsilon > 0$. Since $\lim_{x \to r} f(x) = L$, we can find $\delta_1 > 0$ such that $0 < |x - r| < \delta_1 \to |f(x) - L| < \frac{\epsilon}{2}$. Since $\lim_{x \to r} f(x) = K$, we can find $\delta_2 > 0$ such that $0 < |x - r| < \delta_2 \to |f(x) - K| < \frac{\epsilon}{2}$. Let $\delta = \min\{\delta_1, \delta_2\}$. Suppose that $0 < |x - r| < \delta$. Then $|L - K| = |(f(x) - K) - (f(x) - L)|$ **(SACT)** $\leq |f(x) - K| + |f(x) - L|$ **(TI)** $< \frac{\epsilon}{2} + \frac{\epsilon}{2} = \epsilon$. Since ϵ was an arbitrary positive real number, by Problem 15 from Problem Set 6, we have $|L - K| = 0$. So, $L - K = 0$, and therefore, $L = K$. $\square$

Note: SACT stands for the Standard Advanced Calculus Trick and **TI** stands for the Triangle Inequality.

Limits in $\mathbb{R}$ Involving Infinity

Recall that a **horizontal strip** in $\mathbb{R} \times \mathbb{R}$ is a set of the form $\mathbb{R} \times (c, d) = \{(x, y) \mid c < y < d\}$ and a **vertical strip** is a set of the form $(a, b) \times \mathbb{R} = \{(x, y) \mid a < x < b\}$. If we allow a and/or c to take on the value $-\infty$ (in which case we say that the strip **contains** $-\infty$) and we allow b and/or d to take on the value $+\infty$ (in which case we say that the strip **contains** $+\infty$), we can extend our definition of limit to handle various situations involving infinity.

Example 8.16: Let's take a look at the horizontal strip $\mathbb{R} \times (1, +\infty)$ and the vertical strip $(-\infty, -2) \times \mathbb{R}$ in the xy-plane. These can be visualized as follows:

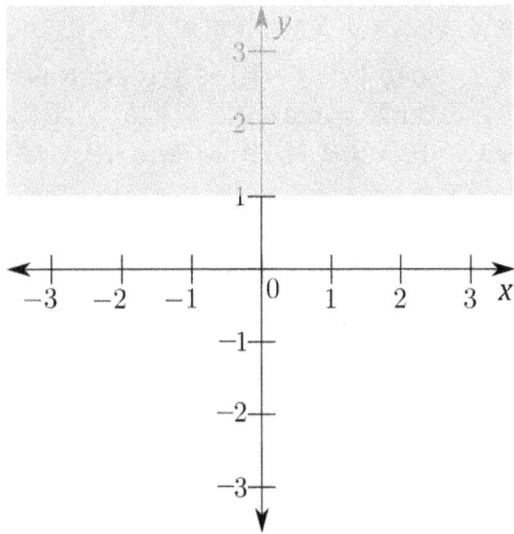

$$\mathbb{R} \times (1, +\infty)$$

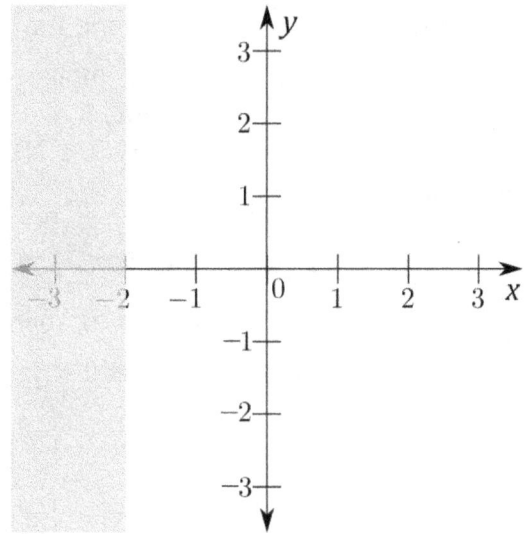

$$(-\infty, -2) \times \mathbb{R}$$

The horizontal strip $\mathbb{R} \times (1, +\infty)$ contains $+\infty$ and the vertical strip $(-\infty, -2) \times \mathbb{R}$ contains $-\infty$.

Note: Strips that contain $+\infty$ or $-\infty$ are usually called **half planes**. Here, we will continue to use the expression "strip" because it allows us to handle all types of limits (finite and infinite) without having to discuss every case individually.

By allowing strips to contain $+\infty$ or $-\infty$, intersections of horizontal and vertical strips can now be **unbounded**. The resulting open rectangles $(a, b) \times (c, d)$ can have a and/or c taking on the value $-\infty$ and b and/or d taking on the value $+\infty$.

Example 8.17: Consider the horizontal strip $H = \mathbb{R} \times (1, +\infty)$ and the vertical strip $V = (-2, -1) \times \mathbb{R}$. The intersection of these strips is the open rectangle $R = (-2, -1) \times (1, +\infty)$. The rectangle R traps $(-1.5, 3)$, whereas $(-1.5, 0)$ escapes from R. This can be seen in the figure below on the left. Also, consider the horizontal strip $H = \mathbb{R} \times (1, +\infty)$ and the vertical strip $V = (-\infty, -2) \times \mathbb{R}$. The intersection of these strips is the open rectangle $S = (-\infty, -2) \times (1, +\infty)$. The rectangle S traps $(-3, 2)$, whereas $(-3, -1)$ escapes from S. This can be seen in the figure below on the right.

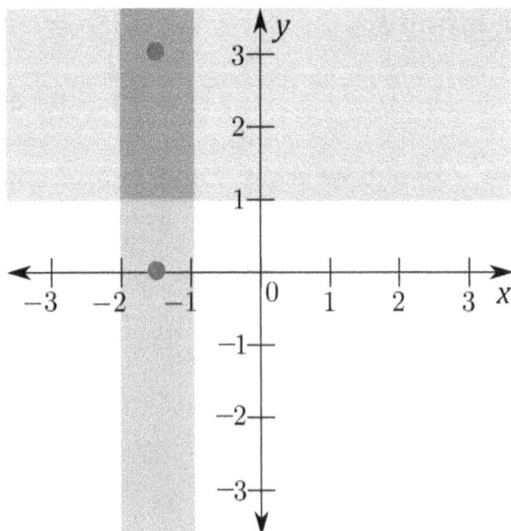

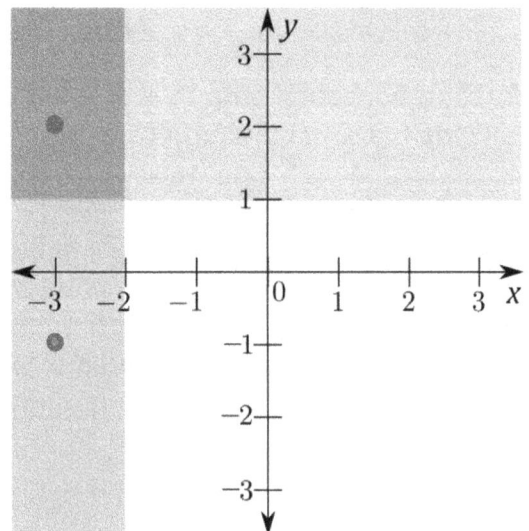

150

When we allow $+\infty$ and $-\infty$, the definitions of "trap" and "escape" are just about the same. We just need to make the following minor adjustment.

Small technicality: If $r = +\infty$ or $r = -\infty$, then we define **R traps f around r** to simply mean that **R traps f**. In other words, when checking a limit that is approaching $+\infty$ or $-\infty$, we do not exclude any point from consideration as we would do if r were a finite real number.

Example 8.18: Let $f : \mathbb{R} \to \mathbb{R}$ be defined by $f(x) = \frac{1}{x^2}$, let $r = 0$, and let $L = +\infty$. Let's show that $\lim_{x \to 0} f(x) = +\infty$. Let $H = \mathbb{R} \times (c, +\infty)$ be a horizontal strip that contains $+\infty$. Next, let $V = \left(-\frac{1}{\sqrt{c}}, \frac{1}{\sqrt{c}}\right) \times \mathbb{R}$. We will show that $H \cap V = \left(-\frac{1}{\sqrt{c}}, \frac{1}{\sqrt{c}}\right) \times (c, +\infty)$ traps f around 0. Let $x \in \left(-\frac{1}{\sqrt{c}}, \frac{1}{\sqrt{c}}\right) \setminus \{0\}$, so that $-\frac{1}{\sqrt{c}} < x < \frac{1}{\sqrt{c}}$ and $x \neq 0$. Then $-\frac{1}{\sqrt{c}} < x < 0$ or $0 < x < \frac{1}{\sqrt{c}}$. In either case, $x^2 < \frac{1}{c}$, and therefore, $c < \frac{1}{x^2} = f(x)$. Since $x \in \left(-\frac{1}{\sqrt{c}}, \frac{1}{\sqrt{c}}\right) \setminus \{0\}$ and $f(x) \in (c, +\infty)$, it follows that $(x, f(x)) \in \left(-\frac{1}{\sqrt{c}}, \frac{1}{\sqrt{c}}\right) \times (c, +\infty) = H \cap V$. So, $H \cap V$ traps f around 0. Thus, $\lim_{x \to 0} f(x) = +\infty$.

Example 8.19: Let $f : \mathbb{R} \to \mathbb{R}$ be defined by $f(x) = -x + 3$, let $r = +\infty$, and let $L = -\infty$. Let's show that $\lim_{x \to +\infty} f(x) = -\infty$. Let $H = \mathbb{R} \times (-\infty, d)$ be a horizontal strip that contains $-\infty$. Next, let $V = (3 - d, +\infty) \times \mathbb{R}$. We will show that $H \cap V = (3 - d, +\infty) \times (-\infty, d)$ traps f around $+\infty$ (or more simply, that $(3 - d, +\infty) \times (-\infty, d)$ traps f). Let $x \in (3 - d, +\infty)$, so that $x > 3 - d$. Then we have $-x < d - 3$, and so, $f(x) = -x + 3 < d$. Since $x \in (3 - d, +\infty)$ and $f(x) \in (-\infty, d)$, it follows that $(x, f(x)) \in (3 - d, +\infty) \times (-\infty, d) = H \cap V$. Therefore, $H \cap V$ traps f. So, $\lim_{x \to +\infty} f(x) = -\infty$.

We can find equivalent definitions for limits involving infinity on a case-by-case basis. We will do one example here and you will look at others in Problem 15 below.

Theorem 8.20: $\lim_{x \to r} f(x) = +\infty$ if and only if $\forall M > 0 \, \exists \delta > 0 \, (0 < |x - r| < \delta \to f(x) > M)$.

Proof: Suppose that $\lim_{x \to r} f(x) = +\infty$ and let $M > 0$. Let $H = \mathbb{R} \times (M, +\infty)$. Since $\lim_{x \to r} f(x) = +\infty$, there is a vertical strip $V = (a, b) \times \mathbb{R}$ that contains r such that the rectangle $(a, b) \times (M, +\infty)$ traps f around r. Let $\delta = \min\{r - a, b - r\}$, and let $0 < |x - r| < \delta$. Then $x \neq r$ and $-\delta < x - r < \delta$. So, $r - \delta < x < r + \delta$. Since $\delta \leq r - a$, we have $a \leq r - \delta$. Since $\delta \leq b - r$, we have $b \geq r + \delta$. Therefore, $a < x < b$, and so, $x \in (a, b)$. Since $x \neq r$, $x \in (a, b)$, and $(a, b) \times (M, +\infty)$ traps f around r, we have $f(x) \in (M, +\infty)$. Thus, $f(x) > M$.

Conversely, suppose that $\forall M > 0 \, \exists \delta > 0 \, (0 < |x - r| < \delta \to f(x) > M)$. Let $H = \mathbb{R} \times (c, +\infty)$ be a horizontal strip containing $+\infty$ and let $M = \max\{c, 1\}$. Then there is $\delta > 0$ such that $0 < |x - r| < \delta$ implies $f(x) > M$. Let $V = (r - \delta, r + \delta) \times \mathbb{R}$ and let $R = H \cap V = (r - \delta, r + \delta) \times (c, +\infty)$. We show that R traps f around r. Indeed, if $x \in (r - \delta, r + \delta)$ and $x \neq r$, then $0 < |x - r| < \delta$ and so, $f(x) > M$. So, $(x, f(x)) \in (r - \delta, r + \delta) \times (M, +\infty) \subseteq (r - \delta, r + \delta) \times (c, +\infty)$ (because $c \leq M$). So, R traps f around r. $\qquad\square$

One-sided Limits in $\mathbb{R}$

Let $A \subseteq \mathbb{R}$, let $f: A \to \mathbb{R}$, and let $r \in \mathbb{R}$ and $L \in \mathbb{R} \cup \{-\infty, +\infty\}$. We say that the **limit of f as x approaches r from the right is L**, written $\lim_{x \to r^+} f(x) = L$, if for every horizontal strip H that contains L there is a vertical strip V of the form $(r, b) \times \mathbb{R}$ such that the rectangle $H \cap V$ traps f.

Example 8.21: Let $f: \mathbb{R} \setminus \{1\} \to \mathbb{R}$ be defined by $f(x) = \frac{1}{x-1}$, let $r = 1$, and let $L = +\infty$. Let's show that $\lim_{x \to 1^+} f(x) = +\infty$. Let $H = \mathbb{R} \times (c, +\infty)$ be a horizontal strip that contains $+\infty$ and let $M = \max\{1, c\}$. Let $V = \left(1, \frac{1}{M} + 1\right) \times \mathbb{R}$. We will show that $H \cap V = \left(1, \frac{1}{M} + 1\right) \times (c, +\infty)$ traps f. Let $x \in \left(1, \frac{1}{M} + 1\right)$, so that $1 < x < \frac{1}{M} + 1$. Then we have $0 < x - 1 < \frac{1}{M}$, and so, $\frac{1}{x-1} > M \geq c$. So, $f(x) > c$. Since $x \in \left(1, \frac{1}{M} + 1\right)$ and $f(x) \in (c, +\infty)$, $(x, f(x)) \in \left(1, \frac{1}{M} + 1\right) \times (c, +\infty) = H \cap V$. Therefore, $H \cap V$ traps f. So, $\lim_{x \to 1^+} f(x) = +\infty$.

Theorem 8.22: Let L be a finite real number. Then we have $\lim_{x \to r^+} f(x) = L$ if and only if $\forall \epsilon > 0 \, \exists \delta > 0 \, (0 < x - r < \delta \to |f(x) - L| < \epsilon)$.

Proof: Suppose that $\lim_{x \to r^+} f(x) = L$ and let $\epsilon > 0$. Let $H = \mathbb{R} \times (L - \epsilon, L + \epsilon)$. Since $\lim_{x \to r^+} f(x) = L$, there is a vertical strip $V = (r, b) \times \mathbb{R}$ such that the rectangle $H \cap V = (r, b) \times (L - \epsilon, L + \epsilon)$ traps f. Let $\delta = b - r$ and let $0 < x - r < \delta$. Then $r < x < b$ and so, $x \in (r, b)$. Since $(r, b) \times (L - \epsilon, L + \epsilon)$ traps f, we have $f(x) \in (L - \epsilon, L + \epsilon)$. Thus, $L - \epsilon < f(x) < L + \epsilon$, or equivalently, $|f(x) - L| < \epsilon$.

Conversely, suppose that $\forall \epsilon > 0 \, \exists \delta > 0 \, (0 < x - r < \delta \to |f(x) - L| < \epsilon)$. Let $H = \mathbb{R} \times (c, d)$ be a horizontal strip containing L and let $\epsilon = \min\{L - c, d - L\}$. Then there is $\delta > 0$ such that $0 < x - r < \delta$ implies $|f(x) - L| < \epsilon$. Let $V = (r, r + \delta) \times \mathbb{R}$ and $R = H \cap V = (r, r + \delta) \times (c, d)$. We show that R traps f. If $x \in (r, r + \delta)$, then $r < x < r + \delta$, or equivalently, $0 < x - r < \delta$. So, $|f(x) - L| < \epsilon$. Therefore, $-\epsilon < f(x) - L < \epsilon$, or equivalently, $L - \epsilon < f(x) < L + \epsilon$. Thus, $(x, f(x)) \in (r, r + \delta) \times (L - \epsilon, L + \epsilon) \subseteq (r, r + \delta) \times (c, d)$ (Check this!). So, R traps f. $\square$

Limits and Continuity in $\mathbb{C}$

Let $A \subseteq \mathbb{C}$, let $f: A \to \mathbb{C}$, let $L \in \mathbb{C}$, and let $a \in \mathbb{C}$ be a point such that A contains some deleted neighborhood of a. We say that the **limit of f as z approaches a is L**, written $\lim_{z \to a} f(z) = L$, if for every positive number ϵ, there is a positive number δ such that $0 < |z - a| < \delta \to |f(z) - L| < \epsilon$.

Notes: (1) The statement of this definition of limit is essentially the same as the statement of the $\epsilon - \delta$ definition of a limit of a real-valued function. However, the geometry looks very different.

For a real-valued function, a deleted neighborhood of a has the form $N_\epsilon^\odot(a) = (a - \epsilon, a) \cup (a, a + \epsilon)$ and we can visualize this neighborhood as follows:

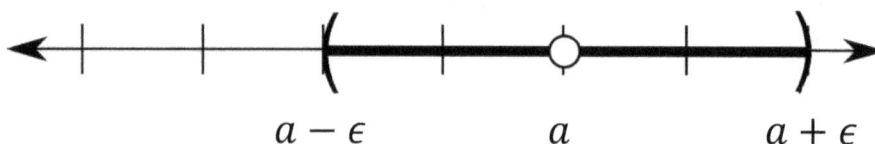

$$a - \epsilon \qquad\qquad a \qquad\qquad a + \epsilon$$

For a complex-valued function, a deleted neighborhood of a, say $N_\epsilon^\odot(a) = \{z \in \mathbb{C} \mid 0 < |z - a| < \epsilon\}$, is a punctured disk with center a. We can see a visualization of such a neighborhood to the right.

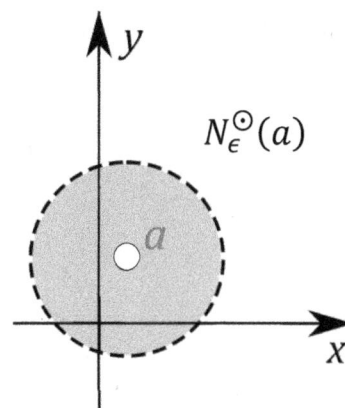

(2) In $\mathbb{R}$, there is a simple one to one correspondence between neighborhoods (open intervals) and (vertical or horizontal) strips.

In $\mathbb{C}$ there is no such correspondence. Therefore, for complex-valued functions, we start right away with the $\epsilon - \delta$ definition.

(3) Recall that in $\mathbb{R}$, the expression $|x - a| < \delta$ is equivalent to $a - \delta < x < a + \delta$, or $x \in (a - \delta, a + \delta)$.

Also, the expression $0 < |x - a|$ is equivalent to $x - a \neq 0$, or $x \neq a$.

Therefore, $0 < |x - a| < \delta$ is equivalent to $x \in (a - \delta, a) \cup (a, a + \delta)$.

In $\mathbb{C}$, if we let $z = x + yi$ and $a = b + ci$, then

$$|z - a| = |(x + yi) - (b + ci)| = |(x - b) + (y - c)i| = \sqrt{(x - b)^2 + (y - c)^2}.$$

So, $|z - a| < \delta$ is equivalent to $(x - b)^2 + (y - c)^2 < \delta^2$. In other words, (x, y) is inside the disk with center (b, c) and radius δ.

Also, we have

$$0 < |z - a| \Leftrightarrow (x - b)^2 + (y - c)^2 \neq 0 \Leftrightarrow x - b \neq 0 \text{ or } y - c \neq 0 \Leftrightarrow x \neq b \text{ or } x \neq c \Leftrightarrow z \neq a.$$

Therefore, $0 < |z - a| < \delta$ is equivalent to "z is in the punctured disk with center a and radius δ."

(4) Similarly, in $\mathbb{R}$, we have that $|f(x) - L| < \epsilon$ is equivalent to $f(x) \in (L - \epsilon, L + \epsilon)$, while in $\mathbb{C}$, we have $|f(z) - L| < \epsilon$ is equivalent to "$f(z)$ is in the disk with center L and radius ϵ."

(5) Just like for real-valued functions, we can think of determining if $\lim_{z \to a} f(z) = L$ as the result of an $\epsilon - \delta$ game. Player 1 "attacks" by choosing a positive number ϵ. This is equivalent to Player 1 choosing the disk $N_\epsilon(L) = \{w \in \mathbb{C} \mid |w - L| < \epsilon\}$.

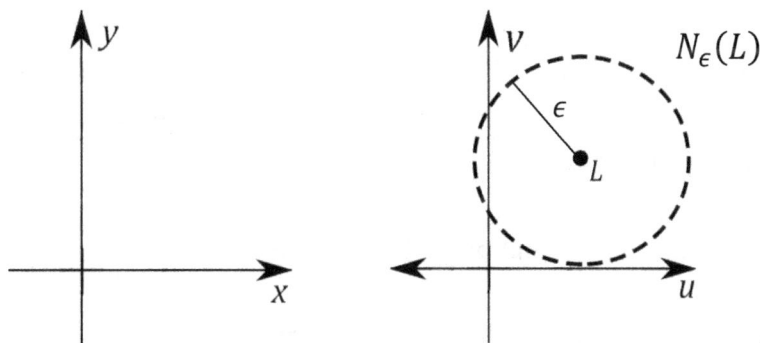

Player 2 then tries to "defend" by finding a positive number δ. This is equivalent to Player 2 choosing the punctured disk $N_\delta^\odot(a) = \{z \in \mathbb{C} \mid 0 < |z - a| < \delta\}$.

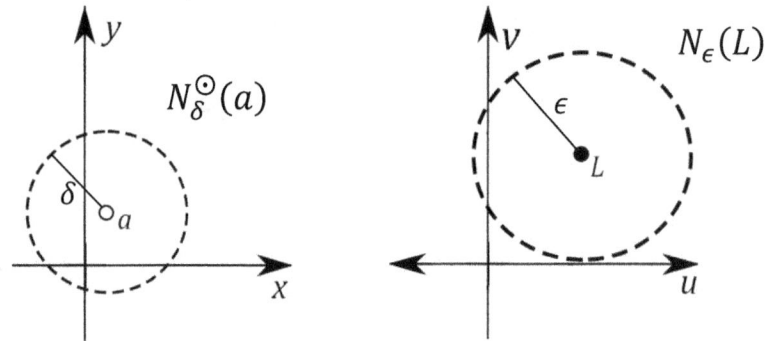

The defense is successful if $z \in N_\delta^\odot(a)$ implies $f(z) \in N_\epsilon(L)$, or equivalently, $f[N_\delta^\odot(a)] \subseteq N_\epsilon(L)$.

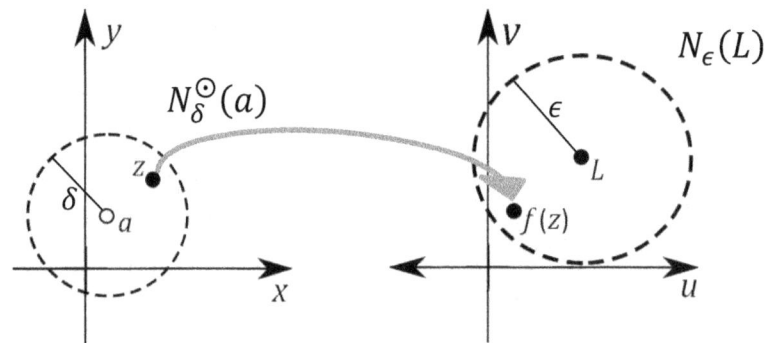

If Player 2 defends successfully, then Player 1 chooses a new positive number ϵ', or equivalently, a new neighborhood $N_{\epsilon'}(L) = \{w \in \mathbb{C} \mid |w - L| < \epsilon'\}$. If Player 1 is smart, then he/she will choose ϵ' to be less than ϵ (otherwise, Player 2 can use the same δ). The smaller the value of ϵ', the smaller the neighborhood $N_{\epsilon'}(L)$, and the harder it will be for Player 2 to defend. Player 2 once again tries to choose a positive number δ' so that $f[N_{\delta'}^\odot(a)] \subseteq N_\epsilon'(L)$. This process continues indefinitely. Player 1 wins the $\epsilon - \delta$ game if at some stage, Player 2 cannot defend successfully. Player 2 wins the $\epsilon - \delta$ game if he or she defends successfully at every stage.

Player 1 has a winning strategy for the $\epsilon - \delta$ game if and only if $\lim\limits_{z \to a} f(z) \neq L$, while Player 2 has a winning strategy for the $\epsilon - \delta$ game if and only if $\lim\limits_{z \to a} f(z) = L$.

(6) If for a given $\epsilon > 0$, we have found a $\delta > 0$ such that $f[N_\delta^\odot(a)] \subseteq N_\epsilon(L)$, then any positive number smaller than δ works as well. Indeed, if $0 < \delta' < \delta$, then $N_{\delta'}^\odot(a) \subseteq N_\delta^\odot(a)$. It then follows that $f[N_{\delta'}^\odot(a)] \subseteq f[N_\delta^\odot(a)] \subseteq N_\epsilon(L)$.

Example 8.23: Let's use the $\epsilon - \delta$ definition of limit to prove that $\lim\limits_{z \to 3+6i} \left(\frac{iz}{3} + 2\right) = i$.

Analysis: Given $\epsilon > 0$, we will find $\delta > 0$ so that $0 < |z - (3 + 6i)| < \delta$ implies $\left|\left(\frac{iz}{3} + 2\right) - i\right| < \epsilon$. First note that

$$\left|\left(\tfrac{iz}{3}+2\right)-i\right| = \left|\tfrac{1}{3}(iz+6)-\tfrac{1}{3}(3i)\right| = \left|\tfrac{1}{3}i(z-6i-3)\right| = \left|\tfrac{1}{3}i\right|\,|z-3-6i| = \tfrac{1}{3}|z-(3+6i)|.$$

So, $\left|\left(\tfrac{iz}{3}+2\right)-i\right| < \epsilon$ is equivalent to $|z-(3+6i)| < 3\epsilon$. Therefore, $\delta = 3\epsilon$ should work.

Proof: Let $\epsilon > 0$ and let $\delta = 3\epsilon$. Suppose that $0 < |z-(3+6i)| < \delta$. Then we have

$$\left|\left(\tfrac{iz}{3}+2\right)-i\right| = \tfrac{1}{3}|z-(3+6i)| < \tfrac{1}{3}\delta = \tfrac{1}{3}(3\epsilon) = \epsilon.$$

Since $\epsilon > 0$ was arbitrary, we have $\forall \epsilon > 0\,\exists \delta > 0\,\left(0 < |z-(3+6i)| < \delta \rightarrow \left|\left(\tfrac{iz}{3}+2\right)-i\right| < \epsilon\right)$.

Therefore, $\displaystyle\lim_{z\to 3+6i}\left(\tfrac{iz}{3}+2\right) = i$. $\qquad\square$

Example 8.24: Let's use the $\epsilon - \delta$ definition of limit to prove that $\displaystyle\lim_{z\to i} z^2 = -1$.

Analysis: Given $\epsilon > 0$, we need to find $\delta > 0$ so that $0 < |z-i| < \delta$ implies $|z^2 - (-1)| < \epsilon$. First note that $|z^2 - (-1)| = |z^2 + 1| = |(z-i)(z+i)| = |z-i||z+i|$. Therefore, $|z^2 - (-1)| < \epsilon$ is equivalent to $|z-i||z+i| < \epsilon$.

As in Example 8.11, $|z-i|$ is not an issue because we're going to be choosing δ so that this expression is small enough. But to make the argument work we need to make $|z+i|$ small too. Remember from Note 6 above that if we find a value for δ that works, then any smaller positive number will work too. This allows us to start by assuming that δ is smaller than any positive number we choose. So, let's just assume that $\delta \leq 1$ and see what effect that has on $|z+i|$.

Well, if $\delta \leq 1$ and $0 < |z-i| < \delta$, then $|z+i| = |(z-i)+2i| \leq |z-i|+|2i| < 1+2 = 3$. Here we used the Standard Advanced Calculus Trick (SACT) from Note 7 following the proof of Theorem 5.11, followed by the Triangle Inequality (Theorem 7.4), and then the computation $|2i| = |2||i| = 2\cdot 1 = 2$.

So, if we assume that $\delta \leq 1$, then $|z^2 - (-1)| = |z-i||z+i| < \delta \cdot 3 = 3\delta$. Therefore, if we want to make sure that $|z^2 - (-1)| < \epsilon$, then is suffices to choose δ so that $3\delta \leq \epsilon$, as long as we also have $\delta \leq 1$. So, we will let $\delta = \min\left\{1,\tfrac{\epsilon}{3}\right\}$.

Proof: Let $\epsilon > 0$ and let $\delta = \min\left\{1,\tfrac{\epsilon}{3}\right\}$. Suppose that $0 < |z-i| < \delta$. Then since $\delta \leq 1$, we have $|z+i| = |(z-i)+2i| \leq |z-i|+|2i| = |z-i|+|2||i| = |z-i|+2 < 1+2 = 3$, and therefore, $|z^2 - (-1)| = |z^2 + 1| = |(z-i)(z+i)| = |z-i||z+i| < \delta \cdot 3 \leq \tfrac{\epsilon}{3}\cdot 3 = \epsilon$.

Since $\epsilon > 0$ was arbitrary, we have $\forall \epsilon > 0\,\exists \delta > 0\,(0 < |z-i| < \delta \rightarrow |z^2 - (-1)| < \epsilon)$. Therefore, $\displaystyle\lim_{z\to i} z^2 = -1$. $\qquad\square$

Just like for real-valued functions, if $f\colon \mathbb{C}\to\mathbb{C}$, then we have the following:

Theorem 8.25: If $\displaystyle\lim_{z\to a} f(z)$ exists, then it is unique.

The proof is exactly the same as the proof of Theorem 8.15 (you might just want to replace x with z and r with a).

Example 8.26: Let's show that $\lim\limits_{z \to 0} \left(\frac{z}{\bar{z}}\right)^2$ does not exist.

Proof: If we consider complex numbers of the form $x + 0i$, $\left(\frac{z}{\bar{z}}\right)^2 = \left(\frac{x+0i}{x-0i}\right)^2 = \left(\frac{x}{x}\right)^2 = 1^2 = 1$. Since every deleted neighborhood of 0 contains points of the form $x + 0i$, we see that if $\lim\limits_{z \to 0} \left(\frac{z}{\bar{z}}\right)^2$ exists, it must be equal to 1.

Next, let's consider complex numbers of the form $x + xi$. In this case, $\left(\frac{z}{\bar{z}}\right)^2 = \left(\frac{x+xi}{x-xi}\right)^2 = \frac{2x^2 i}{-2x^2 i} = -1$. Since every deleted neighborhood of 0 contains points of the form $x + xi$, we see that if $\lim\limits_{z \to 0} \left(\frac{z}{\bar{z}}\right)^2$ exists, it must be equal to -1.

By Theorem 8.25, the limit does not exist. $\qquad\square$

We have the following definition of continuity for complex-valued functions:

Let $A \subseteq \mathbb{C}$, let $f : A \to \mathbb{C}$, and let $a \in A$ be a point such that A contains some neighborhood of a. f is **continuous at a** if and only if for every positive number ϵ, there is a positive number δ such that

$$|z - a| < \delta \to |f(z) - f(a)| < \epsilon.$$

Example 8.27: Let $f : \mathbb{C} \to \mathbb{C}$ be defined by $f(z) = \frac{iz}{3} + 2$. In Example 8.23, we showed that $\lim\limits_{z \to 3+6i} f(z) = i$. Since $f(3 + 6i) = \frac{i(3+6i)}{3} + 2 = \frac{3i-6}{3} + 2 = \frac{3(i-2)}{3} + 2 = i - 2 + 2 = i$, we see from the proof in Example 8.23 that if $|z - (3 + 6i)| < \delta$, then $|f(z) - f(3 + 6i)| = \left|\left(\frac{iz}{3} + 2\right) - i\right| < \epsilon$. It follows that f is continuous at $3 + 6i$.

More generally, let's show that for all $a \in \mathbb{C}$, f is continuous at a.

Proof: Let $a \in \mathbb{C}$, let $\epsilon > 0$ and let $\delta = 3\epsilon$. Suppose that $|z - a| < \delta$. Then we have

$$|f(z) - f(a)| = \left|\left(\frac{iz}{3} + 2\right) - \left(\frac{ia}{3} + 2\right)\right| = \left|\frac{i}{3}(z - a)\right| = \left|\frac{i}{3}\right| |z - a| < \frac{1}{3}\delta = \frac{1}{3}(3\epsilon) = \epsilon.$$

Since $\epsilon > 0$ was arbitrary, we have $\forall \epsilon > 0 \, \exists \delta > 0 \, (|z - a| < \delta \to |f(z) - f(a)| < \epsilon)$.

Therefore, f is continuous at a. $\qquad\square$

Notes: (1) We proved $\forall a \in \mathbb{C} \, \forall \epsilon > 0 \, \exists \delta > 0 \, \forall z \in \mathbb{C} (|z - a| < \delta \to |f(z) - f(a)| < \epsilon)$. In words, we proved that for every complex number a, given a positive real number ϵ, we can find a positive real number δ such that whenever the distance between z and a is less than δ, the distance between $f(z)$ and $f(a)$ is less than ϵ. And of course, a simpler way to say this is "for every complex number a, f is continuous at a," or $\forall a \in \mathbb{C} \, (f \text{ is continuous at } a)$."

(2) If we move the expression $\forall a \in \mathbb{C}$ next to $\forall z \in \mathbb{C}$, we get a concept that is stronger than continuity. We say that a function $f: A \rightarrow \mathbb{C}$ is **uniformly continuous** on A if

$$\forall \epsilon > 0 \, \exists \delta > 0 \, \forall a, z \in A \, (|z - a| < \delta \rightarrow |f(z) - f(a)| < \epsilon).$$

(3) As a quick example of uniform continuity, let's prove that the function $f: \mathbb{C} \rightarrow \mathbb{C}$ defined by $f(z) = \frac{iz}{3} + 2$ is uniformly continuous on $\mathbb{C}$.

New proof: Let $\epsilon > 0$ and let $\delta = 3\epsilon$. Let $a, z \in \mathbb{C}$ and suppose that $|z - a| < \delta$. Then we have

$$|f(z) - f(a)| = \left|\left(\frac{iz}{3} + 2\right) - \left(\frac{ia}{3} + 2\right)\right| = \left|\frac{i}{3}(z - a)\right| = \left|\frac{i}{3}\right| |z - a| < \frac{1}{3}\delta = \frac{1}{3} \cdot 3\epsilon = \epsilon.$$

Since $\epsilon > 0$ was arbitrary, we have $\forall \epsilon > 0 \, \exists \delta > 0 \, \forall a, z \in \mathbb{C} \, (|z - a| < \delta \rightarrow |f(z) - f(a)| < \epsilon)$. Therefore, f is uniformly continuous on $\mathbb{C}$.

(4) The difference between continuity and uniform continuity on a set A can be described as follows: In both cases, an ϵ is given and then a δ is chosen. For continuity, for each value of a, we are allowed to choose a different δ. For uniform continuity, once we choose a δ for some value of a, we need to be able to use the **same** δ for every other value of a in A.

In terms of disks, once a disk of radius ϵ is given, we need to be more careful how we choose our disk of radius δ. As we check different z-values, we can translate our chosen disk as much as we like around the xy-plane. However, we are not allowed to decrease the radius of the disk.

(5) The function $f: \mathbb{C} \rightarrow \mathbb{C}$ defined by $f(z) = \frac{iz}{3} + 2$ is a homeomorphism. The argument is similar to the argument given in Note 5 following Example 8.12. I leave the details to the reader.

The Riemann Sphere

We have used the symbols $-\infty$ and ∞ (or $+\infty$) to describe unbounded intervals of real numbers, as well as certain limits of real-valued functions. These symbols are used to express a notion of "infinity." If we pretend for a moment that we are standing on the real line at 0, and we begin walking to the right, continuing indefinitely, then we might say we are walking toward ∞. Similarly, if we begin walking to the left instead, continuing indefinitely, then we might say we are walking toward $-\infty$.

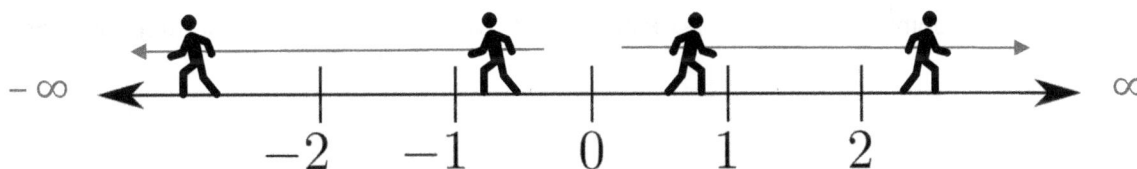

We would like to come up with a coherent notion of infinity with respect to the Complex Plane. There is certainly more than one way to do this. A method that is most analogous to the picture described above would be to define a set of infinities $\{\infty_\theta | 0 \leq \theta < 2\pi\}$, the idea being that for each angle θ made with the positive x-axis, we have an infinity, ∞_θ, describing where we would be headed if we were to start at the origin and then begin walking along the ray determined by θ, continuing indefinitely.

The method in the previous paragraph, although acceptable, has the disadvantage of having to deal with uncountably many "infinities." Instead, we will explore a different notion that involves just a single **point at infinity**. The idea is relatively simple. Pretend you have a large sheet of paper balancing on the palm of your hand. The sheet of paper represents the Complex Plane with the origin right at the center of your palm. The palm of your hand itself represents the unit circle together with its interior.

Now, imagine using the pointer finger on your other hand to press down on the origin of that sheet of paper (the Complex Plane), forcing your hand to form a unit sphere (reshaping the Complex Plane into a unit sphere as well). Notice that the origin becomes the "south pole" of the sphere, while all the "infinities" described in the last paragraph are forced together at the "north pole" of the sphere. Also, notice that the unit circle stays fixed, the points interior to the unit circle form the lower half of the sphere, and the points exterior to the unit circle form the upper half of the sphere with the exception of the "north pole."

When we visualize the unit sphere in this way, we refer to it as the **Reimann Sphere**.

Let's let $\mathbb{S}^2$ be the Reimann Sphere and let's officially define the **north pole** and **south pole** of $\mathbb{S}^2$ to be the points $N = (0, 0, 1)$ and $S = (0, 0, -1)$, respectively.

Also, since $\mathbb{S}^2$ is a subset of three-dimensional space $\mathbb{R}^3$, while $\mathbb{C}$ is only two dimensional, let's identify $\mathbb{C}$ with $\mathbb{C} \times \{0\}$ so that we write points in the Complex Plane as $(a, b, 0)$ instead of (a, b). We can then visualize the Complex Plane as intersecting the Reimann sphere in the unit circle. To the right we have a picture of the Reimann Sphere together with the Complex Plane.

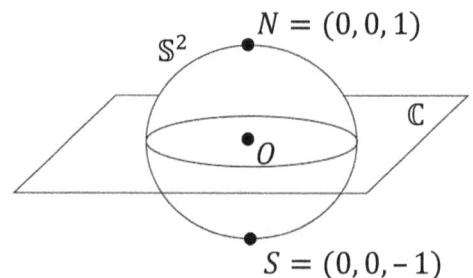

For each point z in the Complex Plane, consider the line passing through the points N and z. This line intersects $\mathbb{S}^2$ in exactly one point P_z. This observation allows us to define a bijection $f: \mathbb{C} \to \mathbb{S}^2 \setminus N$ defined by $f(z) = P_z$. An explicit definition of f can be given by

$$f(z) = \left(\frac{z + \bar{z}}{1 + |z|^2}, \frac{z - \bar{z}}{i(1 + |z|^2)}, \frac{|z|^2 - 1}{|z|^2 + 1} \right)$$

Below is a picture of a point z in the Complex Plane and its image $f(z) = P_z$ on the Riemann Sphere.

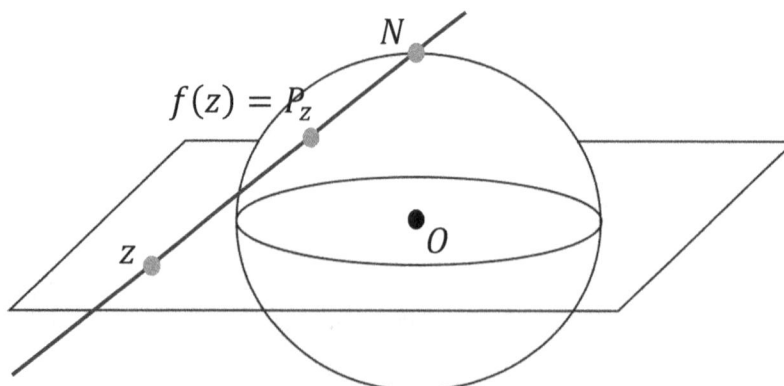

In Problem 25 below, you will be asked to verify that f is a homeomorphism. If we let $\overline{\mathbb{C}} = \mathbb{C} \cup \{\infty\}$, then we can extend f to a function $\overline{f}: \overline{\mathbb{C}} \to \mathbb{S}^2$ by defining $\overline{f}(\infty) = N$. $\overline{\mathbb{C}}$ is called the **Extended Complex Plane**. We consider $U \subseteq \overline{\mathbb{C}}$ to be open in $\overline{\mathbb{C}}$ if and only if either U is open in $\mathbb{C}$ or $U = V \cup \{\infty\}$, where V is the complement of a closed, bounded set in $\mathbb{C}$. In this way, all the basic properties of "openness" are still satisfied and $\overline{f}$ is a homeomorphism from $\overline{\mathbb{C}}$ to $\mathbb{S}^2$ in a way that will be made precise in Lesson 13.

If ϵ is a small positive number, then $\frac{1}{\epsilon}$ is a large positive number. We see that the set $N_{\frac{1}{\epsilon}} = \left\{ z \mid |z| > \frac{1}{\epsilon} \right\}$ is a neighborhood of ∞ in the following sense. Notice that $N_{\frac{1}{\epsilon}}$ consists of all points outside of the circle of radius $\frac{1}{\epsilon}$ centered at the origin. The image of this set under the function f is a deleted neighborhood of the north pole, N.

We can now extend our definition of limit to include various infinite cases. We will do one example here and you will look at others in Problem 22 below.

$$\lim_{z \to a} f(z) = \infty \text{ if and only if } \forall \epsilon > 0 \, \exists \delta > 0 \, \left(0 < |z - a| < \delta \to |f(z)| > \frac{1}{\epsilon} \right).$$

Theorem 8.28: $\lim_{z \to a} f(z) = \infty$ if and only $\lim_{z \to a} \frac{1}{f(z)} = 0$.

Proof: Suppose $\lim_{z \to a} f(z) = \infty$ and let $\epsilon > 0$. There is $\delta > 0$ so that $0 < |z - a| < \delta \to |f(z)| > \frac{1}{\epsilon}$. But, $|f(z)| > \frac{1}{\epsilon}$ is equivalent to $\left| \frac{1}{f(z)} - 0 \right| < \epsilon$. So, $\lim_{z \to a} \frac{1}{f(z)} = 0$.

Now, suppose $\lim_{z \to a} \frac{1}{f(z)} = 0$ and let $\epsilon > 0$. There is $\delta > 0$ so that $0 < |z - a| < \delta \to \left| \frac{1}{f(z)} - 0 \right| < \epsilon$. But, $\left| \frac{1}{f(z)} - 0 \right| < \epsilon$ is equivalent to $|f(z)| > \frac{1}{\epsilon}$. So, $\lim_{z \to a} f(z) = \infty$. $\square$

Limits and Continuity in Euclidean Spaces

Let $A \subseteq \mathbb{R}^n$, let $f: A \to \mathbb{R}^m$, let $\mathbf{b} \in \mathbb{R}^m$, and let $\mathbf{a} \in \mathbb{R}^n$ be a point such that A contains some deleted neighborhood of $\mathbf{a}$. We say that the **limit of f as x approaches a is b**, written $\lim_{\mathbf{x} \to \mathbf{a}} f(\mathbf{x}) = \mathbf{b}$, if for every positive number ϵ, there is a positive number δ such that $0 < |\mathbf{x} - \mathbf{a}| < \delta \to |f(\mathbf{x}) - \mathbf{b}| < \epsilon$.

Notes: (1) Recall from Lesson 7 that $\mathbf{x} \in \mathbb{R}^n$ can be written as $(x_1, x_2, \ldots, x_n)$, where $x_1, x_2, \ldots, x_n \in \mathbb{R}$. Also, the norm of $\mathbf{x}$ is $|\mathbf{x}| = \sqrt{x_1^2 + x_2^2 + \cdots + x_n^2}$. Therefore,

$$|\mathbf{x} - \mathbf{a}| = \sqrt{(x_1 - a_1)^2 + (x_2 - a_2)^2 + \cdots + (x_n - a_n)^2}.$$

(2) If $\mathbf{x} \in \mathbb{R}^n$ and f is a function with domain a subset of $\mathbb{R}^n$, then we will write $f(x_1, x_2, \ldots, x_n)$ instead of $f\big((x_1, x_2, \ldots, x_n)\big)$.

(3) Although we will be working only in Euclidean spaces ($\mathbb{R}^n$), the definitions and results in this section work equally as well for unitary spaces ($\mathbb{C}^n$).

Example 8.29: Let $f: \mathbb{R}^4 \to \mathbb{R}$ be defined by $f(x, y, z, w) = x + 2y - z + 3w$. Let's use the $\epsilon - \delta$ definition of limit to prove that $\displaystyle\lim_{\mathbf{x} \to (1,1,1,1)} f(\mathbf{x}) = 5$.

Analysis: Given $\epsilon > 0$, we will find $\delta > 0$ so that $0 < |(x, y, z, w) - (1, 1, 1, 1)| < \delta$ implies $|(x + 2y - z + 3w) - 5| < \epsilon$. First note that

$$|(x, y, z, w) - (1, 1, 1, 1)| = |(x - 1, y - 1, z - 1, w - 1)|$$
$$= \sqrt{(x-1)^2 + (y-1)^2 + (z-1)^2 + (w-1)^2}.$$

Also, note that $|x - 1| \leq \sqrt{(x-1)^2 + (y-1)^2 + (z-1)^2 + (w-1)^2}$, and similarly for $|y - 1|$, $|z - 1|$, and $|w - 1|$.

Finally, note that $|(x + 2y - z + 3w) - 5| = |(x - 1) + 2(y - 1) - (z - 1) + 3(w - 1)|$.

By the Triangle Inequality, we have

$$|(x - 1) + 2(y - 1) - (z - 1) + 3(w - 1)| \leq |x - 1| + 2|y - 1| + |z - 1| + 3|w - 1|$$

So, $|(x - 1) + 2(y - 1) - (z - 1) + 3(w - 1)| \leq 7\sqrt{(x-1)^2 + (y-1)^2 + (z-1)^2 + (w-1)^2}$.

It follows that $\delta = \dfrac{\epsilon}{7}$ should work.

Proof: Let $\epsilon > 0$ and let $\delta = \dfrac{\epsilon}{7}$. Suppose that $0 < |(x, y, z, w) - (1, 1, 1, 1)| < \delta$. Then we have

$$|(x + 2y - z + 3w) - 5| = |(x - 1) + 2(y - 1) - (z - 1) + 3(w - 1)|$$
$$\leq |x - 1| + 2|y - 1| + |z - 1| + 3|w - 1| \leq 7\sqrt{(x-1)^2 + (y-1)^2 + (z-1)^2 + (w-1)^2}$$
$$= 7|(x, y, z, w) - (1, 1, 1, 1)| < 7\delta = \frac{7\epsilon}{7} = \epsilon.$$

Since $\epsilon > 0$ was arbitrary, we have

$$\forall \epsilon > 0 \, \exists \delta > 0 \, (0 < |(x, y, z, w) - (1, 1, 1, 1)| < \delta \to |(x + 2y - z + 3w) - 5| < \epsilon).$$

Therefore, $\displaystyle\lim_{(x,y,z,w) \to (1,1,1,1)} f(x, y, z, w) = 5$ (or equivalently, $\displaystyle\lim_{\mathbf{x} \to (1,1,1,1)} f(\mathbf{x}) = 5$). $\qquad\square$

Example 8.30: Let $f: \mathbb{R}^2 \to \mathbb{R}^3$ be defined by $f(x, y) = (y, x, y - x)$. Let's use the $\epsilon - \delta$ definition of limit to prove that $\displaystyle\lim_{\mathbf{x} \to (3,2)} f(\mathbf{x}) = (2, 3, -1)$.

Analysis: Given $\epsilon > 0$, we will find $\delta > 0$ so that

$$0 < |(x, y) - (3, 2)| < \delta \text{ implies } |(y, x, y - x) - (2, 3, -1)| < \epsilon.$$

Note that $|(x, y) - (3, 2)| = \sqrt{(x - 3)^2 + (y - 2)^2}$.

Now, we have

$$|y - 2| \leq \sqrt{(x - 3)^2 + (y - 2)^2},$$
$$|x - 3| \leq \sqrt{(x - 3)^2 + (y - 2)^2},$$
$$|y - x - (-1)| = |(y - 2) - (x - 3)| \leq |y - 2| + |x - 3| \leq 2\sqrt{(x - 3)^2 + (y - 2)^2}.$$

Therefore, if $0 < |(x, y) - (3, 2)| < \delta$ (or equivalently, $0 < \sqrt{(x-3)^2 + (y-2)^2} < \delta$), then

$$|(y, x, y - x) - (2, 3, -1)| = |(y - 2, x - 3, y - x + 1)| = \sqrt{(y-2)^2 + (x-3)^2 + (y-x+1)^2}$$
$$\leq \sqrt{\delta^2 + \delta^2 + (2\delta)^2} = \sqrt{6\delta^2} = \sqrt{6}\delta.$$

It follows that $\delta = \dfrac{\epsilon}{\sqrt{6}}$ should work.

Proof: Let $\epsilon > 0$ and let $\delta = \dfrac{\epsilon}{\sqrt{6}}$. Suppose that $0 < |(x, y) - (3, 2)| < \delta$. We have

$$|(x, y) - (3, 2)| = |(x - 3, y - 2)| = \sqrt{(x-3)^2 + (y-2)^2}.$$

Note that

$$|y - 2| \leq \sqrt{(x-3)^2 + (y-2)^2}, \ |x - 3| \leq \sqrt{(x-3)^2 + (y-2)^2},$$
$$\text{and } |y - x - (-1)| = |(y-2) - (x-3)| \leq |y - 2| + |x - 3| \leq 2\sqrt{(x-3)^2 + (y-2)^2}.$$

It follows that

$$|y - 2| \leq \sqrt{(x-3)^2 + (y-2)^2} < \frac{\epsilon}{\sqrt{6}}$$

$$|x - 3| \leq \sqrt{(x-3)^2 + (y-2)^2} < \frac{\epsilon}{\sqrt{6}}$$

$$|y - x + 1| = |y - x - (-1)| \leq 2\sqrt{(x-3)^2 + (y-2)^2} < \frac{2\epsilon}{\sqrt{6}}$$

Therefore,

$$|(y, x, y - x) - (2, 3, -1)| = |(y - 2, x - 3, y - x + 1)| = \sqrt{(y-2)^2 + (x-3)^2 + (y-x+1)^2}$$

$$\leq \sqrt{\left(\frac{\epsilon}{\sqrt{6}}\right)^2 + \left(\frac{\epsilon}{\sqrt{6}}\right)^2 + \left(\frac{2\epsilon}{\sqrt{6}}\right)^2} = \sqrt{6 \cdot \frac{\epsilon^2}{6}} = \sqrt{\epsilon^2} = \epsilon.$$

So, $\lim\limits_{(x,y)\to(3,2)} f(x, y) = (2, 3, -1)$ (or equivalently, $\lim\limits_{\mathbf{x}\to(3,2)} f(\mathbf{x}) = (2, 3, -1)$). $\qquad\square$

Let $A \subseteq \mathbb{R}^n$, let $f: A \to \mathbb{R}^m$, and let $\mathbf{a} \in A$ be a point such that A contains some neighborhood of $\mathbf{a}$. f is **continuous at a** if and only if

$$\forall \epsilon > 0 \ \exists \delta > 0 \ (|\mathbf{x} - \mathbf{a}| < \delta \to |f(\mathbf{x}) - f(\mathbf{a})| < \epsilon).$$

Furthermore, we say that f is **uniformly continuous on A** if and only if

$$\forall \epsilon > 0 \ \exists \delta > 0 \ \forall \mathbf{a}, \mathbf{x} \in A \ (|\mathbf{x} - \mathbf{a}| < \delta \to |f(\mathbf{x}) - f(\mathbf{a})| < \epsilon).$$

The proofs given in Examples 8.29 and 8.30 can be modified slightly to prove that the functions in those examples are uniformly continuous on $\mathbb{R}^4$ and $\mathbb{R}^2$, respectively. I leave the details to the reader.

Problem Set 8

Full solutions to these problems are available for free download here:

www.SATPrepGet800.com/TFBZLF

LEVEL 1

1. Let $f: \mathbb{R} \to \mathbb{R}$ be defined by $f(x) = 5x - 1$.

   (i)   Prove that $\lim\limits_{x \to 3} f(x) = 14$.

   (ii)   Prove that f is continuous on $\mathbb{R}$.

2. Let $r, c \in \mathbb{R}$ and let $f: \mathbb{R} \to \mathbb{R}$ be defined by $f(x) = c$. Prove that $\lim\limits_{x \to r}[f(x)] = c$.

3. Let $A \subseteq \mathbb{R}$, let $f: A \to \mathbb{R}$, let $r, k \in \mathbb{R}$, and suppose that $\lim\limits_{x \to r}[f(x)]$ is a finite real number. Prove that $\lim\limits_{x \to r}[kf(x)] = k \lim\limits_{x \to r}[f(x)]$.

LEVEL 2

4. Let $A \subseteq \mathbb{R}$, let $f: A \to \mathbb{R}$, and let $r \in \mathbb{R}$. Prove that f is continuous at r if and only if $\lim\limits_{x \to r}[f(x)] = f(r)$.

5. Prove that every polynomial function $p: \mathbb{R} \to \mathbb{R}$ is continuous on $\mathbb{R}$.

LEVEL 3

6. Let $g: \mathbb{R} \to \mathbb{R}$ be defined by $g(x) = 2x^2 - 3x + 7$.

   (i)   Prove that $\lim\limits_{x \to 1} g(x) = 6$.

   (ii)   Prove that g is continuous on $\mathbb{R}$.

7. Suppose that $f, g: \mathbb{R} \to \mathbb{R}$, $a \in \mathbb{R}$, f is continuous at a, and g is continuous at $f(a)$. Prove that $g \circ f$ is continuous at a.

LEVEL 4

8. Let $h: \mathbb{R} \to \mathbb{R}$ be defined by $h(x) = \frac{x^3 - 4}{x^2 + 1}$. Prove that $\lim\limits_{x \to 2} h(x) = \frac{4}{5}$.

9. Let $k: (0, \infty) \to \mathbb{R}$ be defined by $k(x) = \sqrt{x}$.

   (i)   Prove that $\lim\limits_{x \to 25} k(x) = 5$.

   (ii)   Prove that f is continuous on $(0, \infty)$.

   (iii) Is f uniformly continuous on $(0, \infty)$?

10. Let $f: \mathbb{R} \to \mathbb{R}$ be defined by $f(x) = x^2$. Prove that f is continuous on $\mathbb{R}$, but not uniformly continuous on $\mathbb{R}$.

11. Let $A \subseteq \mathbb{R}$, let $f: A \to \mathbb{R}$, let $r \in \mathbb{R}$, and suppose that $\lim\limits_{x \to r}[f(x)] > 0$. Prove that there is a deleted neighborhood N of r such that $f(x) > 0$ for all $x \in N$.

12. Let $A \subseteq \mathbb{R}$, let $f: A \to \mathbb{R}$, let $r \in \mathbb{R}$, and suppose that $\lim\limits_{x \to r}[f(x)]$ is a finite real number. Prove that there is $M \in \mathbb{R}$ and an open interval (a, b) containing r such that $|f(x)| \le M$ for all $x \in (a, b) \setminus \{r\}$.

13. Let $A \subseteq \mathbb{R}$, let $f, g, h: A \to \mathbb{R}$, let $r \in \mathbb{R}$, let $f(x) \le g(x) \le h(x)$ for all $x \in A \setminus \{r\}$, and suppose that $\lim\limits_{x \to r}[f(x)] = \lim\limits_{x \to r}[h(x)] = L$. Prove that $\lim\limits_{x \to r}[g(x)] = L$.

LEVEL 5

14. Let $A \subseteq \mathbb{R}$, let $f, g: A \to \mathbb{R}$ such that $g(x) \ne 0$ for all $x \in A$, let $r \in \mathbb{R}$, and suppose that $\lim\limits_{x \to r}[f(x)]$ and $\lim\limits_{x \to r}[g(x)]$ are both finite real numbers such that $\lim\limits_{x \to r}[g(x)] \ne 0$. Prove that
$$\lim_{x \to r}\left[\frac{f(x)}{g(x)}\right] = \frac{\lim\limits_{x \to r} f(x)}{\lim\limits_{x \to r} g(x)}.$$

15. Give a reasonable equivalent definition for each of the following limits (like what was done in Theorem 8.20). r and L are finite real numbers.

   (i) $\lim\limits_{x \to r} f(x) = -\infty$

   (ii) $\lim\limits_{x \to +\infty} f(x) = L$

   (iii) $\lim\limits_{x \to -\infty} f(x) = L$

   (iv) $\lim\limits_{x \to +\infty} f(x) = +\infty$

   (v) $\lim\limits_{x \to +\infty} f(x) = -\infty$

   (vi) $\lim\limits_{x \to -\infty} f(x) = +\infty$

   (vii) $\lim\limits_{x \to -\infty} f(x) = -\infty$

16. Let $f(x) = -x^2 + x + 1$. Use the $M - K$ definition of an infinite limit (that you came up with in Problem 15) to prove $\lim\limits_{x \to +\infty} f(x) = -\infty$.

17. Give a reasonable definition for each of the following limits (like what was done in Theorem 8.22). r and L are finite real numbers.

    (i) $\lim\limits_{x \to r^-} f(x) = L$

    (ii) $\lim\limits_{x \to r^+} f(x) = +\infty$

    (iii) $\lim\limits_{x \to r^+} f(x) = -\infty$

    (iv) $\lim\limits_{x \to r^-} f(x) = +\infty$

    (v) $\lim\limits_{x \to r^-} f(x) = -\infty$

18. Use the $M - \delta$ definition of a one-sided limit (that you came up with in Problem 17) to prove that $\lim\limits_{x \to 3^-} \dfrac{1}{x-3} = -\infty$.

19. Let $f(x) = \dfrac{x+1}{(x-1)^2}$. Prove that

    (i) $\lim\limits_{x \to +\infty} f(x) = 0$.

    (ii) $\lim\limits_{x \to 1^+} f(x) = +\infty$.

20. Let $f: \mathbb{R} \to \mathbb{R}$ be defined by $f(x) = \begin{cases} 0 & \text{if } x \text{ is rational.} \\ 1 & \text{if } x \text{ is irrational.} \end{cases}$ Prove that for all $r \in \mathbb{R}$, $\lim\limits_{x \to r}[f(x)]$ does not exist.

21. Let $A \subseteq \mathbb{C}$, let $f: A \to \mathbb{C}$, let $L = j + ki \in \mathbb{C}$, and let $a = b + ci \in \mathbb{C}$ be a point such that A contains some deleted neighborhood of a. Suppose that $f(x + yi) = u(x,y) + iv(x,y)$. Prove that $\lim\limits_{z \to a} f(z) = L$ if and only if $\lim\limits_{(x,y) \to (b,c)} u(x,y) = j$ and $\lim\limits_{(x,y) \to (b,c)} v(x,y) = k$.

22. Give a reasonable definition for each of the following limits (like what was done right before Theorem 8.28). L is a finite real number.

    (i) $\lim\limits_{z \to \infty} f(z) = L$

    (ii) $\lim\limits_{z \to \infty} f(z) = \infty$

23. Prove each of the following:

    (i) $\lim\limits_{z \to \infty} f(z) = L$ if and only $\lim\limits_{z \to 0} f\left(\dfrac{1}{z}\right) = L$

    (ii) $\lim\limits_{z \to \infty} f(z) = \infty$ if and only $\lim\limits_{z \to 0} \dfrac{1}{f\left(\frac{1}{z}\right)} = 0$.

24. Let $f: \mathbb{R}^2 \to \mathbb{R}$ be defined by $f(x, y) = xy$. Prove that f is continuous. Then generalize this result to the function $f: \mathbb{R}^n \to \mathbb{R}$ defined by $f(\mathbf{x}) = x_1 \cdot x_2 \cdots x_n$.

CHALLENGE PROBLEM

25. Let $f: \mathbb{C} \to \mathbb{S}^2 \setminus N$ be defined as follows:

$$f(z) = \left(\frac{z + \bar{z}}{1 + |z|^2}, \frac{z - \bar{z}}{i(1 + |z|^2)}, \frac{|z|^2 - 1}{|z|^2 + 1} \right)$$

Note that $\mathbb{S}^2 \setminus N$ is a subset of $\mathbb{R}^3$. Prove that f is a homeomorphism.

LESSON 9
TOPOLOGICAL SPACES

Definitions and Examples

A topological space consists of a set S together with a collection of "open" subsets of S. Before we give the formal definition of "open," let's quickly review one of the standard examples from Lesson 7.

Consider the set $\mathbb{R}$ of real numbers and call a subset X of $\mathbb{R}$ open if for every real number $x \in X$, there is an open interval (a, b) with $x \in (a, b)$ and $(a, b) \subseteq X$. We were first introduced to this definition of an open set in Lesson 7. In that same lesson, we proved that $\emptyset$ and $\mathbb{R}$ are both open in $\mathbb{R}$ (Theorem 7.11), we proved that an arbitrary union of open sets in $\mathbb{R}$ is open in $\mathbb{R}$ (Theorem 7.15), and we proved that a finite intersection of open sets in $\mathbb{R}$ is open in $\mathbb{R}$ (Theorem 7.19 and part (iii) of Problem 5 from Problem Set 7). As it turns out, with this definition of open, every nonempty open set in $\mathbb{R}$ can be expressed as a countable union of pairwise disjoint open intervals (Theorem 7.18).

In this lesson we will move to a more general setting and explore arbitrary sets together with various collections of "open" subsets of these sets. Let's begin by giving the formal definition of a topological space.

Let S be a set and let $\mathcal{T}$ be a collection of subsets of S. $\mathcal{T}$ is said to be a **topology** on S if the following three properties are satisfied:

1. $\emptyset \in \mathcal{T}$ and $S \in \mathcal{T}$.

2. If $X \subseteq \mathcal{T}$, then $\bigcup X \in \mathcal{T}$ ($\mathcal{T}$ is closed under taking arbitrary unions).

3. If $Y \subseteq \mathcal{T}$ and Y is finite, then $\bigcap Y \in \mathcal{T}$ ($\mathcal{T}$ is closed under taking finite intersections).

A **topological space** is a pair $(S, \mathcal{T})$, where S is a set and $\mathcal{T}$ is a topology on S. We will call the elements of $\mathcal{T}$ **open sets**. Complements of elements of $\mathcal{T}$ will be called **closed sets** (A is closed if and only if $S \setminus A$ is open).

We may sometimes refer to the topological space S. When we do, there is a topology $\mathcal{T}$ on S that we are simply not mentioning explicitly.

Example 9.1:

1. Let $S = \{a\}$ be a set consisting of just the one element a. There is just one topology on S. It is the topology $\mathcal{T} = \{\emptyset, \{a\}\}$.

    Note that the power set of S is $\mathcal{P}(S) = \{\emptyset, \{a\}\}$ and $\mathcal{P}(\mathcal{P}(S)) = \{\emptyset, \{\emptyset\}, \{\{a\}\}, \{\emptyset, \{a\}\}\}$. Notice that the topology $\mathcal{T} = \{\emptyset, \{a\}\}$ is an element of $\mathcal{P}(\mathcal{P}(S))$. However, the other three elements of $\mathcal{P}(\mathcal{P}(S))$ are **not** topologies on $S = \{a\}$.

    In general, for any set S, a topology on S is a subset of $\mathcal{P}(S)$, or equivalently, an element of $\mathcal{P}(\mathcal{P}(S))$. If $S \neq \emptyset$, then not every element of $\mathcal{P}(\mathcal{P}(S))$ will be a topology on S.

For example, if $S = \{a\}$, Then $\emptyset, \{\emptyset\}$, and $\{\{a\}\}$ are all elements of $\mathcal{P}\big(\mathcal{P}(S)\big)$ that are **not** topologies on S. $\emptyset$ and $\{\emptyset\}$ fail to be topologies on $\{a\}$ because they do not contain $\{a\}$, while $\{\{a\}\}$ fails to be a topology on $\{a\}$ because it does not contain $\emptyset$.

2. Let $S = \{a, b\}$ be a set consisting of the two distinct elements a and b. There are four topologies on S: $\mathcal{T}_1 = \{\emptyset, \{a, b\}\}$, $\mathcal{T}_2 = \{\emptyset, \{a\}, \{a, b\}\}$, $\mathcal{T}_3 = \{\emptyset, \{b\}, \{a, b\}\}$, and $\mathcal{T}_4 = \{\emptyset, \{a\}, \{b\}, \{a, b\}\}$. We can visualize these topologies as follows.

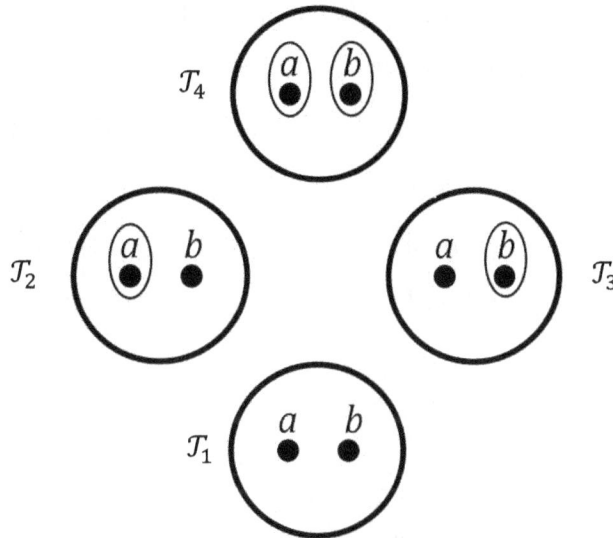

Notice that all four topologies in the figure have the elements a and b inside a large circle because $S = \{a, b\}$ is in all four topologies. Also, it is understood that $\emptyset$ is in all the topologies.

$\mathcal{T}_1$ is called the **trivial topology** (or **indiscrete topology**) on S because it contains only $\emptyset$ and S. $\mathcal{T}_4$ is called the **discrete topology** on S, as it contains every subset of S. The discrete topology is just $\mathcal{P}(S)$ (the power set of S).

The topologies $\mathcal{T}_2$, $\mathcal{T}_3$, and $\mathcal{T}_4$ are **finer** than the topology $\mathcal{T}_1$ because $\mathcal{T}_1 \subseteq \mathcal{T}_2$, $\mathcal{T}_1 \subseteq \mathcal{T}_3$ and $\mathcal{T}_1 \subseteq \mathcal{T}_4$. We can also say that $\mathcal{T}_1$ is **coarser** than $\mathcal{T}_2$, $\mathcal{T}_3$, and $\mathcal{T}_4$. Similarly, $\mathcal{T}_4$ is finer than $\mathcal{T}_2$ and $\mathcal{T}_3$, or equivalently, $\mathcal{T}_2$ and $\mathcal{T}_3$ are coarser than $\mathcal{T}_4$. The topologies $\mathcal{T}_2$ and $\mathcal{T}_3$ are **incomparable**. Neither one is finer than the other. To help understand the terminology "finer" and "coarser," we can picture the open sets as a pile of rocks. If we were to smash that pile of rocks (the open sets) with a hammer, the rocks will break into smaller pieces (creating more open sets), and the pile of rocks (the topology) will have been made "finer."

Note that for any set S, the discrete topology is always the finest topology and the trivial topology is always the coarsest.

3. Let $S = \{a, b, c\}$ be a set consisting of the three distinct elements a, b, and c. There are 29 topologies on S. Let's look at a few of them.

We have the trivial topology $\mathcal{T}_1 = \{\emptyset, \{a, b, c\}\}$.

If we throw in just a singleton set (a set consisting of just one element), we get the three topologies $\mathcal{T}_2 = \{\emptyset, \{a\}, \{a, b, c\}\}$, $\mathcal{T}_3 = \{\emptyset, \{b\}, \{a, b, c\}\}$, $\mathcal{T}_4 = \{\emptyset, \{c\}, \{a, b, c\}\}$.

Note that we can't throw in just two singleton sets. For example, $\{\emptyset, \{a\}, \{b\}, \{a,b,c\}\}$ is **not** a topology on S. Do you see the problem? It's not closed under taking unions: $\{a\}$ and $\{b\}$ are there, but $\{a,b\} = \{a\} \cup \{b\}$ is not! However, $\mathcal{T}_5 = \{\emptyset, \{a\}, \{b\}, \{a,b\}, \{a,b,c\}\}$ **is** a topology on S.

Here are a few **chains** of topologies on S written in order from the coarsest to the finest topology (chains are linearly ordered subsets of $\{\mathcal{T} \mid \mathcal{T}$ is a topology on $S\}$).

$$\{\emptyset, \{a,b,c\}\} \subseteq \{\emptyset, \{a\}, \{a,b,c\}\} \subseteq \{\emptyset, \{a\}, \{a,b\}, \{a,b,c\}\} \subseteq \{\emptyset, \{a\}, \{b\}, \{a,b\}, \{a,b,c\}\}$$
$$\subseteq \{\emptyset, \{a\}, \{b\}, \{a,b\}, \{a,c\}, \{a,b,c\}\} \subseteq \{\emptyset, \{a\}, \{b\}, \{c\}, \{a,b\}, \{b,c\}, \{a,c\}, \{a,b,c\}\}$$

$$\{\emptyset, \{a,b,c\}\} \subseteq \{\emptyset, \{a\}, \{a,b,c\}\} \subseteq \{\emptyset, \{a\}, \{b,c\}, \{a,b,c\}\} \subseteq \{\emptyset, \{a\}, \{b\}, \{a,b\}, \{b,c\}, \{a,b,c\}\}$$
$$\subseteq \{\emptyset, \{a\}, \{b\}, \{c\}, \{a,b\}, \{b,c\}, \{a,c\}, \{a,b,c\}\}$$

$$\{\emptyset, \{a,b,c\}\} \subseteq \{\emptyset, \{b,c\}, \{a,b,c\}\} \subseteq \{\emptyset, \{c\}, \{b,c\}, \{a,b,c\}\} \subseteq \{\emptyset, \{b\}, \{c\}, \{b,c\}, \{a,b,c\}\}$$
$$\subseteq \{\emptyset, \{b\}, \{c\}, \{a,b\}, \{b,c\}, \{a,b,c\}\} \subseteq \{\emptyset, \{a\}, \{b\}, \{c\}, \{a,b\}, \{b,c\}, \{a,c\}, \{a,b,c\}\}$$

Below is a picture of all 29 topologies on $\{a,b,c\}$ (to avoid clutter we left out the names of the elements). Again, a large circle surrounds a, b, and c in all cases because $S = \{a,b,c\}$ is in all 29 topologies. Also, it is understood that the empty set is in all these topologies.

I organized these topologies by the number of sets in each topology. The lowest row consists of just the trivial topology. The next row up consists of the topologies with just one additional set (three sets in total because $\emptyset$ and S are in every topology), and so on.

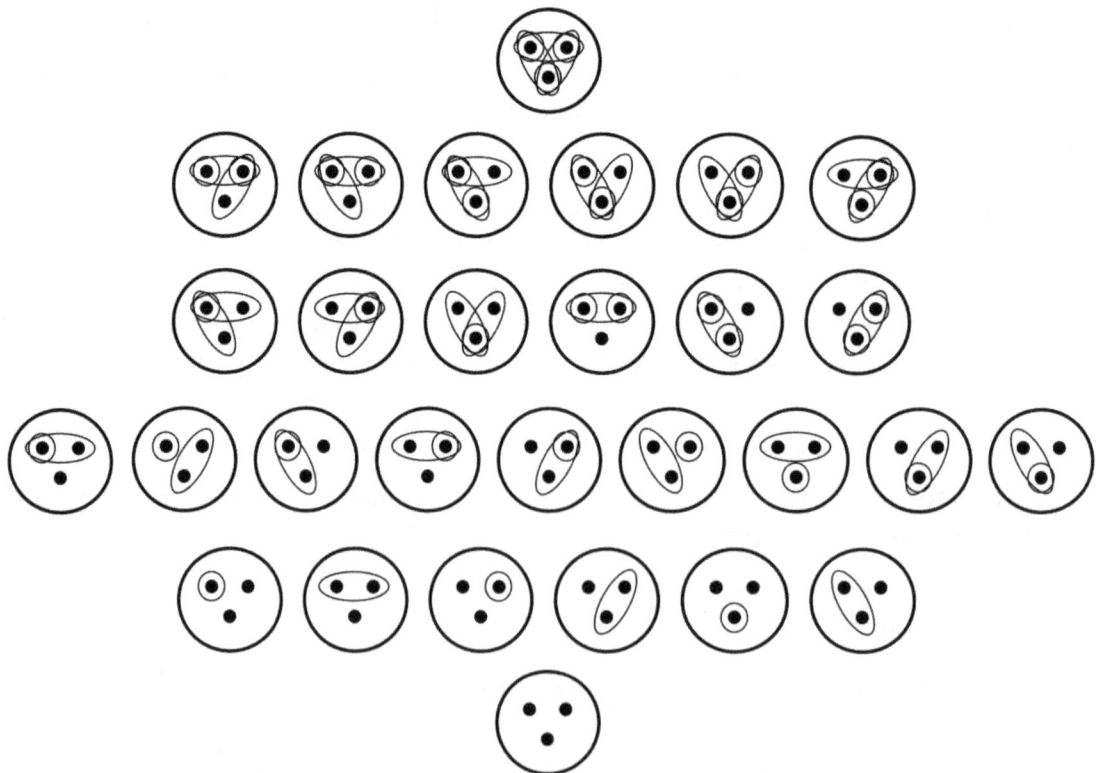

Below we see a visual representation of the three chains described above. As each path moves from the bottom to the top of the picture, we move from coarser to finer topologies.

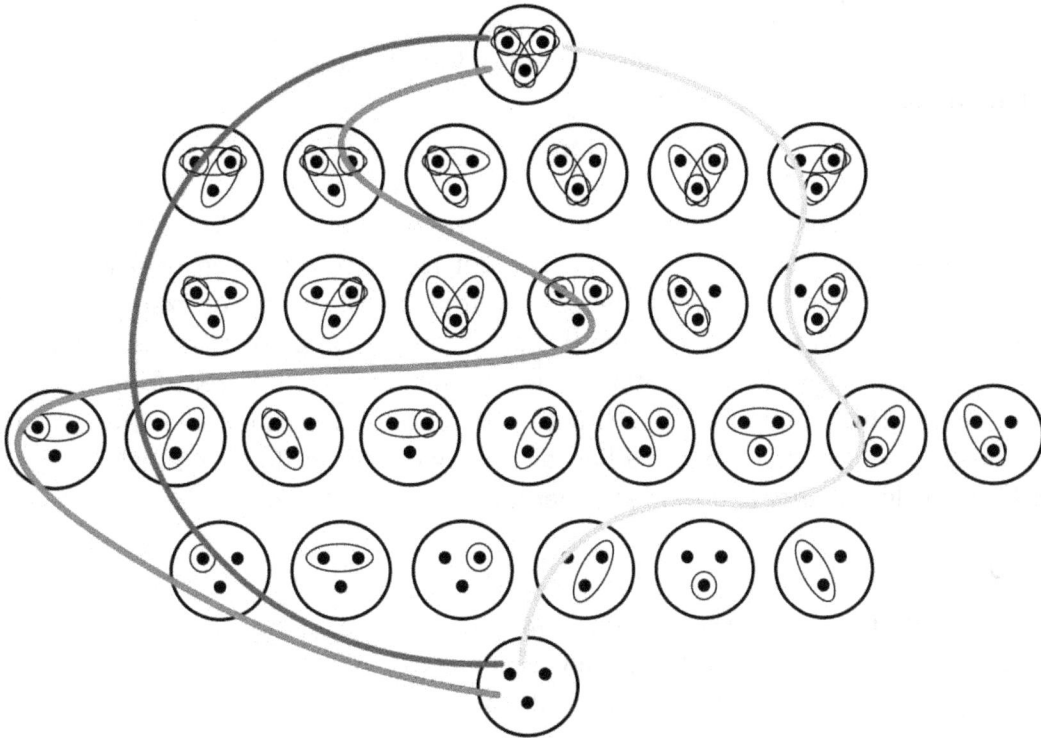

4. Let $S = \mathbb{R}$ and let $\mathcal{T} = \{X \subseteq \mathbb{R} \mid \forall x \in X \; \exists a, b \in \mathbb{R}(x \in (a, b) \land (a, b) \subseteq X\}$. In other words, we are defining a subset of $\mathbb{R}$ to be open as we did in Lesson 7. That is, a subset X of $\mathbb{R}$ is open if for every real number $x \in X$, there is an open interval (a, b) with $x \in (a, b)$ and $(a, b) \subseteq X$. By Theorem 7.11, $\emptyset, \mathbb{R} \in \mathcal{T}$. By Theorem 7.15, $\mathcal{T}$ is closed under taking arbitrary unions. By Problem 5 from Problem Set 7 (part (iii)), $\mathcal{T}$ is closed under taking finite intersections. It follows that $\mathcal{T}$ is a topology on $\mathbb{R}$. This topology is called the **standard topology on $\mathbb{R}$**.

5. Let $S = \mathbb{C}$ and let $\mathcal{T} = \{X \subseteq \mathbb{C} \mid \forall z \in X \; \exists a \in \mathbb{C} \; \exists r \in \mathbb{R}^{+}(z \in N_r(a) \land N_r(a) \subseteq X\}$, where $N_r(a) = \{z \in \mathbb{C} \mid |z - a| < r\}$ is the r-neighborhood of a (see Lesson 7). In other words, we are defining a subset of $\mathbb{C}$ to be open as we did in Lesson 7. That is, a subset X of $\mathbb{C}$ is open if for every complex number $z \in X$, there is an open disk (or neighborhood) $N_r(a)$ with $z \in N_r(a)$ and $N_r(a) \subseteq X$. By Example 7.29 (part 3), $\emptyset, \mathbb{C} \in \mathcal{T}$. By Problem 13 in Problem Set 7 (parts (i) and (ii)), $\mathcal{T}$ is closed under taking arbitrary unions and finite intersections. It follows that $\mathcal{T}$ is a topology on $\mathbb{C}$. This topology is called the **standard topology on $\mathbb{C}$**.

6. Let $S = \mathbb{R}^n$ and let $\mathcal{T} = \{X \subseteq \mathbb{R}^n \mid \forall \mathbf{x} \in X \; \exists \mathbf{a} \in \mathbb{R}^n \; \exists r \in \mathbb{R}^{+}(\mathbf{x} \in B_r(\mathbf{a}) \land B_r(\mathbf{a}) \subseteq X\}$, where $B_r(\mathbf{a}) = \{\mathbf{x} \in \mathbb{R}^n \mid |\mathbf{x} - \mathbf{a}| < r\}$ is the r-neighborhood of $\mathbf{a}$ (or open ball with center $\mathbf{a}$ and radius r—see Lesson 7). Parts 2, 3, and 4 of Theorem 7.38 guarantee that $\mathcal{T}$ is a topology on $\mathbb{R}^n$. This topology is called the **standard topology** on the Euclidean space $\mathbb{R}^n$. The standard topology on the unitary space $\mathbb{C}^n$ can be defined similarly.

Note: In a general topological space $(S, \mathcal{T})$, if $U \in \mathcal{T}$ and $x \in U$, we may call U a neighborhood of x.

We would now like to give an alternate definition of a closed set in a topological space $(S, \mathcal{T})$ in terms of "accumulation points," as we did for $\mathbb{R}$ and $\mathbb{C}$ (as well as $\mathbb{R}^n$ and $\mathbb{C}^n$) in Lesson 7.

Let $(S, \mathcal{T})$ be a topological space and let $A \subseteq S$. Then $x \in S$ is called an **accumulation point** of A if every $U \in \mathcal{T}$ containing x contains at least one point of A different from x. Symbolically, we have that x is an accumulation point of A if and only if

$$\forall U \in \mathcal{T}\, (x \in U \to \exists y \in A(y \in U \wedge y \neq x))$$

If $\mathcal{T}$ is the standard topology on either $\mathbb{R}$ or $\mathbb{C}$ (or more generally, on $\mathbb{R}^n$ or $\mathbb{C}^n$), then an accumulation point of a subset A is really, really close to the set A, as we discussed in Lesson 7. This notion of "closeness" doesn't necessarily apply to a general topological space. For example, let $\mathcal{T}$ be the trivial topology on $\mathbb{R}$ and let $A = (10, 20)$. Then 0 is an accumulation point of A. Indeed, the only open set containing 0 is $\mathbb{R}$, and certainly $\mathbb{R}$ contains at least one point of A different from 0 (in fact, it contains the whole set A). This is an undesirable situation. After all, 0 seems far away from A to me. This type of anomalous behavior may happen when a topological space doesn't satisfy enough "separation axioms." We will learn more about these separation axioms in Lesson 10.

Example 9.2:

1. If $\mathcal{T}$ is the trivial topology on a set S with at least 2 elements, and $A \subseteq S$ with $|A| \geq 2$, then every point of S is an accumulation point of A. Indeed, if $x \in S$, then the only open set containing x is S. Since A has at least 2 elements, there is $y \in A$ with $y \neq x$. Since $A \subseteq S$, we have $y \in S$.

    If $A \subseteq S$ with $|A| = 1$, say $A = \{a\}$, then a is **not** an accumulation point of A because there is no element of A different from a. However, if $y \in S$ with $y \neq a$, then y is an accumulation point of A. Indeed, the only open set containing y is S and a is an element of A with $y \neq a$ and $a \in S$.

2. If $\mathcal{T}$ is the discrete topology on a nonempty set S and $A \subseteq S$, then A has no accumulation points. Indeed, if $x \in S$, then $\{x\}$ is open and there is no $y \in \{x\}$ with $y \neq x$.

3. If $(S, \mathcal{T})$ is any topological space and $x \in S$, then x is **not** an accumulation point of $\{x\}$ because S is an open set containing x and there is no $y \in \{x\}$ with $y \neq x$.

4. Let $S = \{a, b\}$ with $a \neq b$ and let $\mathcal{T}_2 = \{\emptyset, \{a\}, \{a, b\}\}$ (see part 2 of Example 9.1 above). The set $\{a\}$ has one accumulation point, namely b. To see this, note that the only open set containing b is S and a is also in S. The set $\{b\}$ has no accumulation points. b is not an accumulation point of $\{b\}$ by 3 above and a is not an accumulation point of $\{b\}$ because $\{a\}$ is an open set containing a and $b \notin \{a\}$.

5. Examples of accumulation points in the standard topologies on $\mathbb{R}$ and $\mathbb{C}$ can be found in Examples 7.23 and 7.30, respectively. Examples of accumulation points in the standard topology of $\mathbb{R}^n$ can be found in Example 7.37.

Theorem 9.3: Let $(S, \mathcal{T})$ be a topological space with $C \subseteq S$. C is closed in S if and only if C contains each of its accumulation points.

Proof: First assume that C is closed in S. Then $S \setminus C$ is open in S. Let x be any point **not** in C. Then $x \in S \setminus C$. Since $S \setminus C$ is open, there is $U \in \mathcal{T}$ with $x \in U$ and $U \subseteq S \setminus C$. So, if $y \in U$, then $y \notin C$. It follows that x is **not** an accumulation point of C. So, we have shown that $x \notin C$ implies that x is not an accumulation point of C. The contrapositive of this statement is "if x is an accumulation point of C, then $x \in C$." So, C contains each of its accumulation points.

Conversely, assume that C contains each of its accumulation points. We need to show that $S \setminus C$ is open in S. Since $\emptyset$ is open, we can assume that $S \setminus C \neq \emptyset$. Let $x \in S \setminus C$. Then x is **not** an accumulation point of C. So, there is $U \in \mathcal{T}$ with $x \in U$ and $U \cap C = \emptyset$. Therefore, $U \subseteq S \setminus C$. Since $x \in S \setminus C$ was arbitrary, we have shown that $S \setminus C$ is open in S. □

Let $(S, \mathcal{T})$ be a topological space. If $A \subseteq S$, then the **closure** of A in S, written $\overline{A}$, is equal to $\bigcap \{ C \mid A \subseteq C \wedge C$ is closed in $S \}$.

Since $\overline{A}$ is an intersection of closed sets in S, by Problem 11 below, $\overline{A}$ is closed in S. We now state a few basic facts about the closure of a set A.

Theorem 9.4: Let $(S, \mathcal{T})$ be a topological space and let $A \subseteq S$.

1. $A \subseteq \overline{A}$.

2. If C is closed in S with $A \subseteq C$, then $\overline{A} \subseteq C$.

3. $\overline{A} = A \cup \{ x \in S \mid x$ is an accumulation point of $A \}$.

4. A is closed in S if and only if $A = \overline{A}$.

5. $x \in \overline{A}$ if and only if every open set containing x contains at least one point of A.

The proof of Theorem 9.4 is similar to the proof Theorem 7.25, and so, I leave it to the reader.

Example 9.5:

1. If $\mathcal{T}$ is the trivial topology on a nonempty set S, and $A \subseteq S$, then by part 1 of Example 9.2 and part 3 of Theorem 9.4, $\overline{A} = S$.

2. If $\mathcal{T}$ is the discrete topology on a nonempty set S and $A \subseteq S$, then by part 2 of Example 9.2 and part 3 of Theorem 9.4, $\overline{A} = A$.

3. Let $S = \{a, b\}$ with $a \neq b$ and let $\mathcal{T}_2 = \{\emptyset, \{a\}, \{a, b\}\}$ (see part 2 of Example 9.1 and part 4 of Example 9.2 above). By part 4 of Example 9.2 and part 3 of Theorem 9.4, $\overline{\{a\}} = S$ and $\overline{\{b\}} = \{b\}$.

4. For examples involving closures of subsets of real and complex numbers, see Examples 7.26 and 7.33, respectively. For examples in $\mathbb{R}^n$, see Example 7.37.

Bases

If $(S, \mathcal{T})$ is a topological space, then a **basis** for the topology $\mathcal{T}$ is a subset $\mathcal{B} \subseteq \mathcal{T}$ such that every element of $\mathcal{T}$ can be written as a union of elements from $\mathcal{B}$. We say that $\mathcal{T}$ is **generated** by $\mathcal{B}$ or $\mathcal{B}$ **generates** $\mathcal{T}$.

Notes: (1) Given a topological space $\mathcal{T}$, it can be cumbersome to describe all the open sets in $\mathcal{T}$. However, it is usually not too difficult to describe a topology in terms of its basis elements.

(2) If $(S, \mathcal{T})$ is a topological space and $\mathcal{B}$ is a basis for $\mathcal{T}$, then $\mathcal{T} = \{ \bigcup X \mid X \subseteq \mathcal{B} \}$.

(3) More generally, if $\mathcal{X}$ is any collection of subsets of S, then we can say that $\mathcal{X}$ generates $\{\bigcup X \mid X \subseteq \mathcal{X}\}$. However, this set will not always be a topology on S.

Example 9.6:

1. Let $S = \{a, b, c\}$ and $\mathcal{T} = \{\emptyset, \{a\}, \{b\}, \{a, b\}, \{b, c\}, \{a, b, c\}\}$. The set $\mathcal{B} = \{\{a\}, \{b\}, \{b, c\}\}$ is a basis for $\mathcal{T}$. Indeed, we have $\{a\} = \bigcup\{\{a\}\}$, $\{b\} = \bigcup\{\{b\}\}$, $\{a, b\} = \bigcup\{\{a\}, \{b\}\} = \{a\} \cup \{b\}$, $\{b, c\} = \bigcup\{\{b, c\}\}$, $\{a, b, c\} = \bigcup\{\{a\}, \{b\}, \{c\}\} = \{a\} \cup \{b, c\}$, and $\emptyset = \bigcup \emptyset$.

    (Note that $\bigcup \emptyset = \{y \mid \text{there is } Y \in \emptyset \text{ with } y \in Y\} = \emptyset$. It follows that $\emptyset$ does **not** need to be included in a basis.)

    We can visualize the basis $\mathcal{B}$ and the topology $\mathcal{T}$ that is generated by $\mathcal{B}$ as follows.

    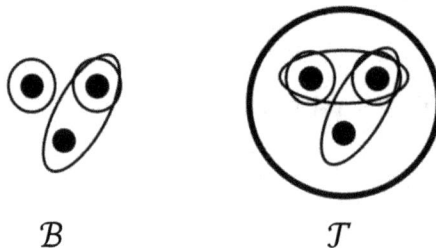

    $\mathcal{B}$ $\qquad\qquad$ $\mathcal{T}$

    We know $\emptyset \in \mathcal{T}$ (even though $\emptyset$ is not indicated in the picture of $\mathcal{T}$) because $\mathcal{T}$ is a topology. On the other hand, it is unclear from the picture of $\mathcal{B}$ whether $\emptyset \in \mathcal{B}$. However, it doesn't really matter. Since $\emptyset$ is equal to an empty union, $\emptyset$ will always be generated from $\mathcal{B}$ anyway.

    There can be more than one basis for the same topology. Here are a few more bases for the topology $\mathcal{T}$ just discussed (are there any others?):

    $$\mathcal{B}_1 = \{\{a\}, \{b\}, \{a, b\}, \{b, c\}\} \qquad \mathcal{B}_2 = \{\{a\}, \{b\}, \{a, b\}, \{b, c\}, \{a, b, c\}\}$$
    $$\mathcal{B}_3 = \{\emptyset, \{a\}, \{b\}, \{b, c\}\} \qquad \mathcal{B}_4 = \mathcal{T} = \{\emptyset, \{a\}, \{b\}, \{a, b\}, \{b, c\}, \{a, b, c\}\}$$

2. Let $S = \{a, b, c\}$ and let $\mathcal{X} = \{\{a\}, \{b\}\}$. In this case, $\mathcal{X}$ generates $\{\emptyset, \{a\}, \{b\}, \{a, b\}\}$. This set is **not** a topology on S because $S = \{a, b, c\}$ is not in the set. The reason that $\mathcal{X}$ failed to generate a topology on S is that it didn't completely "cover" S. Specifically, c is not in any set in $\mathcal{X}$.

    In general, if an element x from a set S does not appear in any of the sets in a set $\mathcal{X}$, then no matter how large a union we take from $\mathcal{X}$, we will never be able to generate a set from $\mathcal{X}$ with x in it, and therefore, $\mathcal{X}$ will not generate a topology on S (although it **might** generate a topology on a subset of S).

3. Let $S = \{a, b, c\}$ and $\mathcal{X} = \{\{a, b\}, \{b, c\}\}$. In this case, $\mathcal{X}$ generates $\{\emptyset, \{a, b\}, \{b, c\}, \{a, b, c\}\}$. This set is also **not** a topology because $\{a, b\} \cap \{b, c\} = \{b\}$ is not in the set. In other words, the set is not closed under finite intersections.

    In general, if there are two sets A and B in $\mathcal{X}$ with nonempty intersection such that the intersection $A \cap B$ does **not** include some nonempty set in $\mathcal{X}$, then the set generated by $\mathcal{X}$ will not be closed under finite intersections, and therefore, $\mathcal{X}$ will not generate a topology on S. Note that $A \cap B$ itself does not necessarily need to be in $\mathcal{X}$. However, there does need to be a set C with $C \subseteq A \cap B$ and $C \in \mathcal{X}$.

Parts 2 and 3 from Example 9.6 show us that not every collection $\mathcal{X}$ of subsets of a set S is the basis for a topology on S. Let's see if we can find conditions on a collection $\mathcal{X}$ of subsets of S that will guarantee that $\mathcal{X}$ is a basis for a topology on S.

We say that $\mathcal{X}$ **covers** S if every element of S belongs to at least one member of $\mathcal{X}$. Symbolically, we have

$$\forall x \in S \, \exists A \in \mathcal{X}(x \in A).$$

We say that $\mathcal{X}$ has the **intersection containment property** on S if every element of S that is in the intersection of two sets in $\mathcal{X}$ is also in some set in $\mathcal{X}$ that is contained in that intersection.

$$\forall x \in S \, \forall A, B \in \mathcal{X}\big(x \in A \cap B \rightarrow \exists C \in \mathcal{X} \, (x \in C \wedge C \subseteq A \cap B)\big).$$

Example 9.7:

1. Once again, let $S = \{a, b, c\}$. $\mathcal{X}_1 = \{\{b\}, \{b, c\}\}$ does **not** cover S because $a \in S$ does not belong to any member of $\mathcal{X}_1$. $\mathcal{X}_1$ does have the intersection containment property—the only element of $\{b\} \cap \{b, c\} = \{b\}$ is b, and $b \in \{b\} \in \mathcal{X}_1$ and $\{b\}$ is contained in $\{b\} \cap \{b, c\}$. Notice that the set that $\mathcal{X}_1$ generates is $\{\emptyset, \{b\}, \{b, c\}\}$. This set is not a topology on S because $S = \{a, b, c\}$ is not in this set. However, it is a topology on $\{b, c\}$.

    $\mathcal{X}_2 = \{\{a, b\}, \{b, c\}\}$ covers S, but does **not** have the intersection containment property. Indeed, $\{a, b\} \cap \{b, c\} = \{b\}$ and $b \in \{b\}$, but $\{b\} \notin \mathcal{X}_2$. Notice that the set that $\mathcal{X}_2$ generates is $\{\emptyset, \{a, b\}, \{b, c\}, \{a, b, c\}\}$. This set is not a topology on S because $\{a, b\} \cap \{b, c\} = \{b\}$ is not in the set.

    $\mathcal{B} = \{\{b\}, \{a, b\}, \{b, c\}\}$ covers S and has the intersection containment property. The set that $\mathcal{B}$ generates is the topology $\mathcal{T} = \{\emptyset, \{b\}, \{a, b\}, \{b, c\}, \{a, b, c\}\}$.

    We can visualize the sets $\mathcal{X}_1$, $\mathcal{X}_2$, $\mathcal{B}$, and $\mathcal{T}$ as follows.

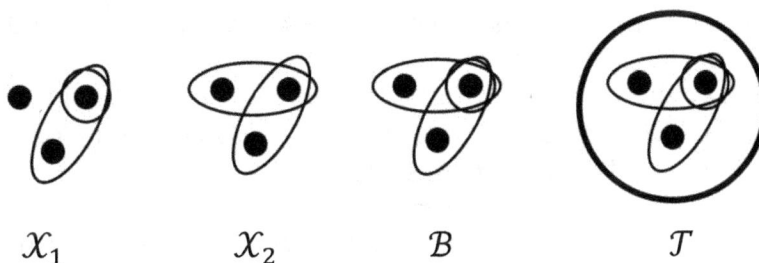

$$\mathcal{X}_1 \qquad \mathcal{X}_2 \qquad \mathcal{B} \qquad \mathcal{T}$$

2. Let $S = \mathbb{R}$ and let $\mathcal{B} = \{(a, b) \mid a, b \in \mathbb{R} \wedge a < b\}$ be the set of open intervals with endpoints in $\mathbb{R}$. $\mathcal{B}$ covers $\mathbb{R}$ because if $x \in \mathbb{R}$, then $x \in (x - 1, x + 1) \in \mathcal{B}$. $\mathcal{B}$ also has the intersection containment property. Indeed, if $x \in \mathbb{R}$ and $(a, b), (c, d) \in \mathcal{B}$ with $x \in (a, b) \cap (c, d)$, then $x \in (a, b) \cap (c, d) = (e, f)$, where $e = \max\{a, c\}$ and $f = \min\{b, d\}$ (see part (ii) of Problem 5 from Problem Set 7) and $(e, f) \in \mathcal{B}$. In fact, $\mathcal{B}$ is a basis on $\mathbb{R}$ that generates the standard topology of $\mathbb{R}$.

To see that $\mathcal{B}$ generates the standard topology on $\mathbb{R}$, let $\mathcal{T}$ be the standard topology on $\mathbb{R}$ and let $\mathcal{T}'$ be the topology generated by $\mathcal{B}$. First, let $X \in \mathcal{T}$ with $X \neq \emptyset$. By Theorem 7.17, X can be expressed as a union of bounded open intervals. So, $X \in \mathcal{T}'$. Since $X \in \mathcal{T}$ was arbitrary, $\mathcal{T} \subseteq \mathcal{T}'$. Now, let $X \in \mathcal{T}'$. Then X is a union of bounded open intervals, say $X = \bigcup Y$. Let $x \in X$. Since $X = \bigcup Y$, $x \in \bigcup Y$. So, $x \in (a, b)$ for some $(a, b) \in Y$. Since $(a, b) \in Y$, $(a, b) \subseteq \bigcup Y = X$. Therefore, $X \in \mathcal{T}$. Since $X \in \mathcal{T}'$ was arbitrary, $\mathcal{T}' \subseteq \mathcal{T}$. Since $\mathcal{T} \subseteq \mathcal{T}'$ and $\mathcal{T}' \subseteq \mathcal{T}$, $\mathcal{T}' = \mathcal{T}$.

3. Let $S = \mathbb{R}$ and let $\mathcal{X} = \{(-\infty, b) \mid b \in \mathbb{R}\} \cup \{(a, \infty) \mid a \in \mathbb{R}\}$. $\mathcal{X}$ covers $\mathbb{R}$ because if $x \in \mathbb{R}$, then $x \in (x - 1, \infty) \in \mathcal{X}$. However, $\mathcal{X}$ does **not** have the intersection containment property. For example, $0 \in (-\infty, 1) \cap (-1, \infty) = (-1, 1)$, but there is no set in $\mathcal{X}$ contained in $(-1, 1)$. The set generated by $\mathcal{X}$ is $\mathcal{X} \cup \{(-\infty, b) \cup (a, \infty) \mid a, b \in \mathbb{R} \wedge b < a\} \cup \{\emptyset, \mathbb{R}\}$. This set is **not** closed under finite intersections, and therefore, it is **not** a topology on $\mathbb{R}$.

Based on the previous examples, the next theorem should come as no surprise.

Theorem 9.8: Let S be a nonempty set and let $\mathcal{B}$ be a collection of subsets of S. $\mathcal{B}$ is a basis for a topology on S if and only if $\mathcal{B}$ covers S and $\mathcal{B}$ has the intersection containment property on S.

Note: The set generated by $\mathcal{B}$ is $\{\bigcup X \mid X \subseteq \mathcal{B}\}$. This set can also be written in the alternative form $\{A \subseteq S \mid \forall x \in A \, \exists B \in \mathcal{B}(x \in B \wedge B \subseteq A)\}$. You will be asked to verify that these two sets are equal in Problem 10 below. We will use this alternative form of the set generated by $\mathcal{B}$ in the proof of Theorem 9.8.

Proof of Theorem 9.8: Suppose that $\mathcal{B}$ covers S and $\mathcal{B}$ has the intersection containment property on S. The set generated by $\mathcal{B}$ is $\mathcal{T} = \{A \subseteq S \mid \forall x \in A \, \exists B \in \mathcal{B}(x \in B \wedge B \subseteq A)\}$. Let's check that $\mathcal{T}$ is a topology on S.

Since $A = \emptyset$ vacuously satisfies the condition $\forall x \in A \, \exists B \in \mathcal{B}(x \in B \wedge B \subseteq A)$, we have $\emptyset \in \mathcal{T}$.

To see that $S \in \mathcal{T}$, let $x \in S$. Since $\mathcal{B}$ covers S, there is $B \in \mathcal{B}$ such that $x \in B$ and $B \subseteq S$. So, $S \in \mathcal{T}$.

Let $X \subseteq \mathcal{T}$ and let $x \in \bigcup X$. Then there is $A \in X$ with $x \in A$. Since $X \subseteq \mathcal{T}$, $A \in \mathcal{T}$. So, there is $B \in \mathcal{B}$ such that $x \in B$ and $B \subseteq A$. Since $B \subseteq A$ and $A \subseteq \bigcup X$, $B \subseteq \bigcup X$. It follows that the condition $\forall x \in \bigcup X \, \exists B \in \mathcal{B}(x \in B \wedge B \subseteq \bigcup X)$ is satisfied. So, $\bigcup X \in \mathcal{T}$.

We now prove by induction on $n \in \mathbb{N}$ that for $n \geq 2$, the intersection of n sets in $\mathcal{T}$ is also in $\mathcal{T}$.

Base Case $(n = 2)$: Let $A_1, A_2 \in \mathcal{T}$ and let $x \in A_1 \cap A_2$. Then there are $B_1, B_2 \in \mathcal{B}$ with $x \in B_1$, $x \in B_2$, $B_1 \subseteq A_1$ and $B_2 \subseteq A_2$. Since $x \in B_1$ and $x \in B_2$, $x \in B_1 \cap B_2$. Since $\mathcal{B}$ has the intersection containment property, there is $C \in \mathcal{B}$ such that $x \in C$ and $C \subseteq B_1 \cap B_2$. Since $B_1 \subseteq A_1$ and $B_2 \subseteq A_2$, $C \subseteq A_1 \cap A_2$. Therefore, $A_1 \cap A_2 \in \mathcal{T}$.

Inductive Step: Suppose that the intersection of k sets in $\mathcal{T}$ is always in $\mathcal{T}$. Let $A_1, A_2, \ldots, A_k, A_{k+1} \in \mathcal{T}$. By the inductive hypothesis, $A_1 \cap A_2 \cap \cdots \cap A_k \in \mathcal{T}$. If we let $C = A_1 \cap A_2 \cap \cdots \cap A_k$ and $D = A_{k+1}$, then we have $C, D \in \mathcal{T}$. By the base case, $C \cap D \in \mathcal{T}$. It follows that

$$A_1 \cap A_2 \cap \cdots \cap A_k \cap A_{k+1} = (A_1 \cap A_2 \cap \cdots \cap A_k) \cap A_{k+1} = C \cap D \in \mathcal{T}.$$

Since $\emptyset, S \in \mathcal{T}$, $\mathcal{T}$ is closed under arbitrary unions, and $\mathcal{T}$ is closed under finite intersections, it follows that $\mathcal{T}$ is a topology. By the note following the statement of Theorem 9.8, $\mathcal{B}$ generates $\mathcal{T}$.

Conversely, suppose that $\mathcal{B}$ is a basis for a topology $\mathcal{T}$ on S. Since $\mathcal{T}$ is a topology on S, $S \in \mathcal{T}$. Since $\mathcal{B}$ is a basis for $\mathcal{T}$, $S = \bigcup \mathcal{X}$ for some $\mathcal{X} \subseteq \mathcal{B}$. Let $x \in S$. Then $x \in \bigcup \mathcal{X}$. So, there is $A \in \mathcal{X}$ with $x \in A$. Since $\mathcal{X} \subseteq \mathcal{B}$, $A \in \mathcal{B}$. Since $x \in S$ was arbitrary, $\mathcal{B}$ covers S.

Let $x \in A_1 \cap A_2$, where $A_1, A_2 \in \mathcal{B}$. Then $A_1, A_2 \in \mathcal{T}$, and since $\mathcal{T}$ is a topology on S, $A_1 \cap A_2 \in \mathcal{T}$. Since $\mathcal{B}$ is a basis for $\mathcal{T}$, $A_1 \cap A_2 = \bigcup \mathcal{X}$ for some $\mathcal{X} \subseteq \mathcal{B}$. It follows that $x \in \bigcup \mathcal{X}$. So, there is $C \in \mathcal{X}$ with $x \in C$. Since $\mathcal{X} \subseteq \mathcal{B}$, $C \in \mathcal{B}$. Also, $C \subseteq \bigcup \mathcal{X} = A_1 \cap A_2$. Since $A_1, A_2 \in \mathcal{B}$ and $x \in S$ were arbitrary, $\mathcal{B}$ has the intersection containment property. $\qquad \square$

Example 9.9:

1. If S is any set, then $\mathcal{B} = \{S\}$ is a basis for the trivial topology on S. Note that $\{S\}$ covers S and $\{S\}$ has the intersection containment property on S (there is just one instance to check: $S \cap S = S$ and $S \in \{S\}$).

2. If S is any set, then $\mathcal{B} = \{\{x\} \mid x \in S\}$ is a basis for the discrete topology on S. $\mathcal{B}$ covers S because if $x \in S$, then $\{x\} \in \mathcal{B}$ and $x \in \{x\}$. $\mathcal{B}$ vacuously has the intersection containment property because $\mathcal{B}$ is pairwise disjoint.

3. Let $S = \mathbb{R}$ and let $\mathcal{B} = \{(a, b) \mid a, b \in \mathbb{R} \wedge a < b\}$. We saw in Example 9.7 (part 2) that $\mathcal{B}$ covers $\mathbb{R}$ and that $\mathcal{B}$ has the intersection containment property on $\mathbb{R}$. It follows that $\mathcal{B}$ is a basis for a topology on $\mathbb{R}$. In fact, we already saw in the same Example that $\mathcal{B}$ generates the standard topology on $\mathbb{R}$.

    The basis $\mathcal{B}$ just described is uncountable because $\mathbb{R}$ is uncountable and the function $f \colon \mathbb{R} \to \mathcal{B}$ defined by $f(r) = (r, r + 1)$ is injective. Does $\mathbb{R}$ with the standard topology have a countable basis? In fact, it does! Let $\mathcal{B}' = \{(a, b) \mid a, b \in \mathbb{Q} \wedge a < b\}$. In Problem 12 below you will be asked to show that $\mathcal{B}'$ is countable and that $\mathcal{B}'$ is a basis for $\mathbb{R}$ with the standard topology.

4. We saw in part 3 of Example 9.7 that $\mathcal{X} = \{(-\infty, b) \mid b \in \mathbb{R}\} \cup \{(a, \infty) \mid a \in \mathbb{R}\}$ does **not** have the intersection containment property. It follows from Theorem 9.8 that $\mathcal{X}$ is **not** a basis for a topology on $\mathbb{R}$.

    However, $\mathcal{B}^* = \{(a, \infty) \mid a \in \mathbb{R}\}$ covers $\mathbb{R}$ and has the intersection containment property. Therefore, $\mathcal{B}^*$ is a basis for a topology $\mathcal{T}^*$ on $\mathbb{R}$. Since every set of the form (a, ∞) is open in the standard topology, $\mathcal{T}^*$ is coarser than the standard topology on $\mathbb{R}$. Since no bounded open interval is in $\mathcal{T}^*$, we see that $\mathcal{T}^*$ is **strictly** coarser than the standard topology on $\mathbb{R}$.

Although the set $\mathcal{X} = \{(-\infty, b) \mid b \in \mathbb{R}\} \cup \{(a, \infty) \mid a \in \mathbb{R}\}$ is not a basis for a topology on $\mathbb{R}$, if we let $\mathcal{B}$ be the collection of all finite intersections of sets in $\mathcal{X}$, then $\mathcal{B}$ **does** form a basis for $\mathbb{R}$ (because $\mathcal{X}$ covers $\mathbb{R}$). In this case, we call $\mathcal{X}$ a **subbasis** for the topology generated by $\mathcal{B}$. Since every bounded open interval is in $\mathcal{B}$, it is not hard to see that $\mathcal{B}$ generates the standard topology on $\mathbb{R}$. We can also say that the standard topology on $\mathbb{R}$ is generated by the subbasis $\mathcal{X}$.

Note: If $\mathcal{X}$ is **any** set of sets, then although $\mathcal{X}$ will not necessarily be a basis for a topology, $\mathcal{X}$ will always be a subbasis for a topology on $\bigcup \mathcal{X}$.

Subspaces

Let's begin our discussion of subspaces with a standard example.

Example 9.10: Let $A = [0, 1]$ be the unit closed interval consisting of all real numbers between 0 and 1, inclusive, and let

$$\mathcal{B}_A = \{(a, b) \mid 0 \le a < b \le 1\} \cup \{[0, b) \mid 0 < b \le 1\} \cup \{(a, 1] \mid 0 \le a < 1\} \cup \{A, \emptyset\}.$$

Some of the sets in $\mathcal{B}_A$ are $(0, 1)$, $\left(\frac{1}{3}, \frac{5}{7}\right)$, $\left[0, \frac{1}{2}\right)$, $\left(\frac{1}{4}, 1\right]$, $[0, 1]$, and $\emptyset$.

Since $A \in \mathcal{B}_A$, it follows that $\mathcal{B}_A$ covers A. Notice that even if we removed A from $\mathcal{B}_A$, $\mathcal{B}_A$ would still cover A. To see this, let $x \in A$. If $x = 0$, then $x \in [0, 1)$ and $[0, 1) \in \mathcal{B}_A$. If $x = 1$, then $x \in (0, 1]$ and $(0, 1] \in \mathcal{B}_A$. Otherwise, $0 < x < 1$, and so, $x \in (0, 1)$ and $(0, 1) \in \mathcal{B}_A$.

Furthermore, $\mathcal{B}_A$ has the intersection containment property. In fact, the intersection of any two sets in $\mathcal{B}_A$ is either empty or in $\mathcal{B}_A$. This is easy to check but there are several cases and so, I leave the details to the reader.

By Theorem 9.8, $\mathcal{B}_A$ is a basis for a topology on A.

Note that if we were to remove A and $\emptyset$ from $\mathcal{B}_A$, then $\mathcal{B}_A$ would still be a basis for the same topology. I included these two sets only to simplify the discussion in the first observation below.

Let $\mathcal{T}_A$ be the topology on A generated by $\mathcal{B}_A$. At first glance, the topology $\mathcal{T}_A$ might seem a little strange. Since $\mathcal{T}_A$ is a topology on A, $A = [0, 1]$ itself is open. So, in this topology, we have a closed interval that is an open set. However, $[0, 1]$ is the *only* closed interval that is open in this topology. We also have many half-open intervals that are open in this topology. For example, $\left[0, \frac{1}{2}\right)$ and $\left(\frac{1}{4}, 1\right]$ are both open in this topology.

Observe that if we take an arbitrary bounded open interval I in $\mathbb{R}$, then $I \cap A \in \mathcal{B}_A$. To see this, let $a, b \in \mathbb{R}$ with $a < b$ and let $I = (a, b)$. If $a < 0$ and $b > 1$, then $I \cap A = [0, 1] = A \in \mathcal{B}_A$ (for example, we have $(-3, 5) \cap [0, 1] = [0, 1]$). If $0 \le a < 1$ and $b > 1$, then $I \cap A = (a, 1] \in \mathcal{B}_A$ (for example, we have $(0, 3) \cap [0, 1] = (0, 1]$). If $a < 0$ and $0 < b \le 1$, then $I \cap A = [0, b) \in \mathcal{B}_A$ (for example, we have $\left(-2, \frac{1}{2}\right) \cap [0, 1] = \left[0, \frac{1}{2}\right)$). If $0 \le a < b \le 1$, then $I \cap A = (a, b) \in \mathcal{B}_A$ (for example, we have $\left(\frac{2}{7}, \frac{8}{9}\right) \cap [0, 1] = \left(\frac{2}{7}, \frac{8}{9}\right)$). Finally, if $b \le 0$ or $a \ge 1$, then $I \cap A = \emptyset$ (for example, $(-2, -1) \cap [0, 1] = \emptyset$ and $(1, 7) \cap [0, 1] = \emptyset$).

Conversely, observe that every element of $\mathcal{B}_A$ is equal to $I \cap A$ for some bounded open interval I in $\mathbb{R}$: if $0 \le a < b \le 1$, then $(a, b) = (a, b) \cap [0, 1]$ (for example, we have $(0, 1) = (0, 1) \cap [0, 1]$), if $0 < b \le 1$, then $[0, b) = (-1, b) \cap [0, 1]$ (for example, we have $\left[0, \frac{2}{3}\right) = \left(-1, \frac{2}{3}\right) \cap [0, 1]$), if $0 \le a < 1$, then $(a, 1] = (a, 2) \cap [0, 1]$ (for example, we have $\left(\frac{1}{6}, 1\right] = \left(\frac{1}{6}, 2\right) \cap [0, 1]$), $A = (-1, 2) \cap [0, 1]$, and $\emptyset = (1, 2) \cap [0, 1]$.

Since the bounded open intervals of reals form a basis for the standard topology of $\mathbb{R}$, by the previous observations, the open sets in $\mathcal{T}_A$ are precisely the sets of the form $U \cap A$, where U is some open set in the standard topology of $\mathbb{R}$. To see this, let U be open in $\mathbb{R}$. Then $U = \bigcup \mathcal{X}$ for some set $\mathcal{X}$ of open intervals in $\mathbb{R}$. It follows that $U \cap A = \bigcup \mathcal{X} \cap A = \bigcup\{I \cap A \mid I \in \mathcal{X}\}$. By the first observation above, $U \cap A$ is open in $\mathcal{T}_A$. Conversely, let V be open in $\mathcal{T}_A$. Since $\mathcal{B}_A$ is a basis for $\mathcal{T}_A$, we have $V = \bigcup \mathcal{X}$ for some $\mathcal{X} \subseteq \mathcal{B}_A$. By the second observation above, every element of $\mathcal{X}$ has the form $I \cap A$ for some bounded open interval I in $\mathbb{R}$. If we let $\mathcal{Y} = \{I \mid I \cap A \in \mathcal{X}\}$, we have $V = \bigcup \mathcal{X} = \bigcup \mathcal{Y} \cap A$ (Check this!). Since the bounded open intervals form a basis for the standard topology on $\mathbb{R}$, $\bigcup \mathcal{Y}$ is open in $\mathbb{R}$.

Example 9.10 leads to the following theorem.

Theorem 9.11: If $(S, \mathcal{T})$ is a topological space with basis $\mathcal{B}$ and $A \subseteq S$, then $\mathcal{B}_A = \{U \cap A \mid U \in \mathcal{B}\}$ is a basis for a topology $\mathcal{T}_A$ on A.

Proof: Let $x \in A$. Since $A \subseteq S$, $x \in S$. Since $\mathcal{B}$ is a basis for $(S, \mathcal{T})$, $\mathcal{B}$ covers S, and so, there is $U \in \mathcal{B}$ with $x \in U$. Since we chose x to be in A, we have $x \in U \cap A$. So, x is in a set in $\mathcal{B}_A$. Since $x \in A$ was arbitrary, it follows that $\mathcal{B}_A$ covers A.

Now, let $U, V \in \mathcal{B}$ and let $x \in (U \cap A) \cap (V \cap A)$. Then $x \in A$ and $x \in U \cap V$. Since $\mathcal{B}$ has the intersection containment property, there is $C \in \mathcal{B}$ with $x \in C$ and $C \subseteq U \cap V$. Since $x \in C$ and $x \in A$, we have $x \in C \cap A$. Also, $C \cap A \subseteq (U \cap V) \cap A = (U \cap A) \cap (V \cap A)$. Therefore, $\mathcal{B}_A$ has the intersection containment property.

Since $\mathcal{B}_A$ covers A and $\mathcal{B}_A$ has the intersection containment property, by Theorem 9.8, $\mathcal{B}_A$ is a basis for a topology on A. $\qquad \square$

The topology $\mathcal{T}_A$ given by Theorem 9.11 is called the **subspace topology relative to A**. We will usually just refer to this topology as the **subspace topology**. It should be clear from the context what A is.

Example 9.12: Let $\mathcal{T}$ be the standard topology on $\mathbb{R}$ and let $\mathcal{B} = \{(a, b) \mid a, b \in \mathbb{R} \wedge a < b\}$ be the set of open intervals with endpoints in $\mathbb{R}$. We saw in part 2 of Example 9.7 that $\mathcal{B}$ is a basis for $\mathbb{R}$. If we let $A = [0, 1]$, then by Theorem 9.11, $\mathcal{B}_A = \{(a, b) \cap [0, 1] \mid a, b \in \mathbb{R} \wedge a < b\}$ is a basis for the subspace topology $([0, 1], \mathcal{T}_{[0,1]})$. By the discussion given in Example 9.10, $\mathcal{B}_A$ can also be described as follows:

$$\mathcal{B}_A = \{(a, b) \mid 0 \leq a < b \leq 1\} \cup \{[0, b) \mid 0 < b \leq 1\} \cup \{(a, 1] \mid 0 \leq a < 1\}$$

Notice that this time we left out $[0, 1]$ and $\emptyset$ from the basis. Recall that those were included in Example 9.10 to make the discussion easier to follow, but they weren't necessary.

Theorem 9.13: Let $(S, \mathcal{T})$ be a topological space and let $A \subseteq S$. Then $\mathcal{T}_A = \{U \cap A \mid U \in \mathcal{T}\}$.

Note that this theorem implies that if $\mathcal{B}$ and $\mathcal{B}'$ are bases for the topology $\mathcal{T}$, then $\mathcal{B}_A$ and $\mathcal{B}'_A$ generate the same topology. Therefore, for $A \subseteq S$, the subspace topology $\mathcal{T}_A$ is well-defined.

Proof: Let $\mathcal{B}$ be a basis for $(S, \mathcal{T})$, let $\mathcal{B}_A = \{U \cap A \mid U \in \mathcal{B}\}$, let $\mathcal{T}_A$ be the subspace topology (the topology generated by $\mathcal{B}_A$), and let $\mathcal{U} = \{U \cap A \mid U \in \mathcal{T}\}$. We need to show that $\mathcal{T}_A = \mathcal{U}$.

First, let $V \in \mathcal{T}_A$. Then $V = \bigcup \mathcal{X}$ for some $\mathcal{X} \subseteq \mathcal{B}_A$. Let $Y = \bigcup \{U \mid U \cap A \in \mathcal{X}\}$. Since Y is a union of sets in $\mathcal{B}$, $Y \in \mathcal{T}$. Now, $V = \bigcup \mathcal{X} = \bigcup \{U \cap A \mid U \cap A \in \mathcal{X}\} = \bigcup \{U \mid U \cap A \in \mathcal{X}\} \cap A = Y \cap A \in \mathcal{U}$. Therefore, $\mathcal{T}_A \subseteq \mathcal{U}$.

Next, let $V \in \mathcal{U}$, say $V = U \cap A$ with $U \in \mathcal{T}$. Since $\mathcal{B}$ is a basis for $\mathcal{T}$, $U = \bigcup \mathcal{X}$ for some $\mathcal{X} \subseteq \mathcal{B}$. Therefore, $V = \bigcup \mathcal{X} \cap A = \bigcup \{W \mid W \in \mathcal{X}\} \cap A = \bigcup \{W \cap A \mid W \in \mathcal{X}\}$. This is a union of elements of $\mathcal{B}_A$, and so, $V \in \mathcal{T}_A$. Therefore, $\mathcal{U} \subseteq \mathcal{T}_A$.

Since $\mathcal{T}_A \subseteq \mathcal{U}$ and $\mathcal{U} \subseteq \mathcal{T}_A$, we have $\mathcal{T}_A = \mathcal{U}$. $\qquad\square$

Example 9.14: Let $\mathcal{T}$ be the standard topology on $\mathbb{R}$.

1. The subspace topology $\mathcal{T}_{\mathbb{Z}}$ is the discrete topology on $\mathbb{Z}$. To see this, just observe that if $n \in \mathbb{Z}$, then $\{n\} = (n-1, n+1) \cap \mathbb{Z}$, and so, every singleton set $\{n\}$ is open in $\mathcal{T}_{\mathbb{Z}}$.

2. Let's take a look at $\mathbb{Q}$ as a subspace of $\mathbb{R}$. Any set of the form $(a, b) \cap \mathbb{Q}$ with $a, b \in \mathbb{R}$ is open in $\mathbb{Q}$ by the definition of the subspace topology. If $a, b \in \mathbb{R} \setminus \mathbb{Q}$, then $(a, b) \cap \mathbb{Q}$ is also closed in $\mathbb{Q}$ because in this case, $\mathbb{Q} \setminus ((a, b) \cap \mathbb{Q}) = ((-\infty, a) \cap \mathbb{Q}) \cup ((b, \infty) \cap \mathbb{Q})$, a union of two sets that are each open in $\mathbb{Q}$. So, as a subspace of $\mathbb{R}$, $\mathbb{Q}$ has uncountably many sets that are both open and closed.

   Is every subset of $\mathbb{Q}$ open in the subspace topology?

   Well, if $a, b \in \mathbb{Q}$, then $[a, b] \cap \mathbb{Q}$ is **not** open in $\mathbb{Q}$. To see this, suppose toward contradiction that $[a, b] \cap \mathbb{Q} = U \cap \mathbb{Q}$ for some open set U in $\mathbb{R}$. Then $a \in U$. Since U is open in $\mathbb{R}$, there is an open interval (c, d) containing a with $(c, d) \subseteq U$. By the Density Theorem, there is a rational number q with $c < q < a$. Then $q \in U \cap \mathbb{Q}$, but $q \notin [a, b] \cap \mathbb{Q}$. So, $[a, b] \cap \mathbb{Q} \neq U \cap \mathbb{Q}$, a contradiction.

   Since we have found sets that are not open in $\mathbb{Q}$, it follows that the subspace topology $\mathcal{T}_{\mathbb{Q}}$ is **not** the discrete topology.

3. Consider $[0, 1] \cup [9, 10]$ as a subspace of $\mathbb{R}$. $[0, 1]$ is open in the subspace topology because $(-1, 2) \cap ([0, 1] \cup [9, 10]) = [0, 1]$. Similarly, $[9, 10]$ is open in the subspace topology. Since $[0, 1]$ and $[9, 10]$ are complements of each other, they are also both closed in the subspace topology.

4. Consider $\{0\} \cup (1, 2]$ as a subspace of $\mathbb{R}$. $\{0\}$ is open in the subspace topology because $(-1, 1) \cap (\{0\} \cup (1, 2]) = \{0\}$. Also, $(1, 2]$ is open in the subspace topology because $(1, 3) \cap (\{0\} \cup (1, 2]) = (1, 2]$. Since $\{0\}$ and $(1, 2]$ are complements of each other, they are also both closed in the subspace topology.

5. Let A be a closed disk in $\mathbb{C}$, let B be an open disk in $\mathbb{C}$, let $z \in \mathbb{C}$, and suppose that A, B, and $\{z\}$ are pairwise disjoint. Consider $X = A \cup B \cup \{z\}$ as a subspace of $\mathbb{C}$. Then A, B, and $\{z\}$ are each both open and closed in the subspace topology.

Products

Recall from Lesson 3 that the Cartesian product of two sets S_1 and S_2 consists of all ordered pairs (x, y) with $x \in S_1$ and $y \in S_2$. Symbolically, we have the following:

$$S_1 \times S_2 = \{(x, y) \mid x \in S_1 \land y \in S_2\}.$$

Theorem 9.15: Suppose that $\mathcal{T}_1$ is a topology on S_1 and that $\mathcal{T}_2$ is a topology on S_2. Then $\mathcal{B} = \{U \times V \mid U \in \mathcal{T}_1 \land V \in \mathcal{T}_2\}$ is a basis for a topology $\mathcal{T}$ on $S_1 \times S_2$.

Proof: Since $S_1 \in \mathcal{T}_1$ and $S_2 \in \mathcal{T}_2$, $S_1 \times S_2 \in \mathcal{B}$. So, $\mathcal{B}$ covers $S_1 \times S_2$.

Now, let $(x, y) \in (U_1 \times V_1) \cap (U_2 \times V_2)$, where $U_1 \times V_1, U_2 \times V_2 \in \mathcal{B}$. Since $U_1, U_2 \in \mathcal{T}_1$, $U_1 \cap U_2 \in \mathcal{T}_1$. Since $V_1, V_2 \in \mathcal{T}_2$, $V_1 \cap V_2 \in \mathcal{T}_2$. Therefore, $(U_1 \cap U_2) \times (V_1 \cap V_2) \in \mathcal{B}$. By part (i) of Problem 13 from Problem Set 3, $(U_1 \times V_1) \cap (U_2 \times V_2) = (U_1 \cap U_2) \times (V_1 \cap V_2)$. Thus, $(U_1 \times V_1) \cap (U_2 \times V_2) \in \mathcal{B}$. Therefore, $\mathcal{B}$ has the intersection containment property.

Since $\mathcal{B}$ covers $S_1 \times S_2$ and $\mathcal{B}$ has the intersection containment property, it follows that $\mathcal{B}$ is a basis for a topology on $S_1 \times S_2$. $\qquad\square$

The topology $\mathcal{T}$ that we get from Theorem 9.15 is called the **product topology** on $S_1 \times S_2$.

Notes: (1) The set $\mathcal{X} = \{U \times S_2 \mid U \in \mathcal{T}_1\} \cup \{S_1 \times V \mid V \in \mathcal{T}_2\}$ is a **subbasis** for the product topology on $S_1 \times S_2$. To see this, simply observe that if $U \in \mathcal{T}_1$ and $V \in \mathcal{T}_2$, then $U \times V = (U \times S_2) \cap (S_1 \times V)$, $U \times S_2 \in \mathcal{X}$, and $S_1 \times V \in \mathcal{X}$. The image below on the left provides a visualization of $U \times V$ in the $S_1 \times S_2$ plane. In this particular visualization, we are "imagining" that U and V are open intervals parallel to the S_1 and S_2-axes, respectively. Then $U \times S_2$ can be represented by an open vertical strip and $S_1 \times V$ can be represented by an open horizontal strip. $U \times V$ is then represented by the open rectangle that is the intersection of the two strips.

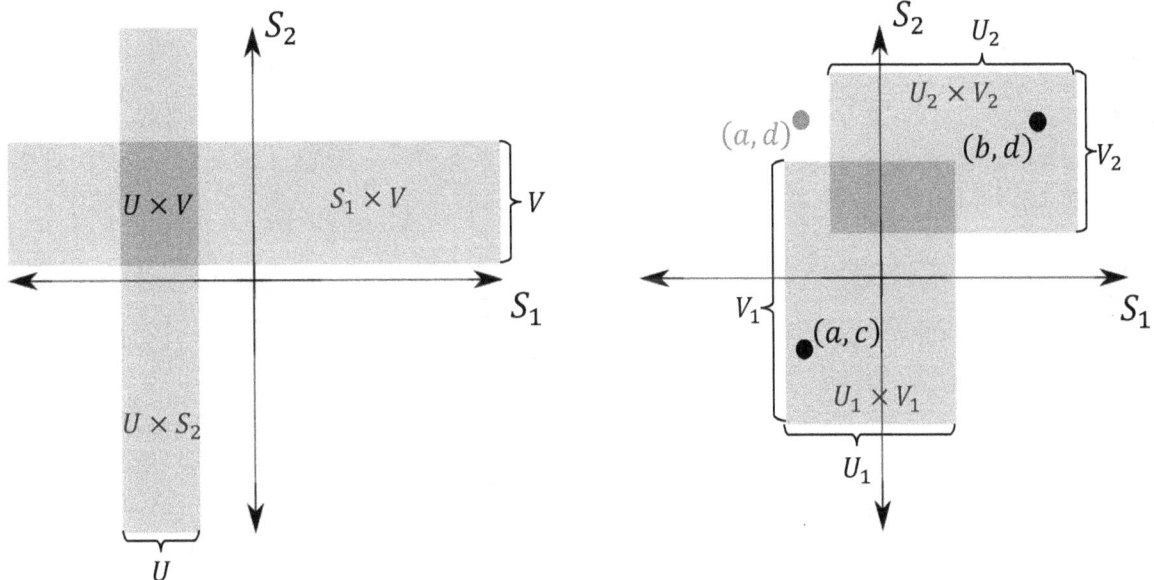

(2) $\mathcal{B} = \{U \times V \mid U \in \mathcal{T}_1 \wedge V \in \mathcal{T}_2\}$ is usually **not** a topology on $S_1 \times S_2$. To see this, let S_1 and S_2 be sets, each with at least two elements and let $\mathcal{T}$ be the topology generated by $\mathcal{B}$. Let a and b be distinct elements in S_1, let c and d be distinct elements in S_2, let $U_1, U_2 \in \mathcal{T}_1$ with $a \in U_1, b \notin U_1, a \notin U_2$, $b \in U_2$ and let $V_1, V_2 \in \mathcal{T}_2$ with $c \in V_1, d \notin V_1, c \notin V_2, d \in V_2$. Then the set $X = (U_1 \times V_1) \cup (U_2 \times V_2)$ is in $\mathcal{T}$. However, $X \neq U \times V$ for any $U \in \mathcal{T}_1$ and $V \in \mathcal{T}_2$ because $(a, c) \in X, (b, d) \in X$, but $(a, d) \notin X$. The image above on the right provides a visualization of this situation.

Since topologies are often defined in terms of bases, it would be nice if we could define a basis for the product topology on $S_1 \times S_2$ in terms of bases for the topologies on S_1 and S_2. Luckily, this is easy to do, as we now show.

Theorem 9.16: Suppose that $\mathcal{T}_1$ is a topology on S_1 with basis $\mathcal{B}_1$ and that $\mathcal{T}_2$ is a topology on S_2 with basis $\mathcal{B}_2$. Then $\mathcal{C} = \{U \times V \mid U \in \mathcal{B}_1 \wedge V \in \mathcal{B}_2\}$ is a basis for the product topology $\mathcal{T}$ on $S_1 \times S_2$.

Proof: Let $\mathcal{B} = \{U \times V \mid U \in \mathcal{T}_1 \wedge V \in \mathcal{T}_2\}$. By Theorem 9.15, $\mathcal{B}$ is a basis for the product topology.

Since $\mathcal{B}_1 \subseteq \mathcal{T}_1$ and $\mathcal{B}_2 \subseteq \mathcal{T}_2$, $\mathcal{C} \subseteq \mathcal{B}$. Therefore, the set generated by $\mathcal{C}$ is contained in the set generated by $\mathcal{B}$, which is $\mathcal{T}$.

Now, let $U \times V \in \mathcal{B}$ and let $(x, y) \in U \times V$. Then $x \in U$ and $y \in V$. Since $U \in \mathcal{T}_1$, there is $U_x \in \mathcal{B}_1$ with $U_x \subseteq U$. Similarly, since $V \in \mathcal{T}_2$, there is $V_y \in \mathcal{B}_2$ with $V_y \subseteq V$. Let $W = \cup\{U_x \times V_y \mid x \in U \wedge y \in V\}$. We will show that $U \times V = W$. First, let $(x, y) \in U \times V$. Then $x \in U_x$ and $y \in V_y$. So, $(x, y) \in U_x \times V_y$. Therefore, $(x, y) \in W$. Since $(x, y) \in U \times V$ was arbitrary, $U \times V \subseteq W$. For the reverse inclusion, observe that by construction, for each $x \in U$ and $y \in V$, we have $U_x \times V_y \subseteq U \times V$. It follows that $W \subseteq U \times V$. Since $U \times V \subseteq W$ and $W \subseteq U \times V$, we have $U \times V = W$. This shows that every set in $\mathcal{B}$ is a union of sets in $\mathcal{C}$. Therefore, the set generated by $\mathcal{B}$ (which is $\mathcal{T}$) is contained in the set generated by $\mathcal{C}$.

Since the set generated by $\mathcal{C}$ is contained in $\mathcal{T}$ and $\mathcal{T}$ is contained in the set generated by $\mathcal{C}$, we see that $\mathcal{C}$ generates $\mathcal{T}$. $\qquad \square$

Note: The set $\mathcal{X} = \{U \times S_2 \mid U \in \mathcal{B}_1\} \cup \{S_1 \times V \mid V \in \mathcal{B}_2\}$ is a **subbasis** for the product topology on $S_1 \times S_2$. To see this, simply observe that if $U \in \mathcal{B}_1$ and $V \in \mathcal{B}_2$, then $U \times V = (U \times S_2) \cap (S_1 \times V)$, $U \times S_2 \in \mathcal{X}$, and $S_1 \times V \in \mathcal{X}$.

Example 9.17:

1. Let $S = \{a\}$ and $\mathcal{T} = \{\emptyset, \{a\}\}$. The product topology on $S^2 = S \times S$ is $\mathcal{U} = \{\emptyset, \{(a, a)\}\}$. This is the only topology on $\{a\} \times \{a\}$.

2. Let $S = \{a, b\}$ and $\mathcal{T}_2 = \{\emptyset, \{a\}, \{a, b\}\}$ (see part 2 of Example 9.1 above). We have

   $\emptyset \times \emptyset = \emptyset$ (and the product of each other set with $\emptyset$ is $\emptyset$ as well)

   $\{a\} \times \{a\} = \{(a, a)\}$ $\{a\} \times \{a, b\} = \{(a, a), (a, b)\}$

   $\{a, b\} \times \{a\} = \{(a, a), (b, a)\}$ $\{a, b\} \times \{a, b\} = \{(a, a), (a, b), (b, a), (b, b)\}$

Each of these five sets is open in the product topology on S^2. There is one more set that is open in the product topology. It is $\{(a,a),(a,b)\} \cup \{(a,a),(b,a)\} = \{(a,a),(a,b),(b,a)\}$.

3. Let $S = \mathbb{R}$ and let $\mathcal{T}$ be the standard topology on $\mathbb{R}$. Since the set of all bounded open intervals of the form (a,b) forms a basis for $(\mathbb{R}, \mathcal{T})$, by Theorem 9.16, a basis for the product topology on $\mathbb{R}^2 = \mathbb{R} \times \mathbb{R}$ is $\mathcal{B} = \{(a,b) \times (c,d) \mid a,b,c,d \in \mathbb{R}\}$. This is the set of all bounded open rectangles, as defined in Lesson 8.

It's worth noting that another basis for the product topology on $\mathbb{R}$ is the set of all bounded open disks $\mathcal{C} = \{N_r(a,b) \mid a,b \in \mathbb{R} \wedge r \in \mathbb{R}^+\}$, where $N_r(a,b)$ is the open disk (or ball) with center (a,b) and radius r. That is, $N_r(a,b) = \{(x,y) \in \mathbb{R}^2 \mid \sqrt{(x-a)^2 + (y-b)^2} < r\}$. It is not hard to show that if $(x,y) \in R$ for some bounded open rectangle R, then there is an open disk D containing (x,y) with $D \subseteq R$. It is also not hard to show that if $(x,y) \in D$ for some open disk D, then there is a bounded open rectangle R containing (x,y) with $R \subseteq D$ (see pictures below). It follows that the bounded open rectangles and the open disks generate the same topology. The dedicated reader may want to write out a detailed proof.

We would now like to generalize to products of more than two sets.

For three sets S_1, S_2, and S_2, we have the following Cartesian product (see Lesson 3):

$$S_1 \times S_2 \times S_3 = \{(x,y,z) \mid x \in S_1 \wedge y \in S_2 \wedge z \in S_3\}.$$

Assuming that $\mathcal{T}_1$, $\mathcal{T}_2$, and $\mathcal{T}_3$ are topologies on S_1, S_2, and S_3, respectively, then we have that $\mathcal{B} = \{U \times V \times W \mid U \in \mathcal{T}_1 \wedge V \in \mathcal{T}_2 \wedge W \in \mathcal{T}_3\}$ is a basis for a topology $\mathcal{T}$ on $S_1 \times S_2 \times S_3$. Once again, we call this topology the **product topology** on $S_1 \times S_2 \times S_3$.

Note: If $\mathcal{B}_1$, $\mathcal{B}_2$ and $\mathcal{B}_3$ are bases for $\mathcal{T}_1$, $\mathcal{T}_2$, and $\mathcal{T}_3$, respectively, then we can replace $\mathcal{T}_1$, $\mathcal{T}_2$, and $\mathcal{T}_3$ by $\mathcal{B}_1$, $\mathcal{B}_2$ and $\mathcal{B}_3$ in the definition of $\mathcal{B}$ above.

More generally, for sets $S_1, S_2,...,S_n$ with topologies $\mathcal{T}_1, \mathcal{T}_2,...,\mathcal{T}_n$, respectively, the following is a basis for the **product topology** on $S_1 \times S_2 \times \cdots \times S_n$:

$$\mathcal{B} = \{U_1 \times U_2 \times \cdots \times U_n \mid U_1 \in \mathcal{T}_1 \wedge U_2 \in \mathcal{T}_2 \wedge \cdots \wedge U_n \in \mathcal{T}_n\}$$

Furthermore, if $\mathcal{B}_1, \mathcal{B}_2,...,\mathcal{B}_n$ are bases for $\mathcal{T}_1, \mathcal{T}_2,...,\mathcal{T}_n$, respectively, then the following is also a basis for the **product topology** on $S_1 \times S_2 \times \cdots \times S_n$:

$$\mathcal{B} = \{U_1 \times U_2 \times \cdots \times U_n \mid U_1 \in \mathcal{B}_1 \wedge U_2 \in \mathcal{B}_2 \wedge \cdots \wedge U_n \in \mathcal{B}_n\}$$

Example 9.18:

1. Let $S_n = \{a_n\}$ and $\mathcal{T}_n = \{\emptyset, \{a_n\}\}$ for $n = 1, 2, 3, 4$. The product topology on $S_1 \times S_2 \times S_3 \times S_4$ is $\mathcal{U} = \{\emptyset, \{(a_1, a_2, a_3, a_4)\}\}$. This is the only topology on $\{a_1\} \times \{a_2\} \times \{a_3\} \times \{a_4\}$.

2. If $S_1, S_2,..., S_n$ are each given the discrete topology, then the product topology on $S_1 \times \cdots \times S_n$ is also discrete. To see this, simply observe that if $a_1 \in S_1, a_2 \in S_2,..., a_n \in S_n$, then each of the sets $\{a_1\}, \{a_2\},..., \{a_n\}$ are open in $S_1, S_2,...,S_n$, respectively. So, $\{(a_1, a_2, ..., a_n)\}$ is open in the product topology on $S_1 \times \cdots \times S_n$.

3. Let $S = \mathbb{R}$ and let $\mathcal{T}$ be the standard topology on $\mathbb{R}$. A basis for the product topology on $\mathbb{R}^n$ is $\mathcal{B} = \{(a_1, b_1) \times (a_2, b_2) \times \cdots \times (a_n, b_n) \mid a_1, b_1, a_2, b_2, ..., a_n, b_n \in \mathbb{R}\}$. If $n = 2$, this is the set of all bounded open rectangles. If $n = 3$, this is the set of all bounded open rectangular boxes. If $n > 3$, we call an element of $\mathcal{B}$ a "hyperrectangle." Although we cannot draw an accurate visualization of a hyperrectangle, when proving theorems, it can be useful to visualize these hyperrectangles as ordinary rectangles in the plane.

   As in part 3 of Example 9.17, it's worth noting that another basis for the product topology on $\mathbb{R}^n$ is the set of all open balls $\mathcal{C} = \{B_r(\mathbf{x}) \mid \mathbf{x} \in \mathbb{R}^n \wedge r \in \mathbb{R}^+\}$. Here $B_r(\mathbf{x})$ is the open ball with center $\mathbf{x}$ and radius r. That is, $B_r(\mathbf{x}) = \{\mathbf{y} \in \mathbb{R}^n \mid |\mathbf{y} - \mathbf{x}| < r\}$. I leave the details to the reader.

Let's generalize these definitions even further. Let K be a set and for each $k \in K$, let S_k be a set. Also, let $S = \bigcup\{S_k \mid k \in K\}$. The **Cartesian product** of $\{S_k \mid k \in K\}$ can be expressed as follows:

$$\prod_{k \in K} S_k = \{f : K \to S \mid \forall k \in K(f(k) \in S_k)\}$$

The set K is called an **index set** and we say that $\{S_k \mid k \in K\}$ is indexed by K. So, an element of the Cartesian product of $\{S_k \mid k \in K\}$ is a function whose domain is the index set K with $f(k) \in S_k$ for each $k \in K$. We will often write $s_k = f(k)$, and we may write an element f in the Cartesian product as $(s_k)_{k \in K}$. We call S_k the **kth coordinate set** and we call $f(k) = s_k$ the **kth coordinate** of $f = (s_k)_{k \in K}$.

We may abbreviate the cartesian product as $\prod S_k$ if the indexing set is clear or if it is unnecessary to know the indexing set for the discussion.

Example 9.19:

1. Let $K = \{0, 1\}$, let $S_0 = \{a, b\}$, and let $S_1 = \{c, d, e\}$. Then we have

$$\prod_{k \in K} S_k = \{f : \{0, 1\} \to S_0 \cup S_1 \mid f(0) \in S_0 \wedge f(1) \in S_1\} = \{(s_0, s_1) \mid s_0 \in S_0 \wedge s_1 \in S_1\}$$

$$= S_0 \times S_1 = \{a, b\} \times \{c, d, e\} = \{(a, c), (a, d), (a, e), (b, c), (b, d), (b, e)\}.$$

So, we see that if the index set K contains just two elements, then the general definition of the Cartesian product is the same as our original definition of the Cartesian product of two sets.

2. More generally, if $K = \{0, 1, \ldots, n-1\}$, then

$$\prod_{k \in K} S_k = \{f : \{0, 1, \ldots, n-1\} \to S_0 \cup S_1 \cup \cdots \cup S_{n-1} \mid f(0) \in S_0 \wedge \cdots \wedge f(n-1) \in S_{n-1}\}$$

$$= \{(s_0, s_1, \ldots, s_{n-1}) \mid s_0 \in S_0 \wedge s_1 \in S_1 \wedge \cdots \wedge s_{n-1} \in S_{n-1}\} = S_0 \times S_1 \times \cdots \times S_{n-1}.$$

3. If $K = \mathbb{N}$, then we can visualize the corresponding Cartesian product as follows:

$$\prod_{k \in K} S_k = \{f : \mathbb{N} \to \cup\{S_n \mid n \in \mathbb{N}\} \mid \forall n \in \mathbb{N}(f(n) \in S_n)\}$$

$$= \{(s_0, s_1, \ldots, s_n, \ldots) \mid \forall n \in \mathbb{N}(s_n \in S_n)\} = S_0 \times S_1 \times \cdots \times S_n \times \cdots.$$

If each S_k is the same set, say $S_k = A$ for all $k \in \mathbb{N}$, then

$$\prod_{k \in K} S_k = {}^{\mathbb{N}}A$$

Recall from Lesson 4 that ${}^{\mathbb{N}}A$ is the set of functions from $\mathbb{N}$ to A, or equivalently, the set of infinite sequences with codomain A. We can visualize a typical element of ${}^{\mathbb{N}}A$ as

$$(a_0, a_1, a_2, a_3, \ldots).$$

4. More generally, if K is any index set and $S_k = A$ for all $k \in K$, then

$$\prod_{k \in K} S_k = {}^{K}A$$

Once again, recall from Lesson 4 that ${}^{K}A$ is the set of functions from K to A. If K is uncountable, we cannot give a visualization that is as nice as in the previous examples. However, we can mimic the visualization by writing a general element of ${}^{K}A$ as $(a_k)_{k \in K}$.

As a specific example, let $K = \mathbb{R}$ and $A = \{0, 1\}$. A typical element of ${}^{\mathbb{R}}\{0, 1\}$ has the form $(a_k)_{k \in K}$, where for each $k \in \mathbb{R}$, $a_k = 0$ or $a_k = 1$. If we let $a_k = 1$ for all $k \in \mathbb{R}$, then $(a_k)_{k \in K} = (1)_{k \in K}$ is an element of ${}^{\mathbb{R}}\{0, 1\}$ that is a constant function. A more complicated element of ${}^{\mathbb{R}}\{0, 1\}$ would be the function $(a_k)_{k \in K}$, where $a_k = 0$ if $k \in \mathbb{Q}$ (k is rational) and $a_k = 1$ if $k \in \mathbb{R} \setminus \mathbb{Q}$ (k is irrational).

We now define the **product topology** on a general Cartesian product via the following Theorem.

Theorem 9.20: Let K be an index set and let $(S_k, \mathcal{T}_k)$ be a topological space for each $k \in K$. Then $\mathcal{B} = \{\prod U_k \mid \forall k \in K(U_k \in \mathcal{T}_k) \wedge (U_k = S_k \text{ for all but finitely many } k)\}$ is a basis for a topology $\mathcal{T}$ on $\prod S_k$.

You will be asked to prove Theorem 9.20 in Problem 15 below.

The topology $\mathcal{T}$ that we get from Theorem 9.20 is called the **product topology** on $\prod S_k$.

Notes: (1) The set $\mathcal{X} = \{\prod U_k \mid U_k \in \mathcal{T}_K \wedge (U_k = S_k \text{ for all but one } k)\}$ is a subbasis for the product topology on $\prod S_k$.

(2) $\mathcal{C} = \{\prod U_k \mid \forall k \in K(U_k \in \mathcal{B}_k) \wedge (U_k = S_k \text{ for all but finitely many } k)\}$, where $\mathcal{B}_k$ is a basis for $\mathcal{T}_k$ for each $k \in K$, is also a basis for the product topology on $\prod S_k$.

(3) If we remove "$U_k = S_k$ for all but finitely many k" from the definition of $\mathcal{B}$ in Theorem 9.20, we get a topology finer than the product topology called the **box topology** (you will be asked to prove this in Problem 15 below). If the index set K is finite, then clearly the product topology and box topology are equal. If the index set K is infinite, then the box topology is *strictly* finer than the product topology. The box topology isn't as well-behaved as the product topology, and so, we do not emphasize it here.

Example 9.21:

1. Let $K = \mathbb{N}$ and for each $k \in \mathbb{N}$, let $S_k = \{0, 1\}$ be given the discrete topology. Then the product topology on $\prod S_k = {}^{\mathbb{N}}\{0, 1\}$ is **not** discrete. In fact, no set with a single element is open in ${}^{\mathbb{N}}\{0, 1\}$. To see this, let $\mathcal{B}$ be the basis for the product topology given in Theorem 9.20. Then any basis element $\prod U_k$ has $U_k = \{0, 1\}$ for infinitely many k. Therefore, if $(x_k)_{k \in \mathbb{N}}$ is in a basis element $\prod U_k$, we can choose some $m \in \mathbb{N}$ such that $U_m = \{0, 1\}$. Let $y_m = 1 - x_m$. Then there is $(z_k)_{k \in \mathbb{N}} \in \prod U_k$ with $z_m = y_m$. Since $y_m \neq x_m$, it follows that $(z_k)_{k \in \mathbb{N}} \neq (x_k)_{k \in \mathbb{N}}$. Thus, every open set contains more than one element. In particular, $\{(x_k)_{k \in \mathbb{N}}\}$ is not open.

2. Let $K = \mathbb{N}$ and for each $k \in \mathbb{N}$, let $S_k = \mathbb{R}$ be given the standard topology. Then the product ${}^{\mathbb{N}}(-1, 1)$ is **not** open in the product topology. To see this, we will show that the point $(0)_{k \in \mathbb{N}} \in {}^{\mathbb{N}}(-1, 1)$ but $(0)_{k \in \mathbb{N}}$ has no basic open neighborhood $\prod U_k$ with $\prod U_k \subseteq {}^{\mathbb{N}}(-1, 1)$. That $(0)_{k \in \mathbb{N}} \in {}^{\mathbb{N}}(-1, 1)$ is clear. Let $\prod U_k$ be an open neighborhood with $(0)_{k \in \mathbb{N}} \in \prod U_k$. Since $U_k = \mathbb{R}$ for all but finitely many k, we can choose some $m \in \mathbb{N}$ such that $U_m = \mathbb{R}$. So, $2 \in U_m$. It follows that there is $(x_k)_{k \in \mathbb{N}} \in \prod U_k$ with $x_m = 2$. Therefore, $(x_k)_{k \in \mathbb{N}} \notin {}^{\mathbb{N}}(-1, 1)$. It follows that $\prod U_k \nsubseteq {}^{\mathbb{N}}(-1, 1)$.

Problem Set 9

Full solutions to these problems are available for free download here:

www.SATPrepGet800.com/TFBZLF

LEVEL 1

1. Let $\mathcal{T}$ and $\mathcal{U}$ be topologies on a set S. Prove that $\mathcal{T} \cap \mathcal{U}$ is a topology on S.

2. Let $(S, \mathcal{T})$ be a topological space and let A be a subspace of S.

   (i) Provide an example to show that if B is open in A in the subspace topology, then B need not be open in S.

   (ii) Provide an example to show that if B is closed in A in the subspace topology, then B need not be closed in S.

LEVEL 2

3. Let $K = \left\{ \frac{1}{n} \mid n \in \mathbb{Z}^+ \right\}$, $\mathcal{B} = \{(a, b) \mid a, b \in \mathbb{R} \wedge a < b\} \cup \{(a, b) \setminus K \mid a, b \in \mathbb{R} \wedge a < b\}$. Prove that $\mathcal{B}$ is a basis for a topology $\mathcal{T}_K$ on $\mathbb{R}$.

4. Prove that $\mathcal{B} = \{X \subseteq \mathbb{R} \mid \mathbb{R} \setminus X \text{ is finite}\}$ generates a topology $\mathcal{T}$ on $\mathbb{R}$ that is strictly coarser than the standard topology. $\mathcal{T}$ is called the **cofinite topology** on $\mathbb{R}$.

5. Let S be a nonempty set and let $\mathcal{T} = \{X \subseteq S \mid S \setminus X \text{ is countable}\} \cup \{\emptyset\}$. Is $\mathcal{T}$ a topology on S?

6. Let $\mathcal{X}$ be a set of topologies on a set S. Prove that $\cap \mathcal{X}$ is a topology on S.

LEVEL 3

7. Let $(S, \mathcal{T})$ be a topological space and let A be a subspace of S.

   (i) Prove that if B is open in A in the subspace topology and A is open in S, then B is open in S.

   (ii) Prove that B is closed in A if and only if $B = C \cap A$ for some closed set C of S.

   (iii) Prove that if B is closed in A in the subspace topology and A is closed in S, then B is closed in S.

8. Let $(S, \mathcal{T})$ be a topological space and let X and Y be subsets of S. Prove that $\overline{X \cup Y} = \overline{X} \cup \overline{Y}$.

9. Let $(S, \mathcal{T})$ be a topological space and let $A \subseteq B$ with A and B be subspaces of S. Prove that the subspace topology of A with respect to S is the same as the subspace topology of A with respect to B.

10. Let S be a nonempty set and let $\mathcal{B}$ be a collection of subsets of S. Prove that the set generated by $\mathcal{B}$, $\{\cup X \mid X \subseteq \mathcal{B}\}$, is equal to $\{A \subseteq S \mid \forall x \in A \, \exists B \in \mathcal{B}(x \in B \wedge B \subseteq A)\}$.

LEVEL 4

11. Let $(S, \mathcal{T})$ be a topological space. Prove that an arbitrary intersection of closed sets in S is a closed set in S and a finite union of closed sets in S is a closed set in S.

12. Let $\mathcal{B}' = \{(a, b) \mid a, b \in \mathbb{Q} \wedge a < b\}$. Prove that $\mathcal{B}'$ is countable and that $\mathcal{B}'$ is a basis for a topology on $\mathbb{R}$. Then show that the topology generated by $\mathcal{B}'$ is the standard topology on $\mathbb{R}$.

13. Let $(S, \mathcal{T})$ be a topological space, let A be a subspace of S, let B be a subset of A, and let $\overline{B}$ be the closure of B in S. Prove that the closure of B in A is $\overline{B} \cap A$.

14. Let T_L be the set generated by the half open intervals of the form $[a, b)$ with $a, b \in \mathbb{R}$. Show that T_L is a topology on $\mathbb{R}$ that is strictly finer than the standard topology on $\mathbb{R}$ and incomparable with the topology $\mathcal{T}_K$ from Problem 3 above.

LEVEL 5

15. Let K be an index set and let $(S_k, \mathcal{T}_k)$ be a topological space for each $k \in K$. Prove that the collection $\mathcal{B} = \{\prod U_k \mid \forall k \in K(U_k \in \mathcal{T}_k) \wedge (U_k = S_k \text{ for all but finitely many } k)\}$ is a basis for a topology $\mathcal{T}$ on $\prod S_k$. Furthermore, if $\mathcal{B}_k$ is a basis for $\mathcal{T}_k$ for each $k \in K$, then the collection $\mathcal{C} = \{\prod U_k \mid \forall k \in K(U_k \in \mathcal{B}_k) \wedge (U_k = S_k \text{ for all but finitely many } k)\}$ is a basis for the same topology (this topology is called the **product topology** on $\prod S_k$). Show that in both cases, if we remove "$U_k = S_k$ for all but finitely many k," we get a topology (called the **box topology**) that is finer than the product topology.

16. Let K be an index set, let $(S_k, \mathcal{T}_k)$ be a topological space for each $k \in K$, and let $A_k \subseteq S_k$ for each $k \in K$. Prove that $\overline{\prod A_k} = \prod \overline{A_k}$.

17. Let J and K be index sets. Prove that $\cap\{\prod_{k \in K} B_k^j \mid j \in J\} = \prod_{k \in K}(\cap\{B_k^j \mid j \in J\})$.

CHALLENGE PROBLEM

18. Consider $\mathbb{R}$ with the standard topology. Define $f: \mathcal{P}(\mathbb{R}) \to \mathcal{P}(\mathbb{R})$ by $f(A) = \mathbb{R} \setminus A$ and define $g: \mathcal{P}(\mathbb{R}) \to \mathcal{P}(\mathbb{R})$ by $g(A) = \overline{A}$. For example, we have $f([0, 1]) = (-\infty, 0) \cup (1, \infty)$ and $g([0, 1]) = [0, 1]$. We can also take compositions of these functions to possibly generate new sets. For example, we have the following:

$$(g \circ f)([0, 1]) = g(f([0, 1])) = g((-\infty, 0) \cup (1, \infty)) = (-\infty, 0] \cup [1, \infty)$$
$$(f \circ g \circ f)([0, 1]) = f((g \circ f)([0, 1])) = f((-\infty, 0] \cup [1, \infty)) = (0, 1)$$

Let $\mathcal{X} = \{[0, 1], (0, 1), (-\infty, 0) \cup (1, \infty), (-\infty, 0] \cup [1, \infty)\}$ and observe that whenever we apply f or g to any of the sets in $\mathcal{X}$, we get a set in $\mathcal{X}$. We say that $\mathcal{X}$ is **closed** for the operations f and g. We can also say that $\mathcal{X}$ is the **closure** of $[0, 1]$ under f and g, and we write $\mathcal{X} = cl\{[0, 1]\}$. Observe that $|cl([0, 1])| = 4$.

Determine the maximum value of $|cl(A)|$, where A is a subset of $\mathbb{R}$ and describe a set A whose closure has that maximum value.

LESSON 10
SEPARATION AND COUNTABILITY

T_0- (Kolmogorov) Spaces

A topological space $(S, \mathcal{T})$ is a $\boldsymbol{T_0}$-**space** (or **Kolmogorov space**) if for all $x, y \in S$ with $x \neq y$, there is $U \in \mathcal{T}$ such that either $x \in U$ and $y \notin U$ or $x \notin U$ and $y \in U$.

In other words, in a T_0-space, given any two elements, there is an open set that contains one of the elements and excludes the other. In the picture to the right we see two typical elements a and b in a T_0-space. We have drawn an open set containing a and excluding b. There does not need to be an open set containing b and excluding a (although there can be).

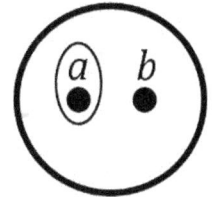

Example 10.1:

1. Let $S = \{a, b\}$ where $a \neq b$. S together with the trivial topology $\{\emptyset, \{a, b\}\}$ is **not** a T_0-space. In fact, the trivial topology on any set with more than one element is not a T_0-space.

   $\{a, b\}$ together with the discrete topology $\{\emptyset, \{a\}, \{b\}, \{a, b\}\}$ **is** a T_0-space because the open set $\{a\}$ satisfies $a \in \{a\}$ and $b \notin \{a\}$. In fact, the discrete topology on any set is a T_0-space.

   The other two topologies on $\{a, b\}$ are also T_0-spaces. For example, $\{\emptyset, \{a\}, \{a, b\}\}$ is a T_0-space because $\{a\}$ is open, $a \in \{a\}$ and $b \notin \{a\}$.

2. Let $S = \mathbb{R}$ and let $\mathcal{T}$ be the topology generated by the basis $\{(a, \infty) \mid a \in \mathbb{R}\}$. Then $(S, \mathcal{T})$ is a T_0-space. If $a, b \in \mathbb{R}$ with $a < b$, then $U = (a, \infty)$ is an open set with $b \in U$ and $a \notin U$.

   It's worth noting that the topology generated by $\{(a, \infty) \mid a \in \mathbb{R}\}$ is $\{(a, \infty) \mid a \in \mathbb{R}\} \cup \{\emptyset, \mathbb{R}\}$.

3. If $(S, \mathcal{T})$ is a T_0-space and $\mathcal{T}'$ is finer than $\mathcal{T}$, then $(S, \mathcal{T}')$ is also a T_0-space. Indeed, if $U \in \mathcal{T}$ with $x \in U$ and $y \notin U$, then since $\mathcal{T}'$ is finer than $\mathcal{T}$, we have $U \in \mathcal{T}'$.

   For example, since the standard topology on $\mathbb{R}$ is finer than the topology generated by $\{(a, \infty) \mid a \in \mathbb{R}\}$, $\mathbb{R}$ together with the standard topology on $\mathbb{R}$ is a T_0-space.

T_1- (Fréchet) Spaces

A topological space $(S, \mathcal{T})$ is a $\boldsymbol{T_1}$-**space** (or **accessible space** or **Fréchet space**) if for all $x, y \in S$ with $x \neq y$, there are $U, V \in \mathcal{T}$ such that $x \in U$ and $y \notin U$ and $x \notin V$ and $y \in V$.

In the picture to the right we see two typical elements a and b in a T_1-space. We have drawn an open set containing a and excluding b and an open set containing b and excluding a. These two open sets **do not** need to be disjoint. The smaller dots in the picture are representing some elements of the space other than a and b.

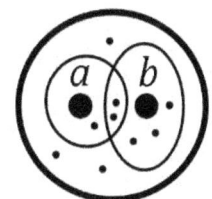

Example 10.2:

1. $S = \{a, b\}$ together with the discrete topology $\{\emptyset, \{a\}, \{b\}, \{a, b\}\}$ **is** a T_1-space because the open sets $\{a\}$ and $\{b\}$ satisfy $a \in \{a\}$, $b \notin \{a\}$, $b \in \{b\}$, and $a \notin \{b\}$. In fact, the discrete topology on any set is a T_1-space.

    It should be clear from the definitions that every T_1-space is a T_0-space. It follows that the trivial topology on any set with more than one element is **not** a T_1-space.

    The other two topologies on $\{a, b\}$ do **not** give us T_1-spaces. For example, $\{\emptyset, \{a\}, \{a, b\}\}$ is not a T_1-space because the only open set containing b also contains a.

    In fact, the only topology on any finite set that is T_1 is the discrete topology. To see this, let $\mathcal{T}$ be a topology on a finite set X that is T_1 and let $x \in X$. For each $y \in X$ with $y \neq x$, there is an open set U_y such that $x \in U_y$ and $y \notin U_y$. It follows that $U = \cap\{U_y \mid y \in X \wedge y \neq x\}$ is open and it is easy to see that $U = \{x\}$. So, $\mathcal{T}$ is generated by the one point sets, and therefore, $\mathcal{T}$ is the discrete topology on X.

2. Let $S = \mathbb{R}$ and let $\mathcal{T}$ be the topology generated by the basis $\{(a, \infty) \mid a \in \mathbb{R}\}$. Then $(S, \mathcal{T})$ is **not** a T_1-space. To see this, let $x, y \in \mathbb{R}$ with $x < y$. Let U be an open set containing x, say $U = (a, \infty)$. Since $x < y$ and $a < x$, we have $a < y$, and so, $y \in U$. Therefore, there is no open set U with $x \in U$ and $y \notin U$.

3. Let $S = \mathbb{R}$ and let $\mathcal{T}$ be the topology generated by the basis $\mathcal{B} = \{X \subseteq \mathbb{R} \mid \mathbb{R} \setminus X \text{ is finite}\}$. $\mathcal{T}$ is called the **cofinite topology** on $\mathbb{R}$. In Problem 4 in Problem Set 9, you were asked to prove that $\mathcal{B}$ generates a topology on $\mathbb{R}$ that is strictly coarser than the standard topology. It's easy to see that $(S, \mathcal{T})$ is a T_1-space. Indeed, if $a, b \in \mathbb{R}$ with $a \neq b$, then let $U = \mathbb{R} \setminus \{b\}$ and $V = \mathbb{R} \setminus \{a\}$.

    It's worth noting that the topology generated by $\mathcal{B}$ is $\{X \subseteq \mathbb{R} \mid \mathbb{R} \setminus X \text{ is finite}\} \cup \{\emptyset\}$.

    Also, there is nothing special about the set $\mathbb{R}$ here. If we let S be any set, then it follows that $\mathcal{B} = \{X \subseteq S \mid S \setminus X \text{ is finite}\}$ is a basis for the **cofinite topology** on S.

4. If $(S, \mathcal{T})$ is a T_1-space and $\mathcal{T}'$ is finer than $\mathcal{T}$, then $(S, \mathcal{T}')$ is also a T_1-space. Indeed, if $U, V \in \mathcal{T}$ with $x \in U$ and $y \notin U$ and $x \notin V$ and $y \in V$, then since $\mathcal{T}'$ is finer than $\mathcal{T}$, we have $U, V \in \mathcal{T}'$.

    For example, since the standard topology on $\mathbb{R}$ is finer than the cofinite topology on $\mathbb{R}$, it follows that $\mathbb{R}$ together with the standard topology on $\mathbb{R}$ is a T_1-space.

In the next Example, we will see that topological spaces that are not T_1-spaces can be a bit "weird."

Example 10.3:

1. Consider the topological space $(S, \mathcal{T})$, where $S = \{a, b\}$ and $\mathcal{T} = \{\emptyset, \{a\}, \{a, b\}\}$. By part 1 of Example 10.1 and part 1 of Example 10.2 this is a T_0-space, but **not** a T_1-space. This topological space has the undesirable property that $\{a\}$ is **not** closed in $(S, \mathcal{T})$. This follows immediately from the fact that $\{a, b\} \setminus \{a\} = \{b\}$ is not open in $(S, \mathcal{T})$.

2. Consider the topological space $(S, \mathcal{T})$, where $S = \mathbb{R}$ and $\mathcal{T}$ is generated by the basis $\{(a, \infty) \mid a \in \mathbb{R}\}$. By part 2 of Example 10.1 and part 2 of Example 10.2 this is a T_0-space, but **not** a T_1-space. This space also has the undesirable property that sets consisting of a single point are not closed. To see this, let $x \in \mathbb{R}$. Then $\mathbb{R} \setminus \{x\} = (-\infty, x) \cup (x, \infty)$ does not have the form (a, ∞). Therefore, $\mathbb{R} \setminus \{x\}$ is not open in $(S, \mathcal{T})$, and so, $\{x\}$ is not closed in $(S, \mathcal{T})$.

Note: Example 10.3 shows that spaces that are not T_1 can be a bit weird, possibly having some counterintuitive properties. After all, in any "reasonable topological space," all sets consisting of a single element should be closed. For example, if we consider $\mathbb{R}$ with the standard topology, the set $\{x\}$ is closed because it is the complement of the open set $(-\infty, x) \cup (x, \infty)$. Luckily, T_1-spaces are more "reasonable" in this sense.

Theorem 10.4: A topological space $(S, \mathcal{T})$ is a T_1-space if and only if for all $x \in S$, $\{x\}$ is closed in $(S, \mathcal{T})$.

Proof: Let $(S, \mathcal{T})$ be a topological space. First, assume that $(S, \mathcal{T})$ is a T_1-space and let $x \in S$. For each $y \in S$ with $y \neq x$, there is an open set U_y with $y \in U_y$ and $x \notin U_y$. Then $U = \bigcup \{U_y \mid y \in S \wedge y \neq x\}$ is open (because U is a union of open sets). Let's check that $\{x\} = S \setminus U$. Since $x \notin U_y$ for all $y \neq x$, $x \notin U$. So, $x \in S \setminus U$. It follows that $\{x\} \subseteq S \setminus U$. If $z \in S \setminus U$, then $z \notin U$. So, for all $y \neq x$, $z \notin U_y$. Thus, for all $y \neq x$, $z \neq y$. Therefore, $z = x$, and so, $z \in \{x\}$. So, $S \setminus U \subseteq \{x\}$. Since $\{x\} \subseteq S \setminus U$ and $S \setminus U \subseteq \{x\}$, we have $\{x\} = S \setminus U$. Since U is open in $(S, \mathcal{T})$, $\{x\} = S \setminus U$ is closed in $(S, \mathcal{T})$.

Conversely, suppose that for all $x \in S$, $\{x\}$ is closed in $(S, \mathcal{T})$. Let $x, y \in S$ with $x \neq y$, let $U = S \setminus \{y\}$, and let $V = S \setminus \{x\}$. Then U and V are open sets such that $x \in U$ and $y \notin U$ and $x \notin V$ and $y \in V$. So, $(S, \mathcal{T})$ is a T_1-space. $\square$

Let $(S, \mathcal{T})$ be a topological space and let $A \subseteq S$. Recall from Lesson 9 that $x \in S$ is called an **accumulation point** of A if every $U \in \mathcal{T}$ containing x contains at least one point of A different from x.

Example 10.5:

1. Consider the T_0-space $(S, \mathcal{T})$, where $S = \{a, b\}$ and $\mathcal{T} = \{\emptyset, \{a\}, \{a, b\}\}$. By part 4 of Example 9.2, b is an accumulation point of the set $\{a\}$. Furthermore, b is the **only** accumulation point of $\{a\}$. Here, $\{a, b\}$ is the only open set containing b. It does contain a point of $\{a\}$ different from b, namely a. But that's it! The open set $\{a, b\}$ does **not** contain infinitely many points of $\{a\}$.

2. let $\mathcal{T}$ be the trivial topology on $\mathbb{R}$ and let $A = (10, 20)$. Then 0 is an accumulation point of A. Indeed, the only open set containing 0 is $\mathbb{R}$, and certainly $\mathbb{R}$ contains at least one point of A different from 0 (in fact, it contains the whole set A). Note that the space $(\mathbb{R}, \mathcal{T})$ is not even a T_0-space.

Note: Example 10.5 provides a little more evidence that spaces that are not T_1 can be weird. We might expect that an open set containing an accumulation point of a set A would contain "lots" of points from A and not just one. In a T_1-space, this is exactly what happens.

Theorem 10.6: Let $(S, \mathcal{T})$ be a T_1-space and let $A \subseteq S$. Then x is an accumulation point of A if and only if every open set containing x contains infinitely many points of A.

189

Analysis: The nontrivial direction of the proof of this theorem is a bit tricky to write down, but the idea is quite simple. Suppose that U is an open set containing x that has only finitely many points in common with A. We form the set V by deleting those finitely many points from U, except for x (if x happens to be one of those points). Then V is an open set containing x that has no points in common with A. Why is V open? This is where we use the fact that $(S, \mathcal{T})$ is a T_1-space. In a T_1-space, finite sets are closed, and therefore, complements of finite sets are open. V is the intersection of U with the complement of a finite set.

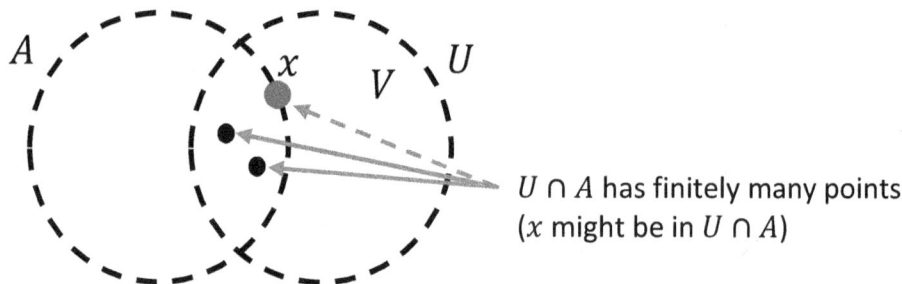

$U \cap A$ has finitely many points (x might be in $U \cap A$)

Proof: Let $x \in U$, where U is an open set containing infinitely many points of A. Then U certainly contains one point of A different from x. So, x is an accumulation point of A.

Conversely, let x be an accumulation point of A and suppose toward contradiction that U is an open set with $x \in U$ such that U contains only finitely many points of A. Let $C = (A \cap U) \setminus \{x\}$. By Theorem 10.4 and Problem 11 from Problem Set 9, C is closed in S (since C is finite, it is a finite union of sets with one element each). Therefore, $S \setminus C$ is open in S. Since the intersection of two open sets in S is open in S, we have that $V = (S \setminus C) \cap U$ is open in S. Since $x \notin C$, $x \in S \setminus C$. So, $x \in V$. Since x is an accumulation point of A, there should be $y \in V \cap A$ with $y \neq x$. But if $y \in V$, then $y \notin C$. So, either $y \notin A$ or $y \notin U$, both of which are impossible. So, there is no $y \in V \cap A$ with $y \neq x$, contradicting our assumption that x is an accumulation point of A. $\qquad \square$

Although Theorems 10.4 and 10.6 provide evidence that T_1-spaces are "not too weird," it turns out that they can still have some undesirable characteristics, as we now show.

Recall from Lesson 4 that an infinite sequence is a function with domain $\mathbb{N}$. As stated in that lesson, if $f: \mathbb{N} \to S$ is defined by $f(n) = s_n$, then we will represent the sequence f using the notation (s_n). We can visualize the sequence (s_n) as the list $s_0, s_1, s_2, \ldots$

Let $(S, \mathcal{T})$ be a topological space and let (s_n) be a sequence with $s_n \in S$ for all $n \in \mathbb{N}$. We say that (s_n) **converges** to $s \in S$, written $s_n \to s$, if for every open set U containing s, there is a natural number K such that $s_n \in U$ for all $n > K$. Symbolically, we can write that $s_n \to s$ if and only if

$$\forall U \in \mathcal{T} \left(s \in U \to \exists K \in \mathbb{N}(n > K \to s_n \in U) \right)$$

If $s_n \to s$, then s is called the **limit** of the sequence.

In words, a sequence converges to a point s in a topological space if every open set containing s contains "almost all" of the sequence. So, if we just chop off a finite piece from the beginning of the sequence, then what's left is trapped in the given open set.

Our intuition seems to tell us that if $s_n \to s$, then the sequence should be getting "closer and closer" to s as we go further out into the sequence. However, for some topological spaces, this intuition can fail us. For example, if S is any set given the trivial topology and (s_n) is any sequence with $s_n \in S$ for all $n \in \mathbb{N}$. Then $s_n \to s$ for every $s \in S$. This is due to the fact that there is only one open set containing s, namely S, and S contains every element of the sequence.

We will see in the next example, that even in T_1-spaces, sequences that converge may converge to multiple limits.

Example 10.7:

1. Let $S = \mathbb{R} \cup \{c\}$, where c is not equal to any real number. Let $\mathcal{B}$ be the collection of all open intervals in $\mathbb{R}$ together will all sets of the form $(-a, 0) \cup \{c\} \cup (0, b)$. We've essentially duplicated the origin. So, there are now two origins (0 and c).

    It is easy to see that $\mathcal{B}$ covers S and that $\mathcal{B}$ has the intersection containment property. So, by Theorem 9.8, $\mathcal{B}$ is a basis for a topology $\mathcal{T}$ on S. The topological space $(S, \mathcal{T})$ is known as the **line with two origins**.

    It is easily seen that $(S, \mathcal{T})$ is a T_1-space. We can separate 0 and c with the open sets $U = (-1, 1)$ and $V = (-1, 0) \cup \{c\} \cup (1, b)$. Observe that $0 \in U$, $c \notin U$, $0 \notin V$, and $c \in V$. The cases where x and y are not 0 and c are also easily verified (this verification is similar to what we will do in part 3 of Example 10.8 below).

    Consider the sequence $\left(\frac{1}{n+1}\right)$. I leave it to the reader to show that this sequence converges to both 0 and c.

2. Consider $(\mathbb{N}, \mathcal{T})$, where $\mathcal{T}$ is the cofinite topology on $\mathbb{N}$. In part 3 of Example 10.2, we showed that $(\mathbb{N}, \mathcal{T})$ is a T_1-space. The sequence (n) converges to **every** natural number m in this space. To see this, let U be an open set containing the natural number m. Then U is the complement of a finite set A. If $A = \emptyset$, then $U = \mathbb{N}$ and so, U contains every element of the sequence. If $A \neq \emptyset$, let K be the greatest element of A. Then $s_n \in U$ for all $n > K$.

Since it is undesirable to work with topological spaces where convergent sequences do not have unique limits, it seems that requiring a topological space to be a T_1-space is usually not quite enough. The next type of topological space that we will discuss is much more pleasant to work with.

T_2- (Hausdorff) Spaces

A topological space $(S, \mathcal{T})$ is a **T_2-space** (or **Hausdorff space**) if for all $x, y \in S$ with $x \neq y$, there are $U, V \in \mathcal{T}$ with $x \in U$, $y \in V$, and $U \cap V = \emptyset$.

In the picture to the right we see two typical elements a and b in a T_2-space. We have drawn disjoint open sets, one including a and the other including b. The smaller dots in the picture represent some elements of the space other than a and b.

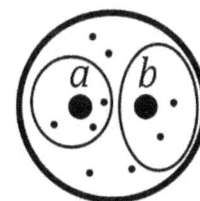

Remark: Many mathematicians will work only with topological spaces that are T_2-spaces. Some even go so far as to include the condition of being a T_2-space in their definition of a topological space. We will not go to that extreme in this book.

Example 10.8:

1. Any set together with the discrete topology on that set is a T_2-space. Indeed, if a and b are distinct points from a T_2-space, then $\{a\}$ and $\{b\}$ are disjoint open sets.

   It should be clear from the definitions that every T_2-space is a T_1-space. It follows that except for the discrete topology, every other topology on a finite set is **not** a T_2-space.

2. The topological space $(\mathbb{R}, \mathcal{T})$, where $\mathcal{T}$ is the cofinite topology on $\mathbb{R}$ (see part 3 of Example 10.2) is **not** a T_2-space. Indeed, if U and V are open sets containing $a, b \in \mathbb{R}$, respectively, then $\mathbb{R} \setminus (U \cap V) = (\mathbb{R} \setminus U) \cup (\mathbb{R} \setminus V)$ (this is **De Morgan's law**), which is finite. So, $U \cap V$ is infinite, and therefore, nonempty.

3. The standard topologies on $\mathbb{R}$ and $\mathbb{C}$ are both T_2. The same argument can be used for both (although the geometry looks very different).

   Let $S = \mathbb{R}$ or $\mathbb{C}$ and let $x, y \in S$. Let $\epsilon = \frac{1}{2}d(x,y) = \frac{1}{2}|x - y|$. Then $U = N_\epsilon(x)$ and $V = N_\epsilon(y)$ are disjoint open sets with $x \in U$ and $y \in V$.

   In the picture below, we have drawn two typical real numbers x and y on the real line and then separated them with the disjoint neighborhoods $U = N_\epsilon(x)$ and $V = N_\epsilon(y)$.

   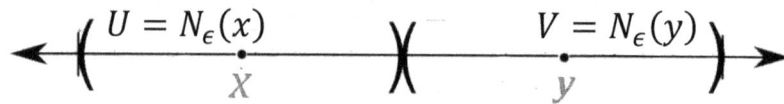

   In the picture to the right, we have drawn two typical complex numbers x and y in the Complex Plane and then separated them with the disjoint neighborhoods $U = N_\epsilon(x)$ and $V = N_\epsilon(y)$.

   Note once again that neighborhoods on the real line are open intervals, whereas neighborhoods in the Complex Plane are open disks.

4. If $(S, \mathcal{T})$ is a T_2-space and $\mathcal{T}'$ is finer than $\mathcal{T}$, then $(S, \mathcal{T}')$ is also a T_2-space. Indeed, if $U, V \in \mathcal{T}$ with $x \in U$, $y \in V$, and $U \cap V = \emptyset$, then since $\mathcal{T}'$ is finer than $\mathcal{T}$, we have $U, V \in \mathcal{T}'$. Let's look at an example of this.

   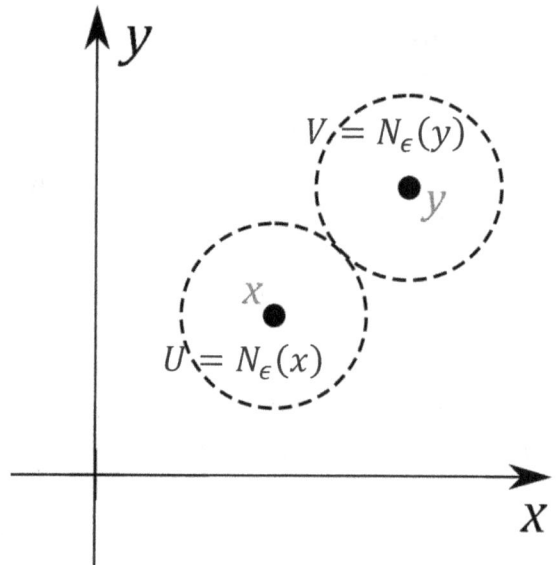

   Let $K = \left\{\frac{1}{n} \mid n \in \mathbb{Z}^+\right\}$, $\mathcal{B} = \{(a,b) \mid a, b \in \mathbb{R} \wedge a < b\} \cup \{(a,b) \setminus K \mid a, b \in \mathbb{R} \wedge a < b\}$. In Problem 3 from Problem Set 9, you were asked to verify that $\mathcal{B}$ is a basis for a topology $\mathcal{T}_K$ on $\mathbb{R}$. Since $\mathcal{T}_K$ contains every basis element of the standard topology on $\mathbb{R}$, we see that $\mathcal{T}_K$ is finer than the standard topology. It follows that $(\mathbb{R}, \mathcal{T}_K)$ is a T_2-space.

As was already mentioned above, many mathematicians like to restrict their study of topological spaces to T_2-spaces. In addition to sets with just one element being closed, T_2-spaces have the desirable property that convergent sequences have a unique limit, as we now prove.

Theorem 10.9: Let $(S, \mathcal{T})$ be a T_2-space and let (s_n) be a convergent sequence. Then the limit of the sequence is unique.

Proof: Suppose that $s_n \to s$ and $s_n \to t$. Assume toward contradiction that $s \neq t$. Since $(S, \mathcal{T})$ is a T_2-space, there are open sets U and V with $s \in U$, $t \in V$, and $U \cap V = \emptyset$. Since $s_n \to s$, there is $K_1 \in \mathbb{N}$ such that $n > K_1$ implies $s_n \in U$. Since $s_n \to t$, there is $K_2 \in \mathbb{N}$ such that $n > K_2$ implies $s_n \in V$. Let $K = \max\{K_1, K_2\}$. Then since $K + 1 > K$, we have $K + 1 > K_1$ and $K + 1 > K_2$. It follows that $s_{K+1} \in U$ and $s_{K+1} \in V$, contradicting that $U \cap V = \emptyset$. Therefore, we must have $s = t$. $\square$

T_3- (Regular) Spaces

A topological space $(S, \mathcal{T})$ is a T_3-**space** (or **Regular space**) if $(S, \mathcal{T})$ is a T_1-space and for every $x \in S$ and closed set X with $X \subseteq S \setminus \{x\}$, there are $U, V \in \mathcal{T}$ with $x \in U$, $X \subseteq V$, and $U \cap V = \emptyset$.

In the picture to the right we see a typical element a and a closed set K in a T_3-space. We have drawn disjoint open sets, one including a and the other containing K. The smaller dots in the picture represent some elements of the space other than a that are not included in K. (Note that we replaced an arbitrary closed set X with the specific closed set K, and similarly, we replaced x with a.)

Example 10.10:

1. Any set together with the discrete topology on that set is a T_3-space. Indeed, if $x \in S$ and A is any subset of $S \setminus \{x\}$ (all subsets of S are closed), simply let $U = \{x\}$ and $V = A$ (all subsets of S are also open).

    Some authors call a set **clopen** if it is both open and closed. If S is given the discrete topology, then all subsets of S are clopen.

2. Every T_3-space is a T_2-space. This follows easily from the fact that a T_3-space is a T_1-space and Theorem 10.4. It follows that except for the discrete topology, every other topology on a finite set is **not** a T_3-space.

3. The standard topologies on $\mathbb{R}$ and $\mathbb{C}$ are T_3. This will follow from Problem 7 in Problem Set 11.

4. Consider the T_2-space $(\mathbb{R}, \mathcal{T}_K)$ from part 4 of Example 10.8. Recall that $K = \{\frac{1}{n} \mid n \in \mathbb{Z}^+\}$ and $(\mathbb{R}, \mathcal{T}_K)$ has basis $\mathcal{B} = \{(a, b) \mid a, b \in \mathbb{R} \wedge a < b\} \cup \{(a, b) \setminus K \mid a, b \in \mathbb{R} \wedge a < b\}$. Let $x = 0$ and $A = K$. $\mathbb{R} \setminus K = (-\infty, 0) \cup [(-1, 1) \setminus K] \cup (1, \infty)$, which is a union of three open sets, thus open. Therefore, K is a closed set in this topology. Let U be an open set containing 0 and let V be an open set containing K. For some $\epsilon > 0$, $(0, \epsilon) \setminus K \subseteq U$. By the Archimedean Property of $\mathbb{R}$, there is $n \in \mathbb{N}$ with $n > \frac{1}{\epsilon}$, or equivalently, $\frac{1}{n} < \epsilon$. There is $0 < \delta \leq \epsilon - \frac{1}{n}$ such that $\left(\frac{1}{n} - \delta, \frac{1}{n} + \delta\right) \subseteq V$. Let r be an irrational number in $\left(\frac{1}{n}, \frac{1}{n} + \delta\right)$. $r \in U \cap V$ and therefore, $U \cap V \neq \emptyset$. Since we cannot separate 0 and K with open sets, $(\mathbb{R}, \mathcal{T}_K)$ is **not** a T_3-space.

    Unlike T_0, T_1, and T_2-spaces, T_3-spaces are not closed under upward refinement. In other words, if $(S, \mathcal{T})$ is a T_3-space and $\mathcal{T}'$ is finer than $\mathcal{T}$, then $(S, \mathcal{T}')$ is **not necessarily** a T_3-space. The topological space $(\mathbb{R}, \mathcal{T}_K)$ proves this. Indeed, $(\mathbb{R}, \mathcal{T}_K)$ is finer than the T_3-space $(\mathbb{R}, \mathcal{T})$, where $\mathcal{T}$ is the standard topology on $\mathbb{R}$, but as we just saw, $(\mathbb{R}, \mathcal{T}_K)$ is **not** a T_3-space.

Also, since $(\mathbb{R}, \mathcal{T})$ is T_3, where $\mathcal{T}$ is the standard topology on $\mathbb{R}$, but $(\mathbb{R}, \mathcal{T}_K)$ is not, the two topological spaces cannot be the same. It follows that $\mathcal{T}_K$ is strictly finer than the standard topology on $\mathbb{R}$.

T_4- (Normal) Spaces

A topological space $(S, \mathcal{T})$ is a **T_4-space** (or **Normal space**) if $(S, \mathcal{T})$ is a T_1-space and for every pair X, Y of disjoint closed subsets of S, there are $U, V \in \mathcal{T}$ with $X \subseteq U, Y \subseteq V$, and $U \cap V = \emptyset$.

In the picture to the right we see two closed sets K and L in a T_4-space. We have drawn disjoint open sets, one containing K and the other containing L. The smaller dots in the picture represent some elements of the space not included in K or L. (Note that we replaced the arbitrary closed sets X and Y with specific closed sets K and L.)

Example 10.11:

1. Any set together with the discrete topology on that set is a T_4-space. Indeed, if A and B are disjoint closed subsets of S, then A and B are also disjoint open subsets of S (because all subsets of S are both open and closed).

   Every T_4-space is a T_3-space. This follows easily from the fact that a T_4-space is a T_1-space and Theorem 10.4\

   . It follows that except for the discrete topology, every other topology on a finite set is **not** a T_4-space.

2. The standard topologies on $\mathbb{R}$ and $\mathbb{C}$ are both T_4. This will follow from Problem 7 in Problem Set 11.

3. In Problem 9 below, you will see a T_3-space that is **not** a T_4-space.

The definitions of T_0, T_1, T_2, T_3, and T_4 are called **separation axioms** because they all involve "separating" points and/or closed sets from each other by open sets.

We will now look at a few additional properties of particular interest that a topological space may have. These next few properties are called **countability axioms**.

Separable Spaces

Let $(S, \mathcal{T})$ be a topological space and let $A \subseteq S$. We say that A is **dense** in S if $\overline{A} = S$.

A topological space $(S, \mathcal{T})$ is **separable** if it has a countable dense subset.

Example 10.12:

1. Let $(S, \mathcal{T})$ be a topological space, where $\mathcal{T}$ is the discrete topology. Then $(S, \mathcal{T})$ is separable if and only if S is countable. If S is countable, then the collection S is a countable dense subset of itself. Now, suppose that S is uncountable and let A be a dense subset of S. By definition, $\overline{A} = S$. Since $\mathcal{T}$ is the discrete topology on S, $\overline{A} = A$ (because every subset of S is closed in S). Therefore, $A = S$, which is uncountable. So, S has no countable dense subset.

2. If $\mathcal{T}$ is the standard topology on $\mathbb{R}$, then $(\mathbb{R}, \mathcal{T})$ is separable because $\overline{\mathbb{Q}} = \mathbb{R}$ (see part 4 of Example 7.26). Similarly, by Problem 20 in Problem Set 7, $\mathbb{C}$ with its standard topology is separable, as is $\mathbb{R}^n$ with the product topology for each $n \in \mathbb{N}$.

3. Let $\mathcal{T}$ be the cofinite topology on $\mathbb{R}$ (or any uncountable set). Recall that this topology is generated by the basis $\mathcal{B} = \{X \subseteq \mathbb{R} \mid \mathbb{R} \setminus X \text{ is finite}\}$. Then $(\mathbb{R}, \mathcal{T})$ is separable. To see this, let A be any countably infinite subset of $\mathbb{R}$. We will prove that $\overline{A} = \mathbb{R}$. To this end, let $x \in \mathbb{R}$ and let U be any basic open set containing x. Then $U = \mathbb{R} \setminus B$ for some finite set B. Therefore, $U \cap A = (\mathbb{R} \setminus B) \cap A = A \setminus B$. Since A is infinite and B is finite, $A \setminus B$ is infinite. In particular, U contains at least one point of A. By part 5 of Theorem 9.4, $x \in \overline{A}$. It follows that $\overline{A} = \mathbb{R}$.

4. A subspace of a separable space need not be separable. You will be asked to verify this in Problem 14 below.

5. If $(S_n, \mathcal{T}_n)$ is a separable topological space for each $n \in \mathbb{N}$, then $\prod S_n$ with the product topology is separable. To see this, for each $n \in \mathbb{N}$, let A_n be a countable dense subset of S_n. Let $\mathbf{x} \in \prod S_n$ and let $A = \bigcup\{\prod_{k \leq n} A_k \times \prod_{k > n} \{x_k\} \mid n \in \mathbb{N}\}$. A is a countable union of countable sets, and therefore, A is countable (see Problem 11 in Problem Set 4). We will show that $\overline{A} = \prod S_n$. To this end, let $\mathbf{y} \in \prod S_n$ and let U be a basic open set in the product topology of $\prod S_n$. Then we have $U = \prod U_n$ and $K \in \mathbb{N}$ such that $U_n = S_n$ for all $n > K$. If $k \leq K$, then since A_k is dense in S_k, there is $a_k \in A_k$ with $a_k \in U_k$. Let $b_k = a_k$ for $k \leq K$ and let $b_k = x_k$ for $x > K$. Then $\mathbf{b} \in A \cap U$. It follows that $\mathbf{y} \in \overline{A}$. Since $\mathbf{y} \in \prod S_n$ was arbitrary, $\overline{A} = \prod S_n$. Therefore, $\prod S_n$ is separable.

First-Countable Spaces

Let $(S, \mathcal{T})$ be a topological space and let $x \in S$. We say that $\mathcal{B}$ is a **countable basis at x** if the following two properties holds:

1. If $V \in \mathcal{B}$, then $x \in V$.

2. For any open set U containing x, there is a $V \in \mathcal{B}$ such that $V \subseteq U$.

A topological space $(S, \mathcal{T})$ is a **first-countable space** if for every $x \in S$ there is a countable basis at x.

Example 10.13:

1. Any set together with the discrete topology on that set is a first-countable space. Indeed, if $x \in S$, then $\{\{x\}\}$ is a countable basis at x. Clearly any open set containing x contains the set $\{x\}$. If S is uncountable, then we get an example of a first-countable space that is **not** separable (by part 1 of Example 10.12).

2. The standard topology on $\mathbb{R}$ is a first-countable space. To see this, let $x \in \mathbb{R}$. Then the set of all open intervals with rational endpoints that contain x is a countable basis at x.

   Similarly, the standard topology on $\mathbb{C}$ is a first-countable space. To see this, let $z \in \mathbb{C}$. Then the set of all open disks with a rational radius and center z is a countable basis at z.

More generally, the Euclidean spaces $\mathbb{R}^n$ and the unitary spaces $\mathbb{C}^n$ are first-countable spaces. Given $\mathbf{x}$ in one of these spaces, the set of open balls with a rational radius and center $\mathbf{x}$ is a countable basis at $\mathbf{x}$.

3. Let $\mathcal{T}$ be the cofinite topology on $\mathbb{R}$. In part 3 of Example 10.12, we showed that $(\mathbb{R}, \mathcal{T})$ is separable. We will now show that $(\mathbb{R}, \mathcal{T})$ is **not** a first-countable space. To see this, let $x \in \mathbb{R}$ and assume toward contradiction that $\mathcal{D} = \{D_n \mid n \in \mathbb{N}\}$ is a countable basis at x. For each $n \in \mathbb{N}$, let $C_n = \mathbb{R} \setminus D_n$. Since each D_n is open, each C_n is finite. Let $C = \cup\{C_n \mid n \in \mathbb{N}\} \cup \{x\}$. Since each C_n is finite, by Problem 11 from Problem Set 4, C is countable. Since $\mathbb{R}$ is uncountable, there is $y \in \mathbb{R} \setminus C$. Let $U = \mathbb{R} \setminus \{y\}$. Then U is the complement of a finite set, and so, U is open. Since $y \notin C$, $y \neq x$ and $y \notin C_n$ for each $n \in \mathbb{N}$. So, $y \in D_n$ for each $n \in \mathbb{N}$. It follows that for each $n \in \mathbb{N}$, $D_n \nsubseteq U$. This contradicts our assumption that $\mathcal{D}$ is a countable basis at x.

This example shows that not every separable space is first-countable. Together with part 1 of this example, we see that neither one of separable nor first-countable imply the other.

4. Every subspace of a first-countable space is also first-countable. To see this, let $(S, \mathcal{T})$ be a first-countable space, let $(A, \mathcal{T}_A)$ be a subspace, and let $x \in A$. Since $(S, \mathcal{T})$ is first-countable, there is a countable basis $\mathcal{B}$ at x in the topological space $(S, \mathcal{T})$. Then $\{V \cap A \mid V \in \mathcal{B}\}$ is a countable basis at x in the subspace $(A, \mathcal{T}_A)$.

5. If $(S_n, \mathcal{T}_n)$ is a first-countable topological space for each $n \in \mathbb{N}$, then $\prod S_n$ with the product topology is first-countable. To see this, let $\mathbf{x} \in \prod S_n$ and for each $n \in \mathbb{N}$, let $\mathcal{D}_n$ be a countable basis at x_n. Let

$$\mathcal{D} = \{\textstyle\prod B_n \mid (B_n = S_n \text{ for all but finitely many } n \in \mathbb{N}) \wedge (B_n \neq S_n \to B_n \in \mathcal{D}_n)\}.$$

$\mathcal{D}$ is a countable collection of open sets containing $\mathbf{x}$. If U is any open set containing $\mathbf{x}$, then we have $\mathbf{x} \in \prod U_n \subseteq U$, where U_n is open in $(S_n, \mathcal{T}_n)$ for all $n \in \mathbb{N}$ and $U_n = S_n$ for all but finitely many $n \in \mathbb{N}$. For all $n \in \mathbb{N}$, $x_n \in U_n$. If $U_n = S_n$, let $B_n = S_n$. If $U_n \neq S_n$, then there is $B_n \in \mathcal{D}_n$ such that $x_n \in B_n \subseteq U_n$. Then $\prod B_n \in \mathcal{D}$ and $\mathbf{x} \in \prod B_n \subseteq \prod U_n \subseteq U$. It follows that $\mathcal{D}$ is a countable basis at $\mathbf{x}$.

Let $(S, \mathcal{T})$ be a topological space, let $A \subseteq S$, let $x \in S$, and let (s_n) be a sequence in A such that $s_n \to x$. Under these conditions, we will always have $x \in \overline{A}$. To see this, let U be an open set containing x. Since $s_n \to x$, there is a natural number K such that $s_n \in U$ for all $n > K$. By part 5 of Theorem 9.4, $x \in \overline{A}$.

In a first-countable space, the converse of the previous statement is true as well.

Theorem 10.14: Let $(S, \mathcal{T})$ be a first-countable topological space and let $x \in \overline{A}$. Then there is a sequence (s_n) in A such that $s_n \to x$.

Proof: Let $(S, \mathcal{T})$ be a first-countable topological space, let $x \in \overline{A}$, and let $\mathcal{B} = \{U_n \mid n \in \mathbb{N}\}$ be a countable basis at x. For each $n \in \mathbb{N}$, let $V_n = \cap\{U_k \mid k \leq n\}$. Let $n \in \mathbb{N}$. Since $x \in U_k$ for each $k = 0, 1, \ldots, n$, $x \in V_n$. Since V_n is a finite intersection of open sets, it is open. For each $n \in \mathbb{N}$, let $x_n \in V_n \cap A$ (we can find x_n by part 5 of Theorem 9.4). To see that $x_n \to x$, let U be an open set containing x. Since $\mathcal{B}$ is a countable basis at x, there is $n \in \mathbb{N}$ with $U_n \subseteq U$. If $m \geq n$, then we have

$$x_m \in V_m \cap A \subseteq V_n \cap A \subseteq U_n \cap A \subseteq U \cap A \subseteq U.$$

Therefore, $x_n \to x$. $\qquad \square$

Second-Countable Spaces

A topological space $(S, \mathcal{T})$ is a **second-countable space** if it has a countable basis.

Theorem 10.15: Every second-countable topological space is separable.

Proof: Let $(S, \mathcal{T})$ be a second-countable space with countable basis $\mathcal{B} = \{U_n \mid n \in \mathbb{N}\}$. We may assume that each $U_n \neq \emptyset$ (if we remove $\emptyset$, we will still have a basis). For each $n \in \mathbb{N}$, let x_n be any element of U_n. The set $A = \{x_n \mid n \in \mathbb{N}\}$ is clearly countable. We need to show that $\overline{A} = S$. To see this, let $x \in S$ and let U_n be a basic open set containing x. Then $x_n \in U_n$. By part 5 of Theorem 9.4, $x \in \overline{A}$. $\qquad \square$

Example 10.16:

1. Let $(S, \mathcal{T})$ be a topological space, where $\mathcal{T}$ is the discrete topology. Then $(S, \mathcal{T})$ is a second-countable space if and only if S is countable. If S is countable, then the collection $\mathcal{B} = \{\{x\} \mid x \in S\}$ is a countable basis for $\mathcal{T}$. If S is uncountable, then by part 1 of Example 10.12, $(S, \mathcal{T})$ is **not** separable. So, by the contrapositive of Theorem 10.15, $(S, \mathcal{T})$ is **not** second-countable.

2. The standard topology on $\mathbb{R}$ is a second-countable space. $\mathcal{B} = \{(a, b) \mid a, b \in \mathbb{Q} \wedge a < b\}$ is a countable basis for this topology. You were asked to prove this in Problem 12 in Problem Set 9.

   Similarly, the Euclidean spaces $\mathbb{R}^n$ and the unitary spaces $\mathbb{C}^n$ are second-countable spaces. The set of open balls with a rational radius and a center whose coordinates are all rational is a countable basis for any of these topologies. I leave the details to the reader.

3. Every second-countable space is first-countable. After all, a countable basis is a countable basis at any point.

   Since the cofinite topology on $\mathbb{R}$ is not first-countable (see part 3 of Example 10.13), it is also not second-countable.

4. Every subspace of a second-countable space is also second-countable. To see this, let $(S, \mathcal{T})$ be a second-countable space with countable basis $\mathcal{B}$ and let $(A, \mathcal{T}_A)$ be a subspace. Then the collection $\mathcal{C} = \{B \cap A \mid B \in \mathcal{B}\}$ is a countable basis for $(A, \mathcal{T}_A)$.

5. If $(S_n, \mathcal{T}_n)$ is a second-countable topological space for each $n \in \mathbb{N}$, then $\prod S_n$ with the product topology is second-countable. To see this, for each $n \in \mathbb{N}$, let $\mathcal{D}_n$ be a countable basis for $\mathcal{T}_n$. By Note 2 following Theorem 9.20,

   $$\mathcal{D} = \{\textstyle\prod B_n \mid (B_n = S_n \text{ for all but finitely many } n \in \mathbb{N}) \wedge (B_n \neq S_n \to B_n \in \mathcal{D}_n)\}$$

   is a countable basis for the product topology on $\prod S_n$.

Problem Set 10

Full solutions to these problems are available for free download here:
www.SATPrepGet800.com/TFBZLF

LEVEL 1

1. Let $(S, \mathcal{T})$ be a T_1-space with $|S| \geq 2$ and let $\mathcal{B}$ be a basis for $(S, \mathcal{T})$. Prove that $\mathcal{B} \setminus \{S\}$ is a basis for $(S, \mathcal{T})$.

2. Prove that a closed subspace of a T_4-space is a T_4-space.

LEVEL 2

3. Let $(S, \mathcal{T})$ be a T_1-space. Prove that $(S, \mathcal{T})$ is a T_3-space if and only if for every point $x \in S$ and every open set U containing x, there is an open set V containing x such that $\overline{V} \subseteq U$.

4. Let $(S, \mathcal{T})$ be a T_1-space. Prove that $(S, \mathcal{T})$ is a T_4-space if and only if for any closed set A in S and every open set U containing A, there is an open set V containing A such that $\overline{V} \subseteq U$.

LEVEL 3

5. Let $(S, \mathcal{T})$ be a T_2-space and $A \subseteq S$. Prove that $(A, \mathcal{T}_A)$ is a T_2-space (Recall that $\mathcal{T}_A$ is the subspace topology on A). Determine if the analogous statement is true for T_3-spaces.

6. Let $(S, \mathcal{T})$ be a topological space and let $D = \{(s, s) \mid s \in S\}$. Prove that $(S, \mathcal{T})$ is a T_2-space if and only if D is closed in $S \times S$ in the product topology.

LEVEL 4

7. Let $(S_1, \mathcal{T}_1)$ and $(S_2, \mathcal{T}_2)$ be T_2-spaces. Prove that $S_1 \times S_2$ with the product topology is also a T_2-space. Determine if the analogous statement is true for T_3-spaces.

8. Prove that a second-countable T_3-space is a T_4-space.

LEVEL 5

9. Consider the topological space $(\mathbb{R}, \mathcal{T}_L)$ (see Problem 14 from Problem Set 9). Prove that $\mathbb{R}^2$ with the corresponding product topology is a T_3-space, but not a T_4-space.

10. Let K be a nonempty set and let S_k be a T_2-space for each $k \in K$. Prove that $\prod S_k$ (with the product topology) is a T_2-space. Determine if the same result is true if we replace T_2 by each of T_3 and T_4.

11. Let K be a nonempty set and let S_k be a nonempty topological space for each $k \in K$. Prove that if $\prod S_k$ (with the product topology) is a T_2-space, then so is S_k for each $k \in K$. Determine if the same result is true if we replace T_2 by each of T_3 and T_4.

12. Prove that an uncountable product of first-countable topological spaces need not be first-countable.

CHALLENGE PROBLEMS

13. Let K be an uncountable set and let $\mathbb{R}$ be given the standard topology. Prove that $^K\mathbb{R}$ with the product topology is not a T_4-space.

14. Find a topological space $(S, \mathcal{T})$ that is separable and a subspace $(A, \mathcal{T}_A)$ of $(S, \mathcal{T})$ that is **not** separable.

15. Find a topological space $(S, \mathcal{T})$ that is separable and not second-countable.

LESSON 11
METRIZABLE SPACES

Metric Spaces

A **metric space** is a pair (S, d), where S is a set and d is a function $d: S \times S \to \mathbb{R}$ with the following properties:

1. For all $x, y \in S$, $d(x, y) = 0$ if and only if $x = y$.

2. For all $x, y \in S$, $d(x, y) = d(y, x)$.

3. For all $x, y, z \in S$, $d(x, z) \leq d(x, y) + d(y, z)$.

The function d is called a **metric** or **distance function**. It is a consequence of the definition that for all $x \in S$, $d(x, x) \geq 0$. You will be asked to prove this in Problem 1 below.

If (S, d) is a metric space, $a \in S$, and $r \in \mathbb{R}^+$, then the **open ball** centered at a with radius r, written $B_r(a)$ (or $B_r(a; d)$ if we need to distinguish this metric from other metrics), is the set of all elements of S whose distance to a is less than r. That is,

$$B_r(a) = \{x \in S \mid d(a, x) < r\}.$$

The collection $\mathcal{B} = \{B_r(a) \mid a \in S \land r \in \mathbb{R}^+\}$ covers S. Indeed, if $a \in S$, then $d(a, a) = 0 < 1$, and so, $a \in B_1(a)$.

Also, the collection $\mathcal{B} = \{B_r(a) \mid a \in S \land r \in \mathbb{R}^+\}$ has the intersection containment property. To see this, let $x \in B_r(a) \cap B_s(b)$ and $k = \min\{r - d(a, x), s - d(b, x)\}$. We have $x \in B_k(x)$ because $d(x, x) = 0 < k$. Now, let $y \in B_k(x)$. Then $d(x, y) < k$. So, we have

$$d(a, y) \leq d(a, x) + d(x, y) < d(a, x) + k$$
$$\leq d(a, x) + r - d(a, x) = r.$$

So, $y \in B_r(a)$. A similar argument shows that $y \in B_s(b)$. So, $y \in B_r(a) \cap B_s(b)$. It follows that $B_k(x) \subseteq B_r(a) \cap B_s(b)$. This verifies that $\mathcal{B}$ has the intersection containment property.

Since the collection of open balls covers S and has the intersection containment property, it follows that this collection is a basis for a topology on S.

Note: Open balls can be visualized as open intervals on the real line $\mathbb{R}$, open disks in the Complex Plane $\mathbb{C}$ (or $\mathbb{R}^2$), or open balls in three-dimensional space $\mathbb{R}^3$.

When proving theorems about metric spaces, it's usually most useful to visualize open balls as open disks in $\mathbb{C}$. This does **not** mean that all metric spaces look like $\mathbb{C}$. The visualization should be used as evidence that a theorem might be true. Of course, a detailed proof still needs to be written.

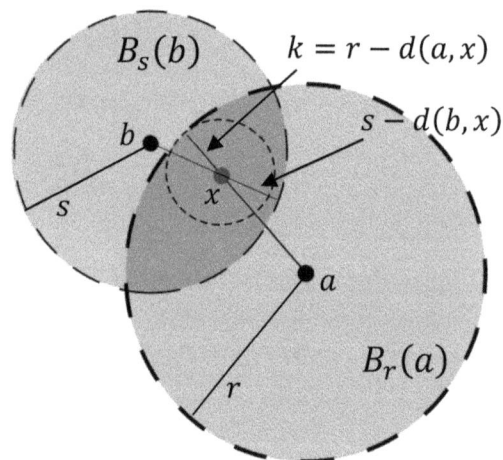

This is exactly what we did when we drew the picture above. That picture represents the open balls $B_r(a)$ and $B_s(b)$ as intersecting open disks. Inside this intersection, we can see the open ball $B_k(x)$. The reader may also want to draw another picture to help visualize the triangle inequality. A picture similar to this is drawn to the right of Note 1 following the proof of Theorem 7.28 in Lesson 7.

A topological space $(S, \mathcal{T})$ is **metrizable** if there is a metric $d: S \times S \to \mathbb{R}$ such that $\mathcal{T}$ is generated from the open balls in (S, d). We also say that the metric d **induces** the topology $\mathcal{T}$. We may refer to $\mathcal{T}$ as the **metric topology** induced by d.

Example 11.1:

1. $(\mathbb{C}, d)$ is a metric space, where $d: \mathbb{C} \times \mathbb{C} \to \mathbb{R}$ is defined by $d(z, w) = |z - w|$. Let's check that the 3 properties of a metric space are satisfied. Property 3 is the Triangle Inequality (Theorem 7.4 and Problem 6 in Problem Set 7). Let's verify the other two properties. Let $z = a + bi$ and $w = c + di$. Then $d(z, w) = |z - w| = \sqrt{(a - c)^2 + (b - d)^2}$. So, $d(z, w) = 0$ if and only if $\sqrt{(a - c)^2 + (b - d)^2} = 0$ if and only if $(a - c)^2 + (b - d)^2 = 0$ if and only if $a - c = 0$ and $b - d = 0$ if and only if $a = c$ and $b = d$ if and only if $z = w$. So, property 1 holds. We have

   $$d(z, w) = |z - w| = |-(w - z)| = |-1(w - z)| = |-1||w - z| = 1|w - z| = d(w, z).$$

   Therefore, property 2 holds.

   If $z \in \mathbb{C}$ and $r \in \mathbb{R}^+$, then the open ball $B_r(z)$ is the set $B_r(z) = \{w \in \mathbb{C} \mid |z - w| < r\}$. This is just an open disk in the Complex Plane, as we defined in Lesson 7.

   Since the collection of open disks in the Complex Plane generates the standard topology on $\mathbb{C}$, we see that $\mathbb{C}$ with the standard topology is a metrizable space.

2. Similarly, $(\mathbb{R}, d)$ is a metric space, where $d: \mathbb{R} \times \mathbb{R} \to \mathbb{R}$ is defined by $d(x, y) = |x - y|$. The proof is similar to the proof above for $(\mathbb{C}, d)$.

   In this case, the open ball $B_r(a)$ is the open interval $(a - r, a + r)$. To see this, observe that we have

   $$B_r(a) = \{x \in \mathbb{R} \mid |x - a| < r\} = \{x \in \mathbb{R} \mid -r < x - a < r\}$$
   $$= \{x \in \mathbb{R} \mid a - r < x < a + r\} = (a - r, a + r).$$

   Since the collection of bounded open intervals of real numbers generates the standard topology on $\mathbb{R}$, we see that $\mathbb{R}$ with the standard topology is a metrizable space.

3. For each $n \in \mathbb{Z}^+$, $(\mathbb{R}^n, d)$ and $(\mathbb{C}^n, k)$ are metric spaces, where $d: \mathbb{R}^n \times \mathbb{R}^n \to \mathbb{R}$ is defined by $d(\mathbf{x}, \mathbf{y}) = |\mathbf{x} - \mathbf{y}|$ (the Euclidean norm of $\mathbf{x} - \mathbf{y}$) and $k: \mathbb{C}^n \times \mathbb{C}^n \to \mathbb{R}$ is defined by $k(\mathbf{z}, \mathbf{w}) = |\mathbf{z} - \mathbf{w}|$ (the unitary norm of $\mathbf{z} - \mathbf{w}$). The proofs for these spaces are similar to the proof above for $(\mathbb{C}, d)$. We use the Generalized Triangle Inequality (Theorem 7.36) for clause 3.

   We call d and k the **Euclidean metric** on $\mathbb{R}^n$ and **unitary metric** on $\mathbb{C}^n$, respectively.

4. Define the functions d_1 and d_2 from $\mathbb{C} \times \mathbb{C}$ to $\mathbb{R}$ by $d_1(z, w) = |\text{Re } z - \text{Re } w| + |\text{Im } z - \text{Im } w|$ and $d_2(z, w) = \max\{|\text{Re } z - \text{Re } w|, |\text{Im } z - \text{Im } w|\}$. In Problem 5 below, you will be asked to verify that $(\mathbb{C}, d_1)$ and $(\mathbb{C}, d_2)$ are metric spaces that induce the standard topology on $\mathbb{C}$.

   So, we see that a metrizable space can be induced by many different metrics.

The open balls $B_r(a; d_1)$ and $B_r(a; d_2)$ are both interiors of squares. For example, the unit open ball in the metric d_1 is $B_1(0; d_1) = \{w \in \mathbb{C} \mid d_1(0, w) < 1\} = \{w \in \mathbb{C} \mid |\text{Re } w| + |\text{Im } w| < 1\}$, which is the interior of a square with vertices $1, i, -1$, and $-i$. Similarly, the unit open ball in the metric d_2 is $B_1(0; d_2) = \{w \in \mathbb{C} \mid d_2(0, w) < 1\} = \{w \in \mathbb{C} \mid \max\{|\text{Re } w|, |\text{Im } w|\} < 1\}$, which is the interior of a square with vertices $1 + i, -1 + i, -1 - i$, and $1 - i$.

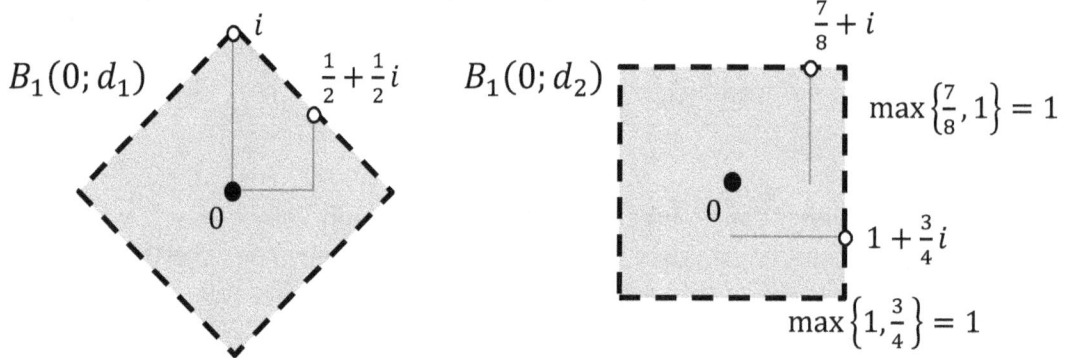

5. Define the function $\rho: \mathbb{R}^n \times \mathbb{R}^n \to \mathbb{R}$ by $\rho(\mathbf{x}, \mathbf{y}) = \max\{|x_1 - y_1|, \ldots, |x_n - y_n|\}$. Let's check that the 3 properties of a metric space are satisfied for $(\mathbb{R}^n, \rho)$. We have $\rho(\mathbf{x}, \mathbf{y}) = 0$ if and only if $|x_k - y_k| = 0$ for each $k = 1, 2, \ldots, n$ if and only if $x_k - y_k = 0$ for each $k = 1, 2, \ldots, n$ if and only if $x_k = y_k$ for each $k = 1, 2, \ldots, n$ if and only if $\mathbf{x} = \mathbf{y}$. This shows that Property 1 holds. Since $|a - b| = |b - a|$ for all real numbers a and b, it is clear that Property 2 holds. For Property 3, we first note that by the Triangle Inequality (and SACT), we have

$$|x_k - z_k| = |(x_k - y_k) + (y_k - z_k)| \leq |x_k - y_k| + |y_k - z_k| \text{ for each } k = 1, 2, \ldots, n.$$

By the definition of ρ, $|x_k - z_k| \leq \rho(\mathbf{x}, \mathbf{y}) + \rho(\mathbf{y}, \mathbf{z})$ for each $k = 1, 2, \ldots, n$. Again, by the definition of ρ, $\rho(\mathbf{x}, \mathbf{z}) \leq \rho(\mathbf{x}, \mathbf{y}) + \rho(\mathbf{y}, \mathbf{z})$. Therefore, property 3 holds.

We call ρ the **square metric** on $\mathbb{R}^n$.

Note that for $n = 1$, the square metric is the same as the Euclidean metric.

In Problem 6 below, you will be asked to prove that ρ induces the same topology as the Euclidean metric $d: \mathbb{R}^n \times \mathbb{R}^n \to \mathbb{R}$ defined by $d(\mathbf{x}, \mathbf{y}) = |\mathbf{x} - \mathbf{y}|$ (see part 3 above) and that this topology is the same as the product topology on $\mathbb{R}^n$.

6. We can turn any nonempty set S into a metric space by defining $d: S \times S \to \mathbb{R}$ by

$$d(x, y) = \begin{cases} 0 & \text{if } x = y \\ 1 & \text{if } x \neq y \end{cases}$$

Properties 1 and 2 are obvious. For Property 3, let $x, y, z \in S$. If $x = z$, then $d(x, z) = 0$, and so, $d(x, z) = 0 \leq d(x, y) + d(y, z)$. If $x \neq z$, then $d(x, z) = 1$. Also, y cannot be equal to both x and z (otherwise $y = x \wedge y = z \to x = z$). So, $d(x, y) = 1$ or $d(y, z) = 1$ (or both). Therefore, $d(x, y) + d(y, z) \geq 1 = d(x, z)$.

If $r > 1$, then $B_r(x) = S$ and if $0 < r \leq 1$, then $B_r(x) = \{x\}$. It follows that every singleton set $\{x\}$ is open and therefore, (S, d) induces the discrete topology on S. Therefore, we will call d the **discrete metric** on S.

Let (S, d) be a metric space. A subset A of S is **bounded** in S if there is $M \in \mathbb{R}^+$ such that for all $x, y \in A$, $d(x, y) \leq M$. If a bounded set A is nonempty, we define the diameter of A to be

$$\text{diam } A = \sup\{d(x, y) \mid x, y \in A\}.$$

If we define $\bar{d} : S \times S \to \mathbb{R}$ by $\bar{d}(x, y) = \min\{1, d(x, y)\}$, then $(S, \bar{d})$ is a metric space such that $\bar{d}$ induces the same topology as d. You will be asked to prove this in Problem 3 below. $\bar{d}$ is called the **standard bounded metric corresponding to d**. We say that $\bar{d}$ is **bounded** because for all $A \subseteq S$, we have that diam A is a finite number. In fact, it is easy to see that for all $A \subseteq S$, diam $A \leq 1$.

Example 11.2: Consider the metric $d : \mathbb{R} \times \mathbb{R} \to \mathbb{R}$ defined by $d(x, y) = |x - y|$ (see part 2 of Example 11.1). The metric d is **not** bounded. Indeed, for each $n \in \mathbb{N}$, $d(0, n) = n$, showing that diam $\mathbb{R} = \infty$. For the metric $\bar{d}$, we have that for all $n \in \mathbb{N}$ with $n \neq 0$, $\bar{d}(0, n) = 1$. By the previous paragraph (and Problem 3 below), we see that we have two metrics that induce the same topology, one bounded and the other not bounded. This shows that boundedness is **not** a topological property. In other words, it doesn't make sense to ask if a topological space is bounded. We will see a more precise definition of topological property in Lesson 13.

Let (S, d) be a metric space and let K be an index set. Then $({}^K S, \rho)$ is a metric space, where $\rho : {}^K S \times {}^K S \to S$ is defined by

$$\rho(\mathbf{x}, \mathbf{y}) = \sup\{\bar{d}(x_k, y_k) \mid k \in K\},$$

where $\bar{d}$ is the standard bounded metric corresponding to d.

Let's check carefully that $({}^K S, \rho)$ is a metric space. We have $\rho(\mathbf{x}, \mathbf{y}) = 0$ if and only if $\bar{d}(x_k, y_k) = 0$ for each $k \in K$ if and only if $x_k = y_k$ for each $k \in K$ (because $\bar{d}$ is a metric) if and only if $\mathbf{x} = \mathbf{y}$. This shows that Property 1 holds. Since $\bar{d}(x_k, y_k) = \bar{d}(y_k, x_k)$ for each $k \in K$ (because $\bar{d}$ is a metric), it is clear that Property 2 holds. For Property 3, we first note that since $\bar{d}$ is a metric, for each $k \in K$, we have

$$\bar{d}(x_k, z_k) \leq \bar{d}(x_k, y_k) + \bar{d}(y_k, z_k).$$

By the definition of ρ, $\bar{d}(x_k, z_k) \leq \rho(\mathbf{x}, \mathbf{y}) + \rho(\mathbf{y}, \mathbf{z})$ for each $k \in K$. Again, by the definition of ρ, $\rho(\mathbf{x}, \mathbf{z}) \leq \rho(\mathbf{x}, \mathbf{y}) + \rho(\mathbf{y}, \mathbf{z})$. Therefore, property 3 holds.

We call ρ the **uniform metric** on ${}^K S$. The metric ρ induces a topology called the **uniform topology** on ${}^K S$. In Problem 10 below, you will be asked to prove that the uniform topology on ${}^K S$ is finer than the product topology on ${}^K S$.

Example 11.3: Let K be an infinite index set and consider ${}^K \mathbb{R}$ with the topologies $\mathcal{T}$ and $\mathcal{U}$, where $\mathcal{T}$ is the product topology on ${}^K \mathbb{R}$ and $\mathcal{U}$ is the uniform topology on ${}^K \mathbb{R}$. By Problem 10 below, $\mathcal{U}$ is finer than $\mathcal{T}$. In this example, $\mathcal{U}$ is strictly finer than $\mathcal{T}$. To see this, let $\mathbf{0} = (0)$ be the element of ${}^K \mathbb{R}$ with all coordinates 0. Then the basic open set $B_1(\mathbf{0}; \rho) \in \mathcal{U}$ satisfies $B_1(\mathbf{0}; \rho) \subseteq \prod(-1, 1)$. Since a basic open set in $\mathcal{T}$ has the form $\prod U_k$, where for all but finitely many $k \in \mathbb{N}$ (and in particular, for at least one $k \in \mathbb{N}$), $U_k = \mathbb{R}$, it follows that there is no basic open set $U \in \mathcal{T}$ with $U \subseteq B_1(\mathbf{0}; \rho)$. So, $B_1(\mathbf{0}; \rho)$ is a set that is open in the uniform topology on ${}^K \mathbb{R}$, but not in the product topology on ${}^K \mathbb{R}$.

Complete Metric Spaces

Recall that for a topological space $(S, \mathcal{T})$ and a sequence (s_n) with $s_n \in S$ for all $n \in \mathbb{N}$, we say that (s_n) **converges** to $s \in S$, written $s_n \to s$, if for every open set U containing s, there is $K \in \mathbb{N}$ such that $n > K$ implies $s_n \in U$.

If the space $(S, \mathcal{T})$ is induced by a metric d, then we have the following.

Theorem 11.4: Let $(S, \mathcal{T})$ be a metrizable topological space such that $\mathcal{T}$ is induced by the metric d, let $s \in S$, and let (s_n) be a sequence with $s_n \in S$ for all $n \in \mathbb{N}$. The following are equivalent.

1. $s_n \to s$.

2. For every open ball $B_r(s)$ with center s, there is $K \in \mathbb{N}$ such that $n > K$ implies $s_n \in B_r(s)$.

3. For every $r \in \mathbb{R}^+$, there is $K \in \mathbb{N}$ such that $n > K$ implies $d(s_n, s) < r$.

Proof: (1→2) Suppose that $s_n \to s$ and let $r \in \mathbb{R}^+$. Since $B_r(s)$ is open, there is $K \in \mathbb{N}$ such that $n > K$ implies $s_n \in B_r(s)$.

(2→3) Suppose 2 holds and let $r \in \mathbb{R}^+$. By 2, there is $K \in \mathbb{N}$ such that $n > K$ implies $s_n \in B_r(s)$. But $s_n \in B_r(s)$ if and only if $d(s_n, s) < r$.

(3→1) Suppose that 3 holds and let U be an open set containing s. Since the open balls form a basis for $\mathcal{T}$, there is $r \in \mathbb{R}^+$ such that $B_r(s) \subseteq U$. By 3, there is $K \in \mathbb{N}$ such that $n > K$ implies $d(s_n, s) < r$. But, if $d(s_n, s) < r$, then $s_n \in B_r(s)$, and since $B_r(s) \subseteq U$, $s_n \in U$. So, $s_n \to s$. $\square$

We now show that if (s_n) is a sequence in a metrizable topological space that converges to a point s in the space, then the coordinates of the sequence are getting closer to each other as we go out further into the sequence. Specifically, we have the following.

Theorem 11.5: Let $(S, \mathcal{T})$ be a metrizable topological space such that $\mathcal{T}$ is induced by the metric d and let (s_n) be a sequence in S such that $s_n \to s$. Then for every $r \in \mathbb{R}^+$, there is $K \in \mathbb{N}$ such that $m \geq n > K$ implies $d(s_m, s_n) < r$.

Proof: Suppose that $s_n \to s$ and let $r \in \mathbb{R}^+$. By Theorem 11.4 (3), there is $K \in \mathbb{N}$ such that $n > K$ implies $d(s_n, s) < \frac{r}{2}$. Let $m \geq n > K$. Then $d(s_m, s_n) \leq d(s_m, s) + d(s, s_n) < \frac{r}{2} + \frac{r}{2} = r$. $\square$

A sequence satisfying the conclusion of Theorem 11.5 is called a **Cauchy sequence**. Note that this definition generalizes the definition of a rational-valued Cauchy sequence that was used to define the real numbers (see Lesson 5).

In general, the converse of Theorem 11.5 is not true. That is, not every Cauchy sequence converges. For example, consider the space $(0, 1)$ as a subspace of $\mathbb{R}$ and let $s_n = \frac{1}{n+1}$ for each $n \in \mathbb{N}$. Then (s_n) is a Cauchy sequence (see part 1 of Example 5.18), but (s_n) does **not** converge. The idea here is that if a sequence converges, then it converges to a point in the space. In this case, the space $(0, 1)$ is missing the point 0 to which the Cauchy sequence would like to converge. We say that the space $(0, 1)$ is not complete.

A metric space (S, d) is a **complete metric space** if every Cauchy sequence converges to a point in the space.

We will next show that for a metric space to be complete, it suffices that every Cauchy sequence have a convergent **subsequence**. First, let us explain what we mean by a subsequence.

The formal definition of a subsequence may seem a bit confusing at first. So, let's begin by looking at a simple example.

Consider the sequence $(s_n) = (\sqrt{n}) = (0, 1, \sqrt{2}, \sqrt{3}, 2, \sqrt{5}, \sqrt{6}, \sqrt{7}, \dots)$.

An example of a subsequence of this sequence is $(0, \sqrt{2}, 2, \sqrt{6}, \dots)$. Notice how we're forming this new sequence by "picking out" every other member of the range of the original sequence.

Using function notation, the original sequence can be described as the function $f: \mathbb{N} \to \mathbb{R}$ defined by $f(n) = \sqrt{n}$.

The subsequence can be described as the function $h: \mathbb{N} \to \mathbb{R}$ defined by $h(k) = \sqrt{2k}$ (I am using the variable k instead of n for the function h only to make the following explanation easier to follow).

Observe that we get from the function f to the function h by replacing n by $2k$. Formally, h is the composition of the functions f and g ($h = f \circ g$), where $g: \mathbb{N} \to \mathbb{N}$ is defined by $g(k) = 2k$. Indeed, we have $(f \circ g)(k) = f(g(k)) = f(2k) = \sqrt{2k} = h(k)$.

Getting back to our original notation, if we let $s_n = f(n) = \sqrt{n}$ and we let $n_k = g(k) = 2k$, then,

$$(s_{n_k}) = (s_{2k}) = \sqrt{2k}.$$

Let (s_n) be a sequence in S. In other words, (s_n) is a function $f: \mathbb{N} \to S$, where $f(n) = s_n$ for each $n \in \mathbb{N}$. Let $g: \mathbb{N} \to \mathbb{N}$ be a strictly increasing function. Then $f \circ g: \mathbb{N} \to S$ is called a **subsequence** of f.

Note: If we let $n_k = g(k)$ for each $k \in \mathbb{N}$, we see that $(f \circ g)(k) = f(g(k)) = f(n_k) = s_{n_k}$. So, we may use the notation (s_{n_k}) to represent a subsequence of (s_n). Since g is a strictly increasing function, it follows that $n_0 < n_1 < n_2 < \cdots$.

Example 11.6:

1. Let $(s_n) = ((-1)^n) = (1, -1, 1, -1, 1, -1, \dots)$. Then (s_n) is a $\mathbb{Z}$-valued sequence. That is, (s_n) is the function $f: \mathbb{N} \to \mathbb{Z}$ given by $f(n) = (-1)^n$. Now, $(s_{n_k}) = (s_{2k}) = (1) = (1, 1, 1, \dots)$ is a subsequence of (s_n). Here, we have $n_k = 2k$. In function notation, n_k is the function $g: \mathbb{N} \to \mathbb{N}$ given by $g(k) = 2k$, and s_{n_k} is the composition $f \circ g: \mathbb{N} \to \mathbb{Z}$, where

    $$(f \circ g)(k) = f(g(k)) = f(2k) = (-1)^{2k} = 1.$$

    Another subsequence is given by $(s_{n_j}) = (s_{2j+1}) = (-1) = (-1, -1, -1, \dots)$. Here, we have $n_j = 2j + 1$. In function notation, n_j is the function $g: \mathbb{N} \to \mathbb{N}$ given by $g(j) = 2j + 1$, and s_{n_j} is the composition $f \circ g: \mathbb{N} \to \mathbb{Z}$, where

    $$(f \circ g)(j) = f(g(j)) = f(2j + 1) = (-1)^{2j+1} = -1.$$

There are many other subsequences of (s_n). For example, (s_{n_t}), where $n_t = \frac{t(t+1)}{2}$ is the subsequence $(1, -1, -1, 1, 1, -1, -1, 1, 1, \ldots)$.

Observe that the sequence (s_n) is not a Cauchy sequence and it does not converge (in any space). The subsequences (s_{n_k}) and (s_{n_j}) converge to 1 and -1, respectively, whereas the subsequence (s_{n_t}) does not converge.

2. Let $(s_n) = (1, 1.4, 1.41, 1.414, 1.4142, \ldots)$ (this is a $\mathbb{Q}$-valued sequence that looks like it's trying to converge to $\sqrt{2}$). Here, (s_n) is a function $f: \mathbb{N} \to \mathbb{Q}$. An example of a subsequence of (s_n) is $(s_{n_k}) = (s_{k^2}) = (1, 1.4, 14142, \ldots)$. Here, we have $n_k = k^2$. In function notation, n_k is the function $g: \mathbb{N} \to \mathbb{N}$ given by $g(k) = k^2$, and s_{n_k} is the composition $f \circ g: \mathbb{N} \to \mathbb{Z}$, where

$$(f \circ g)(k) = f(g(k)) = f(k^2) = s_{k^2}.$$

The sequence (s_n) is a Cauchy sequence, but we cannot say whether or not the sequence converges without knowing what space we are in. For example, in the metric space $(\mathbb{R}, d)$, where d is the Euclidean metric, (s_n) converges to $\sqrt{2}$. However, in the metric space $(\mathbb{Q}, d)$ (with the same metric), (s_n) does not converge. We can say the same of the subsequence $(s_{n_k}) = (s_{k^2})$. It converges in $(\mathbb{R}, d)$ and fails to converge in $(\mathbb{Q}, d)$.

Theorem 11.7: Let (S, d) be a metric space, let (s_n) be a Cauchy sequence in S such that (s_n) has a subsequence (s_{n_k}) such that $s_{n_k} \to s$ for some $s \in S$. Then $s_n \to s$.

Proof: Let (s_n) be a Cauchy sequence in S, let $s \in S$, let (s_{n_k}) be a subsequence of (s_n) such that $s_{n_k} \to s$, and let $r \in \mathbb{R}^+$. Since (s_n) is a Cauchy sequence, there is $K_1 \in \mathbb{N}$ such that $m \geq n > K_1$ implies $d(s_m, s_n) < \frac{r}{2}$. Since $s_{n_k} \to s$, by Theorem 11.4, there is $K_2 \in \mathbb{N}$ so that $k > K_2$ implies $d(s_{n_k}, s) < \frac{r}{2}$. Let $K = \max\{K_1, K_2\} + 1$ and let $m > K$. Since $K > K_2$, $d(s_{n_K}, s) < \frac{r}{2}$. Since $K > K_1$, we have $m > K_1$. We also have $n_K \geq K > K_1$. Therefore, we have either $m \geq n_K > K_1$ or $n_K \geq m > K_1$. In either case, $d(s_m, s_{n_K}) < \frac{r}{2}$. It follows that $d(s_m, s) \leq d(s_m, s_{n_K}) + d(s_{n_K}, s) < \frac{r}{2} + \frac{r}{2} = r$. So, $s_n \to s$. $\quad\square$

Corollary 11.8: A metric space (S, d) is complete if and only if every Cauchy sequence in S has a subsequence that converges to a point in S.

Proof: Let (S, d) be a metric space. First assume that (S, d) is complete and let (s_n) be a Cauchy sequence in S. Then there is $s \in S$ such that $s_n \to s$. Since (s_n) is a subsequence of itself, (s_n) has a subsequence (itself) that converges to a point in S.

Conversely, assume that every Cauchy sequence in S has a subsequence that converges to a point in S. By Theorem 11.7, every Cauchy sequence converges to a point in S. Therefore, (S, d) is complete. $\quad\square$

Let (S, d) be a metric space. A sequence (s_n) is **bounded** in S if the set $\{s_n\}$ is bounded. In other words, (s_n) is bounded if there is $M \in \mathbb{R}^+$ such that for all $n, m \in \mathbb{N}$, $d(s_n, s_m) \leq M$. We will now show that all Cauchy sequences are bounded.

Note: If $S = \mathbb{R}$ and d is the Euclidean metric on $\mathbb{R}$, then a sequence (s_n) is bounded in $\mathbb{R}$ if and only if there is $M \in \mathbb{R}^+$ such that $|s_n| \leq M$ for all $n \in \mathbb{N}$. You will be asked to prove this in Problem 2 below.

Theorem 11.9: Let (S, d) be a metric space and let (s_n) be a Cauchy sequence in S. Then (s_n) is bounded in S.

Proof: Let (S, d) be a metric space and let (s_n) be a Cauchy sequence in S. Then for every $r \in \mathbb{R}^+$, there is $K \in \mathbb{N}$ such that $m \geq n > K$ implies $d(s_m, s_n) < r$. In particular, by letting $r = 1$, we see that there is $K \in \mathbb{N}$ such that $m \geq n > K$ implies $d(s_m, s_n) < 1$. Let

$$M = \max\{d(s_0, s_{K+1}), d(s_1, s_{K+1}), \dots, d(s_K, s_{K+1}), 1\}.$$

Then if $m, n \in \mathbb{N}$, we have $d(s_m, s_n) \leq d(s_m, s_{K+1}) + d(s_{K+1}, s_n) \leq M + M = 2M$. Therefore, (s_n) is bounded in S. $\qquad\square$

We say that a metric space (S, d) has the **Bolzano-Weierstrass Property** if every sequence that is bounded in S has a subsequence that converges to a point of S. By Corollary 11.8 and Theorem 11.9, we see that any metric space with the Bolzano-Weierstrass Property is complete.

Example 11.10:

1. $(\mathbb{R}, d)$, where d is the Euclidean metric, has the Bolzano-Weierstrass Property. You will be asked to prove this in Problem 9 below.

2. $(\mathbb{R}^2, d)$, where d is the Euclidean metric, has the Bolzano-Weierstrass Property. To see this, let $(\mathbf{x}_n)$ be a bounded sequence in $\mathbb{R}^2$, let's say $|\mathbf{x}_n - \mathbf{x}_m| \leq M$ for all $n, m \in \mathbb{N}$. Let's write the coordinates of the sequence in component form, say $\mathbf{x}_n = (x_n, y_n)$. We first show that each of the sequences (x_n) and (y_n) is bounded in $\mathbb{R}$. (x_n) is bounded in $\mathbb{R}$ because $|x_n - x_m| \leq \sqrt{(x_n - x_m)^2 + (y_n - y_m)^2} = |\mathbf{x}_n - \mathbf{x}_m| \leq M$, and similarly for (y_n).

   Since $\mathbb{R}$ has the Bolzano-Weierstrass Property (by 1 above), the sequence (x_n) has a convergent subsequence (x_{n_k}), say $x_{n_k} \to x$. Let $(\mathbf{x}_{n_k}) = (x_{n_k}, y_{n_k})$ be the corresponding subsequence of $(\mathbf{x}_n)$ in $\mathbb{R}^2$. (y_{n_k}) is bounded in $\mathbb{R}$ because $\{y_{n_k}\} \subseteq \{y_n\}$ and (y_n) is bounded in $\mathbb{R}$. It follows once again by the Bolzano-Weierstrass Property for $\mathbb{R}$ that $\{y_{n_k}\}$ has a convergent subsequence $(y_{n_{k_t}})$, say $y_{n_{k_t}} \to y$. I leave it to the reader to verify that $x_{n_{k_t}} \to x$. It follows that $\mathbf{x}_{n_{k_t}} \to (x, y)$.

3. For each $n \in \mathbb{N}$, $(\mathbb{R}^n, d)$ and $(\mathbb{C}^n, k)$ have the Bolzano-Weierstrass Property. Here d is the Euclidean metric and k is the unitary metric. The proofs are analogous to that given in part 2 above, and so, I leave them to the reader.

Theorem 11.11: For each $n \in \mathbb{N}$, $(\mathbb{R}^n, d)$ and $(\mathbb{C}^n, k)$, where d and k are the Euclidean and unitary metrics, respectively, are complete metric spaces.

Proof: By Example 11.10, for each $n \in \mathbb{N}$, $(\mathbb{R}^n, d)$ and $(\mathbb{C}^n, k)$ have the Bolzano-Weierstrass Property. By Theorem 11.9, every Cauchy sequence in each of these spaces is bounded. Using the Bolzano-Weierstrass Property, we see that every Cauchy sequence in each of these spaces has a convergent subsequence. By Corollary 11.8, each of these spaces is a complete metric space. $\qquad\square$

Theorem 11.12: Let $S = {}^{\mathbb{N}}\mathbb{R}$, let $\overline{d}: \mathbb{R} \times \mathbb{R} \to \mathbb{R}$ be the standard bounded metric on $\mathbb{R}$, and define $d^*: {}^{\mathbb{N}}\mathbb{R} \times {}^{\mathbb{N}}\mathbb{R} \to \mathbb{R}$ by $d^*(\mathbf{x}, \mathbf{y}) = \sup\left\{\frac{\overline{d}(x_k, y_k)}{k+1} \,\middle|\, k \in \mathbb{N}\right\}$. Then (S, d^*) is a complete metric space and d^* induces the product topology on $S = {}^{\mathbb{N}}\mathbb{R}$.

Proof: Note that given $\mathbf{x}, \mathbf{y} \in {}^{\mathbb{N}}\mathbb{R}$, we have $\overline{d}(x_k, y_k) \leq 1$ for all $k \in \mathbb{N}$. So, $\frac{\overline{d}(x_k, y_k)}{k+1} \leq \frac{1}{k+1} \leq 1$. By the Completeness Property of $\mathbb{R}$, $\sup\left\{\frac{\overline{d}(x_k, y_k)}{k+1} \,\middle|\, k \in \mathbb{R}\right\}$ exists and is a real number. So, $d^*: {}^{\mathbb{N}}\mathbb{R} \times {}^{\mathbb{N}}\mathbb{R} \to \mathbb{R}$.

We prove that $({}^{\mathbb{N}}\mathbb{R}, d^*)$ is a metric space. We have $d^*(\mathbf{x}, \mathbf{y}) = 0$ if and only if $\frac{\overline{d}(x_k, y_k)}{k+1} = 0$ for each $k \in \mathbb{N}$ if and only if $\overline{d}(x_k, y_k) = 0$ for each $k \in \mathbb{N}$ if and only if $x_k = y_k$ for each $k \in \mathbb{N}$ (because $\overline{d}$ is a metric on $\mathbb{R}$) if and only if $\mathbf{x} = \mathbf{y}$. This shows that Property 1 holds. Since $\overline{d}(a, b) = \overline{d}(b, a)$ for all real numbers a and b, it is clear that Property 2 holds. For Property 3, we first note that for each $k \in \mathbb{N}$,

$$\frac{\overline{d}(x_k, z_k)}{k+1} \leq \frac{\overline{d}(x_k, y_k)}{k+1} + \frac{\overline{d}(y_k, z_k)}{k+1}.$$

This follows from the fact that $\overline{d}$ is a metric on $\mathbb{R}$. By the definition of d^*, for each $k \in \mathbb{N}$,

$$\frac{\overline{d}(x_k, z_k)}{k+1} \leq d^*(\mathbf{x}, \mathbf{y}) + d^*(\mathbf{y}, \mathbf{z}).$$

Again, by the definition of d^*, $d^*(\mathbf{x}, \mathbf{z}) \leq d^*(\mathbf{x}, \mathbf{y}) + d^*(\mathbf{y}, \mathbf{z})$. Therefore, property 3 holds.

We now check that (S, d^*) is complete. Let $(\mathbf{x}^n)$ be a Cauchy sequence in (S, d^*). By part (ii) of Problem 4 below, for each $k \in \mathbb{N}$, the sequence (x_k^n) is a Cauchy sequence in $\mathbb{R}$. By Theorem 11.11, for each $k \in \mathbb{N}$, (x_k^n) converges to an element $c_n \in \mathbb{R}$. Let $\mathbf{x} = (c_n)$. Then $\mathbf{x}^n \to \mathbf{x}$.

Finally, we check that d^* induces the product topology on S. Let U be a basic open set in the product topology on ${}^{\mathbb{N}}\mathbb{R}$ with $\mathbf{x} \in U$, say $U = \prod U_k$, where $U_k = \mathbb{R}$ for all but finitely many $k \in \mathbb{N}$, and if $U_k \neq \mathbb{R}$, then $U_k = B_{r_k}(x_k; \overline{d})$ for some $r_k \in \mathbb{R}$. Let $r = \min\left\{\frac{r_k}{k+1} \,\middle|\, U_k = B_{r_k}(x_k; \overline{d})\right\}$. We will show that $B_r(\mathbf{x}; d^*) \subseteq U$. To see this, let $\mathbf{y} \in B_r(\mathbf{x}; d^*)$, so that $d^*(\mathbf{x}, \mathbf{y}) < r$. Then for each $n \in \mathbb{N}$ with $U_n = B_{r_n}(x_n; \overline{d})$, we have $\frac{\overline{d}(x_n, y_n)}{n+1} \leq \sup\left\{\frac{\overline{d}(x_k, y_k)}{k+1} \,\middle|\, k \in \mathbb{N}\right\} = d^*(\mathbf{x}, \mathbf{y}) < r \leq \frac{r_n}{n+1}$.

So, for each $n \in \mathbb{N}$ with $U_n = B_{r_n}(x_n; \overline{d})$, we have $\overline{d}(x_n, y_n) < r_n$, and thus, $y_n \in B_{r_n}(x_n; \overline{d})$. For each $n \in \mathbb{N}$ with $U_n = \mathbb{R}$, clearly, $y_n \in U_n$. Therefore, $\mathbf{y} \in U$, as desired. Thus, the topology induced by d^* is finer that the product topology on ${}^{\mathbb{N}}\mathbb{R}$.

Conversely, let $B_r(\mathbf{x}; d^*)$ be an arbitrary open ball in $({}^{\mathbb{N}}\mathbb{R}, d^*)$ and let $\mathbf{y} \in B_r(\mathbf{x}; d^*)$. It follows that $d^*(\mathbf{x}, \mathbf{y}) < r$. Since $\frac{1}{n+1} \to 0$, we can find $K \in \mathbb{N}$ such that for all $n > K$, $\frac{1}{n+1} < \frac{r}{2}$. For $n = 0, 1, \dots, K$, we let $U_n = B_{\frac{r}{2}}(x_n; \overline{d})$. For $n > K$, we let $U_n = \mathbb{R}$. We will show that the basic open set $\prod U_k$ in the product topology of $({}^{\mathbb{N}}\mathbb{R}, \overline{d})$ satisfies $\prod U_k \subseteq B_r(\mathbf{x}; d^*)$. To see this, let $\mathbf{y} \in \prod U_k$. Then for $n \leq K$, we have $y_n \in B_{\frac{r}{2}}(x_n; \overline{d})$, and so, $\overline{d}(x_n, y_n) < \frac{r}{2}$, or equivalently, $\frac{\overline{d}(x_n, y_n)}{n+1} < \frac{r}{2(n+1)}$. Since $\frac{r}{2(n+1)} \leq \frac{r}{2}$, we have $\frac{\overline{d}(x_n, y_n)}{n+1} < \frac{r}{2}$. For $n > K$, we have $y_n \in \mathbb{R}$, and so $\overline{d}(x_n, y_n) \leq 1$, or equivalently, $\frac{\overline{d}(x_n, y_n)}{n+1} \leq \frac{1}{n+1} < \frac{r}{2}$. So,

$$d^*(\mathbf{x}, \mathbf{y}) = \sup\left\{\frac{\overline{d}(x_k, y_k)}{k+1} \,\middle|\, k \in \mathbb{N}\right\} \leq \frac{r}{2} < r.$$

Thus, the product topology on ${}^{\mathbb{N}}\mathbb{R}$ is finer than the topology induced by d^*. Since the topology induced by d^* is finer than the product topology on ${}^{\mathbb{N}}\mathbb{R}$ and the product topology on ${}^{\mathbb{N}}\mathbb{R}$ is finer than the topology induced by d^*, d^* induces the product topology on ${}^{\mathbb{N}}\mathbb{R}$. $\qquad\square$

Problem Set 11

Full solutions to these problems are available for free download here:

www.SATPrepGet800.com/TFBZLF

LEVEL 1

1. Let (S, d) be a metric space. Prove that for all $x \in S$, $d(x, x) \geq 0$.

2. Prove that a sequence (s_n) is bounded in $\mathbb{R}$ (with the Euclidean metric) if and only if there is $M \in \mathbb{R}^+$ such that $|s_n| \leq M$ for all $n \in \mathbb{N}$.

LEVEL 2

3. Let (S, d) be a metric space and let $\overline{d}$ be the standard bounded metric corresponding to d. Prove that $(S, \overline{d})$ is a metric space and that $\overline{d}$ induces the same topology as d.

4. Let $d: \mathbb{R} \times \mathbb{R} \to \mathbb{R}$ be the Euclidean metric on $\mathbb{R}$, let $\overline{d}: \mathbb{R} \times \mathbb{R} \to \mathbb{R}$ be the standard bounded metric corresponding to d, and let $d^*: {}^{\mathbb{N}}\mathbb{R} \times {}^{\mathbb{N}}\mathbb{R} \to \mathbb{R}$ be the metric on $\mathbb{R}^n$ defined by $d^*(\mathbf{x}, \mathbf{y}) = \sup\left\{\frac{\overline{d}(x_k, y_k)}{k+1} \;\middle|\; k \in \mathbb{N}\right\}$ (see Theorem 11.12). Prove each of the following:

   (i) (s_n) is a Cauchy sequence in $(\mathbb{R}, d)$ if and only if (s_n) is a Cauchy sequence in $(\mathbb{R}, \overline{d})$

   (ii) If $(\mathbf{x}^n)$ is a Cauchy sequence in $({}^{\mathbb{N}}\mathbb{R}, d^*)$, then for each $k \in K$, the sequence (x_k^n) is a Cauchy sequence in $(\mathbb{R}, d)$.

LEVEL 3

5. Define the functions d_1 and d_2 from $\mathbb{C} \times \mathbb{C}$ to $\mathbb{R}$ by $d_1(z, w) = |\operatorname{Re} z - \operatorname{Re} w| + |\operatorname{Im} z - \operatorname{Im} w|$ and $d_2(z, w) = \max\{|\operatorname{Re} z - \operatorname{Re} w|, |\operatorname{Im} z - \operatorname{Im} w|\}$. Prove that $(\mathbb{C}, d_1)$ and $(\mathbb{C}, d_2)$ are metric spaces such that d_1 and d_2 induce the standard topology on $\mathbb{C}$.

6. Prove that the square metric on $\mathbb{R}^n$ induces the same topology as the Euclidean metric on $\mathbb{R}^n$. Then prove that the topology induced by these metrics is the same as the product topology on $\mathbb{R}^n$.

LEVEL 4

7. Prove that every metrizable space is a T_4-space.

8. Let $(S_1, \mathcal{T}_1)$ and $(S_2, \mathcal{T}_2)$ be metrizable spaces. Prove that $S_1 \times S_2$ with the product topology is metrizable. Use this to show that $(\mathbb{R}, \mathcal{T}_L)$ is not metrizable (see Problem 14 from Problem Set 9 for the definition of $\mathcal{T}_L$).

LEVEL 5

9. Prove that $(\mathbb{R}, d)$, where d is the Euclidean metric, has the Bolzano-Weierstrass Property. You may use the following fact: if $\mathcal{C} = \{[a_k, b_k] \mid k \in \mathbb{N}\}$ is a sequence of closed intervals in $\mathbb{R}$, such that $j < k \rightarrow [a_k, b_k] \subseteq [a_j, b_j]$, then $\cap \mathcal{C} \neq \emptyset$ (Example 12.3 together with the Heine-Borel Theorem (Theorem 12.6) below will explain why this is true).

10. Let (S, d) be a metric space. Prove that for any index set K, the uniform metric on $^K S$ is finer than the product topology on $^K S$.

11. Prove that for any index set K, $(^K\mathbb{R}, \rho)$ is a complete metric space, where ρ is the uniform metric on $^K\mathbb{R}$.

CHALLENGE PROBLEM

12. Prove that the product topology on $^K\mathbb{R}$ is metrizable if and only if K is countable.

LESSON 12
COMPACTNESS

Compact Spaces

Let $(S, \mathcal{T})$ be a topological space. A collection $\mathcal{C}$ of subsets of S is a **covering** of S (or we can say that $\mathcal{C}$ **covers** S) if $\bigcup \mathcal{C} = S$. If $\mathcal{C}$ consists of only open sets, then we will say that $\mathcal{C}$ is an **open covering** of S.

A topological space $(S, \mathcal{T})$ is **compact** if every open covering of S contains a finite subcollection that still covers S (we will call such a finite subcollection a finite **subcover** of the covering).

Example 12.1:

1. If S is a finite set, then for any topology $\mathcal{T}$ on S, $(S, \mathcal{T})$ is compact. After all, any open covering of S is already finite.

2. If S is an infinite set and $\mathcal{T}$ is the discrete topology on S, then $(S, \mathcal{T})$ is **not** compact. Indeed, $\{\{x\} \mid x \in S\}$ is an open covering of S with no finite subcollection covering S.

3. $(\mathbb{R}, \mathcal{T})$, where $\mathcal{T}$ is the standard topology on $\mathbb{R}$, is **not** compact. Indeed, $\{(n, n + 2) \mid n \in \mathbb{Z}\}$ is an open covering of $\mathbb{R}$ with no finite subcollection covering $\mathbb{R}$.

4. Let $(A, \mathcal{T}_A)$ be the topological space where $A \subseteq \mathbb{R}$ is defined by $A = \left\{\frac{1}{n} \mid n \in \mathbb{Z}^+\right\} \cup \{0\}$, and as usual $\mathcal{T}_A$ is the subspace topology. Then $(A, \mathcal{T}_A)$ is compact. To see this, let $\mathcal{C}$ be an open covering of A. Since $0 \in A$, there is $U \in \mathcal{C}$ with $0 \in U$. Then U contains a set of the form $(a, b) \cap A$ with $0 \in (a, b)$. By the Archimedean Property of the real numbers, there is $n \in \mathbb{N}$ such that $n > \frac{1}{b}$, or equivalently, $\frac{1}{n} < b$. So, $\left\{\frac{1}{k} \mid k \geq n\right\} \cup \{0\} \subseteq U$. For each $k \in \mathbb{Z}^+$ with $k < n$, let U_k be an open set in $\mathcal{C}$ such that $\frac{1}{k} \in U_k$. Then $\{U, U_1, U_2, \ldots, U_{k-1}\}$ is a finite subcover of $\mathcal{C}$.

   Note that if we remove 0 from the set A to get the set $B = \left\{\frac{1}{n} \mid n \in \mathbb{Z}^+\right\}$, then we wind up with a subspace of $\mathbb{R}$ that is **not** compact. This follows from the fact that $(B, \mathcal{T}_B)$ is the discrete topology and part 2 above ($\left\{\frac{1}{n}\right\}$ is open because $\left\{\frac{1}{n}\right\} = \left(\frac{1}{n+1}, \frac{1}{n-1}\right) \cap B$ for $n > 1$ and $\{1\} = \left(\frac{1}{2}, 2\right) \cap B$).

5. The open interval $(0, 1)$ considered as a subspace of $\mathbb{R}$ is **not** compact. The collection of open sets $\mathcal{C} = \left\{\left(\frac{1}{n}, 1\right) \mid n \in \mathbb{Z}^+\right\}$ is an open covering of $(0, 1)$ with no finite subcover. The collection $\mathcal{C} = \left\{\left(\frac{1}{n}, 1\right] \mid n \in \mathbb{Z}^+\right\}$ can be used to show that the half-open interval $(0, 1]$ (as a subspace of $\mathbb{R}$) is not compact. Furthermore, it is not hard to show that any open or half-open interval considered as a subspace of $\mathbb{R}$ is not compact. See Problem 1 below.

6. For any set S, the topological space $(S, \mathcal{T})$, where $\mathcal{T}$ is the cofinite topology on S (see part 3 of Example 10.2) is compact. To see this, let $\mathcal{C}$ be an open covering of S, and let A_0 be any set in $\mathcal{C}$. Then $S \setminus A_0$ is finite, say $S \setminus A_0 = \{a_1, a_2, \ldots, a_n\}$. For each $i = 1, 2, \ldots, n$, let $A_i \in \mathcal{C}$ with $a_i \in A_i$. Then the collection $\{A_0, A_1, A_2, \ldots, A_n\}$ is a finite subcollection from $\mathcal{C}$ that covers S.

211

The previous examples might be a bit misleading. Although those examples were quite simple, in general, it can be difficult to determine if a given topological space is compact. For example, consider the closed interval $[0, 1]$ as a subspace of $\mathbb{R}$. As it turns out, this is a compact topological space. However, the proof is not so easy. The compactness of $[0, 1]$ follows immediately from the Heine-Borel Theorem, which we will prove a bit later (Theorem 12.6 below).

We would now like to provide an equivalent definition of compactness in terms of closed sets instead of open sets.

Let $(S, \mathcal{T})$ be a topological space. A collection $\mathcal{C}$ of subsets of S has the **finite intersection property** if for every finite subcollection $\mathcal{D} \subseteq \mathcal{C}$, we have $\bigcap \mathcal{D} \neq \emptyset$.

Theorem 12.2: A topological space $(S, \mathcal{T})$ is compact if and only if for every collection $\mathcal{C}$ of closed sets in S with the finite intersection property, we have $\bigcap \mathcal{C} \neq \emptyset$.

We will prove both directions of Theorem 12.2 by contrapositive (see the analysis after the statement of Theorem 2.7 for an explanation of how this kind of proof works).

The contrapositive of the statement "If a topological space $(S, \mathcal{T})$ is compact, then for every collection $\mathcal{C}$ of closed sets in S with the finite intersection property, we have $\bigcap \mathcal{C} \neq \emptyset$" is the statement "If $(S, \mathcal{T})$ is a topological space and there is a collection $\mathcal{C}$ of closed sets in S with the finite intersection property such that $\bigcap \mathcal{C} = \emptyset$, then $(S, \mathcal{T})$ is not compact."

Similarly, the contrapositive of the statement "If $(S, \mathcal{T})$ is a topological space such that for every collection $\mathcal{C}$ of closed sets in S with the finite intersection property, $\bigcap \mathcal{C} \neq \emptyset$, then $(S, \mathcal{T})$ is compact" is the statement "If a topological space $(S, \mathcal{T})$ is not compact, then there is a collection $\mathcal{C}$ of closed sets in S with the finite intersection property such that $\bigcap \mathcal{C} = \emptyset$."

Let's prove the theorem.

Proof of Theorem 12.2: Let $(S, \mathcal{T})$ be a topological space and suppose there is a collection $\mathcal{C}$ of closed sets in S with the finite intersection property such that $\bigcap \mathcal{C} = \emptyset$. Let $\mathcal{D} = \{S \setminus B \mid B \in \mathcal{C}\}$. Then $\mathcal{D}$ is a collection of open sets in S. Since $\bigcap \mathcal{C} = \emptyset$, by De Morgan's law, $\bigcup \mathcal{D} = S \setminus \bigcap \mathcal{C} = S \setminus \emptyset = S$. Let $\mathcal{E}$ be a finite subcollection of elements in $\mathcal{D}$, say $\mathcal{E} = \{S \setminus B_1, S \setminus B_2, \ldots, S \setminus B_k\}$. Then $\{B_1, B_2, \ldots, B_k\}$ is a finite subcollection of $\mathcal{C}$. Since $\mathcal{C}$ has the finite intersection property, there is $x \in B_1 \cap B_2 \cap \cdots \cap B_k$. So, $x \notin S \setminus (B_1 \cap B_2 \cap \cdots \cap B_k) = (S \setminus B_1) \cup (S \setminus B_2) \cup \cdots \cup (S \setminus B_k)$. It follows that $\mathcal{E}$ does not cover S. Since $\mathcal{E}$ was an arbitrary finite subcollection of $\mathcal{D}$, we have found an open covering of S without a finite subcollection that covers S. Therefore, $(S, \mathcal{T})$ is not compact.

Conversely, assume that the topological space $(S, \mathcal{T})$ is **not** compact. Then there is an open covering $\mathcal{C}$ of S such that no finite subcollection of $\mathcal{C}$ covers S. Let $\mathcal{D} = \{S \setminus U \mid U \in \mathcal{C}\}$. Then $\mathcal{D}$ is a collection of closed sets. Since $\bigcup \mathcal{C} = S$, by De Morgan's law, we have $\bigcap \mathcal{D} = S \setminus \bigcup \mathcal{C} = S \setminus S = \emptyset$. Let $\mathcal{E}$ be a finite subcollection of elements in $\mathcal{D}$, say $\mathcal{E} = \{S \setminus U_1, S \setminus U_2, \ldots, S \setminus U_k\}$. Then $\{U_1, U_2, \ldots, U_k\}$ is a finite subcollection of $\mathcal{C}$. Since $\mathcal{C}$ does not have a finite subcollection that covers S, there is $x \in S$ such that $x \notin U_1 \cup U_2 \cup \cdots \cup U_k$. So, $x \in S \setminus (U_1 \cup U_2 \cup \cdots \cup U_k) = (S \setminus U_1) \cap (S \setminus U_2) \cap \cdots \cap (S \setminus U_k)$. Since $\mathcal{E}$ was an arbitrary finite subcollection of $\mathcal{D}$, we see that $\mathcal{D}$ has the finite intersection property. And yet, we saw that $\bigcap \mathcal{D} = \emptyset$. $\qquad \square$

Example 12.3: Let $(S, \mathcal{T})$ be a compact topological space and let $\mathcal{C} = \{B_k \mid k \in \mathbb{N}\}$ be a collection of nonempty closed sets such that $j < k \rightarrow B_k \subseteq B_j$. $\mathcal{C}$ is called a **nested sequence** of nonempty closed sets in S. We can visualize this sequence as follows.

$$B_0 \supseteq B_1 \supseteq B_2 \supseteq \cdots$$

This collection $\mathcal{C}$ clearly has the finite intersection property. So, by Theorem 12.2, $\cap \mathcal{C} \neq \emptyset$.

Recall from Lesson 9 that a **basis** $\mathcal{B}$ for a topology $\mathcal{T}$ on a set S is a collection of sets such that every element of $\mathcal{T}$ can be written as a union of elements of $\mathcal{B}$. We will call a collection of sets $\mathcal{C}$ a **closed basis** for $\mathcal{T}$ if the set $\mathcal{B} = \{S \setminus B \mid B \in \mathcal{C}\}$ is a basis for $\mathcal{T}$.

Also, recall from Lesson 9 that a **subbasis** $\mathcal{X}$ for a topology $\mathcal{T}$ on a set S is a collection of sets such that the collection of all finite intersections of sets in $\mathcal{X}$ forms a basis for $\mathcal{T}$. We will call a collection of sets $\mathcal{Y}$ a **closed subbasis** for $\mathcal{T}$ if the set $\mathcal{X} = \{S \setminus B \mid B \in \mathcal{Y}\}$ is a subbasis for $\mathcal{T}$.

We have the following theorem.

Theorem 12.4: Let $(S, \mathcal{T})$ be a topological space. The following are equivalent.

1. $(S, \mathcal{T})$ is compact.

2. Every covering of S by basic open sets contains a finite subcollection that still covers S.

3. For every collection $\mathcal{C}$ of basic closed sets in S with the finite intersection property, we have $\cap \mathcal{C} \neq \emptyset$.

4. Every covering of S by subbasic open sets contains a finite subcollection that still covers S.

5. For every collection $\mathcal{C}$ of subbasic closed sets in S with the finite intersection property, we have $\cap \mathcal{C} \neq \emptyset$.

The equivalence of 1 with 2 and 3 in Theorem 12.3 is not difficult to prove. You will be asked to prove this in Problem 5 below. The equivalence of 1 with 4 and 5 requires the **Axiom of Choice** and is much more difficult. You will be asked to prove this in Problem 17 below. You will also be provided with a version of the Axiom of Choice to help you with the proof.

Theorem 12.4 provides us with the machinery we need to prove that an arbitrary product of compact spaces is compact. In particular, we will use clause 5 in the above theorem.

Therorem 12.5 (Tychonoff's Theorem): Let K be an index set and let $(S_k, \mathcal{T}_k)$ be a compact topological space for each $k \in K$. Then $\prod_{k \in K} S_k$ is compact in the product topology. $\quad\square$

Proof: Let K be an index set, let $(S_k, \mathcal{T}_k)$ be a compact topological space for each $k \in K$, and let $\mathcal{C}$ be a collection of subbasic closed sets in $\prod_{k \in K} S_k$ with the finite intersection property, let's say $\mathcal{C} = \{\prod_{k \in K} B_k^j \mid j \in J\}$. So, if $\prod_{k \in K} B_k^j \in \mathcal{C}$, then for each $k \in K$, B_k^j is a closed subset of S_k and $B_k^j = S_k$ for all k except possibly one. For each $k \in K$, we have that $\{B_k^j \mid j \in J\}$ is a collection of closed subsets of S_k with the finite intersection property (Check this!). By the compactness of $(S_k, \mathcal{T}_k)$, there is $x_k \in \cap \{B_k^j \mid j \in J\}$. Then by Problem 17 from Problem Set 9, $\mathbf{x} = (x_k) \in \prod_{k \in K}(\cap \{B_k^j \mid j \in J\}) = \cap \mathcal{C}$. By Theorem 12.4 (5$\rightarrow$ 1), $\prod_{k \in K} S_k$ is compact in the product topology. $\quad\square$

Heine-Borel Theorem

We will now prove the Heine-Borel Theorem, which describes all compact subsets of $\mathbb{R}$. The proof will require a bit of work and we prove it in several steps.

Theorem 12.6: Heine-Borel Theorem: Let $A \subseteq \mathbb{R}$. The topological space $(A, \mathcal{T}_A)$ is compact if and only if A is closed and bounded in $\mathbb{R}$.

We will prove the Heine-Borel Theorem though a series of lemmas.

Lemma 12.7: Let $A \subseteq \mathbb{R}$ and assume that the topological space $(A, \mathcal{T}_A)$ is compact. Then A is bounded in $\mathbb{R}$.

Proof: Let $A \subseteq \mathbb{R}$ and assume that $(A, \mathcal{T}_A)$ is compact. To see that A is bounded in $\mathbb{R}$, consider the following open cover of A: $\mathcal{C} = \{(-k, k) \cap A \mid k \in \mathbb{Z}^+\}$. By compactness, $\mathcal{C}$ has a finite subcover of A, say $\{(-k_0, k_0) \cap A, (-k_1, k_1) \cap A, \dots, (-k_n, k_n) \cap A\}$. Let $N = \max\{k_0, k_1, \dots k_n\}$. Then we have $A = (-N, N) \cap A$. So, for all $x \in A$, we have $|x| \leq N$. Therefore, A is bounded in $\mathbb{R}$. $\square$

Lemma 12.8: Let $A \subseteq \mathbb{R}$ and assume that the topological space $(A, \mathcal{T}_A)$ is compact. Then A is closed in $\mathbb{R}$.

Proof: Let $A \subseteq \mathbb{R}$ and assume that $(A, \mathcal{T}_A)$ is compact. Suppose toward contradiction that A is **not** closed in $\mathbb{R}$. Then by part 4 of Theorem 9.4, we have $\overline{A} \setminus A \neq \emptyset$. Let $x \in \overline{A} \setminus A$. For each $k \in \mathbb{Z}^+$, let $U_k = \left(-\infty, x - \frac{1}{k}\right) \cup \left(x + \frac{1}{k}, \infty\right)$. Then $U_1 \subseteq U_2 \subseteq \cdots$. Using the Archimedean Property of the real numbers, it is easy to see that $\bigcup\{U_k \mid k \in \mathbb{Z}^+\} = \mathbb{R} \setminus \{x\}$ (Prove this!). Since $x \notin A$, we have $A \subseteq \mathbb{R} \setminus \{x\} = \bigcup\{U_k \mid k \in \mathbb{Z}^+\}$. Thus, $\bigcup\{U_k \cap A \mid k \in \mathbb{Z}^+\} = A$. So, $\mathcal{C} = \{U_k \cap A \mid k \in \mathbb{Z}^+\}$ is an open cover of A. By compactness, $\mathcal{C}$ has a finite subcover of A, say $\{U_{k_0} \cap A, U_{k_1} \cap A, \dots, U_{k_n} \cap A\}$. If we let $N = \max\{k_0, k_1, \dots, k_n\}$, then $A = U_N \cap A$. It follows that $A \subseteq U_N = \left(-\infty, x - \frac{1}{N}\right) \cup \left(x + \frac{1}{N}, \infty\right)$. So, $\left(x - \frac{1}{N}, x + \frac{1}{N}\right) \cap A = \emptyset$. By part 5 of Theorem 9.4, $x \notin \overline{A}$, contrary to our assumption. $\square$

Lemma 12.9: Let A be a nonempty bounded subset of $\mathbb{R}$. Then $\sup A$ and $\inf A$ are both in $\overline{A}$.

Proof: Let A be a nonempty bounded subset of $\mathbb{R}$. Since A is bounded, by the Completeness Property of $\mathbb{R}$ (see Lesson 6), $\sup A$ and $\inf A$ both exist. Suppose that $\inf A = a$ and $\sup A = b$. Then $A \subseteq [a, b]$. Let (c, d) be an open interval containing b. Then $(c, d) \cap A \neq \emptyset$ (otherwise, $c < b$ and c is an upper bound of A, contradicting that $b = \sup A$). By part 5 of Theorem 9.4, $b \in \overline{A}$. Similarly, if (e, f) is an open interval containing a, then $(e, f) \cap A \neq \emptyset$, and so, by part 5 of Theorem 9.4, $a \in \overline{A}$. $\square$

Let $(A, \mathcal{T}_A)$ be a subspace of the topological space $(S, \mathcal{T})$. Let $\mathcal{C}$ be a collection of subsets of S such that $A \subseteq \bigcup\mathcal{C}$. In this case, we say that $\mathcal{C}$ is a **covering** of A by sets in S. If all the sets in $\mathcal{C}$ are open in S, then we say that $\mathcal{C}$ is an **open covering** of A by open sets in S.

The following Lemma will allow us to simplify the proof of Lemma 12.11 below. It is also a very useful result in general. It will allow us to save a lot of work when dealing with open coverings and subspaces simultaneously.

Lemma 12.10: Let $(A, \mathcal{T}_A)$ be a subspace of the topological space $(S, \mathcal{T})$. Then $(A, \mathcal{T}_A)$ is compact if and only if every open covering of A by open sets in S contains a finite subcollection that still covers A.

Proof: First suppose that $(A, \mathcal{T}_A)$ is compact and let C be an open covering of A by open sets in S. Then the collection $\{U \cap A \mid U \in C\}$ is an open covering of A. Since $(A, \mathcal{T}_A)$ is compact, there is a finite subcollection $\{U_1 \cap A, U_2 \cap A, ..., U_n \cap A\}$ that is still an open covering of A. So, $\mathcal{D} = \{U_1, U_2, ..., U_n\}$ is a finite subcollection of C such that $A \subseteq \bigcup \mathcal{D}$.

Conversely, suppose that every open covering of A by open sets in S contains a finite subcollection that still covers A. Let C be an open covering of A. Every $U \in C$ has the form $V \cap A$ for some set V that is open in S. Let $\mathcal{D} = \{V \mid V \cap A \in C\}$. Then $\mathcal{D}$ is an open covering of A by open sets in S. Therefore, there is a finite subcollection $\mathcal{E}$ of $\mathcal{D}$ that still covers A. Then $\{V \cap A \mid V \in \mathcal{E}\}$ is a finite subcollection of C that covers A. It follows that $(A, \mathcal{T}_A)$ is compact. $\qquad\square$

Lemma 12.11: Let A be a closed and bounded subset of $\mathbb{R}$. Then $(A, \mathcal{T}_A)$ is compact.

Proof: Let A be a closed and bounded subset of $\mathbb{R}$ and let C be an open covering of A by open sets in $\mathbb{R}$. For each $x \in \mathbb{R}$, let $A_x = (-\infty, x] \cap A$ and let

$$Z = \{x \in \mathbb{R} \mid \text{there exists a finite subcover } \mathcal{D} \text{ of } C \text{ such that } \mathcal{D} \text{ covers } A_x\}.$$

Since A is closed, by part 4 of Theorem 9.4, $A = \overline{A}$. By Lemma 12.9, $\sup A$ and $\inf A$ are both in $\overline{A} = A$. Let $a = \inf A$ and $b = \sup A$.

$A_a = (-\infty, a] \cap A = \{a\}$. Since $\{a\}$ can be covered by one element of C, $a \in Z$. This shows that $Z \neq \emptyset$.

Suppose toward contradiction that Z is bounded above and let $c = \sup Z$. If $c \in A$, then there is $U \in C$ such that $c \in U$ (because C covers A). Since U is open, there is an interval $[e, f] \subseteq U$ with $c \in (e, f)$. Since $e < c$ and $c = \sup Z$, there is a finite subcover $\mathcal{D}$ of C so that $\mathcal{D}$ covers A_e. But then $\mathcal{D} \cup \{U\}$ covers A_f, so that $f \in Z$. This contradicts $c = \sup Z$. So, $c \notin A$. Now, since A is closed, there is $r > 0$ such that $(c - r, c + r) \cap A = \emptyset$. But then $A_{c-r} = A_{c+r}$. Since $c - r \in Z$, we have $c + r \in Z$. This again contradicts $c = \sup Z$. It follows that Z is not bounded from above.

Let $g \in Z$ with $g > b$. Then $A_g = (-\infty, g] \cap A = A$. Therefore, there is a finite subcover of C that covers $A_g = A$ by open sets in $\mathbb{R}$. It follows that $(A, \mathcal{T}_A)$ is compact. $\qquad\square$

Together, Lemmas 12.7 through 12.11 provide a proof of the Heine-Borel Theorem (Theorem 12.6).

Now, we would like to generalize the Heine-Borel Theorem to subspaces of $\mathbb{R}^n$. To do this, we will need two preliminary results. These two results will also be quite useful for proving other theorems throughout the book and in the Problem Sets.

Theorem 12.12: Let $(A, \mathcal{T}_A)$ be a compact subspace of a T_2-space $(S, \mathcal{T})$. Then A is closed in S.

Proof: Let $(A, \mathcal{T}_A)$ be a compact subspace of the T_2-space $(S, \mathcal{T})$. We will show that $S \setminus A$ is open in S. To see this, let $x \in S \setminus A$. For each $y \in A$, let U_y and V_y be disjoint neighborhoods of x and y, respectively (we can do this because $(S, \mathcal{T})$ is a T_2-space space). The collection $\mathcal{C} = \{V_y \mid y \in A\}$ is an open covering of A by open sets in S. By Lemma 12.10, there is a finite subcollection $\mathcal{D}$ of $\mathcal{C}$ such that $\mathcal{D}$ still covers A. Let $V = \cup \mathcal{D}$ and let $U = \cap \{U_y \mid V_y \in \mathcal{D}\}$. Then we have $A \subseteq V$ and $x \in U$. Furthermore, if $z \in V$, then $z \in V_y$ for some $V_y \in \mathcal{D}$. Since $U_y \cap V_y = \emptyset$, $z \notin U_y$. Thus, $z \notin U$. It follows that $U \cap V = \emptyset$. So, U is an open set containing x such that $U \subseteq S \setminus A$. $\quad\square$

Theorem 12.13: Let $(A, \mathcal{T}_A)$ be a subspace of a compact space $(S, \mathcal{T})$ with A closed in S. Then $(A, \mathcal{T}_A)$ is compact.

Proof: Let $(A, \mathcal{T}_A)$ be a subspace of the compact space $(S, \mathcal{T})$ with A closed in S and let $\mathcal{C}$ be an open covering of A by open sets in S. Then $\mathcal{D} = \mathcal{C} \cup \{S \setminus A\}$ is an open covering of S. By the compactness of $(S, \mathcal{T})$, there is a finite subcollection $\mathcal{E}$ of $\mathcal{D}$ such that $\mathcal{E}$ is an open covering of S. It follows that $\mathcal{E} \setminus \{S \setminus A\}$ is an open covering of A by open sets in S. By Lemma 12.10, $(A, \mathcal{T}_A)$ is compact. $\quad\square$

Theorem 12.14 (Generalized Heine-Borel Theorem): Let $A \subseteq \mathbb{R}^n$. The topological space $(A, \mathcal{T}_A)$ is compact if and only A is closed and bounded in $\mathbb{R}^n$.

Note: We should clarify here what we mean by bounded. After all, boundedness is not a topological property (see Example 11.2). This means that the notion of boundedness depends on the metric we decide to use. The most natural choice is the Euclidean metric d defined by $d(\mathbf{x}, \mathbf{y}) = |\mathbf{x} - \mathbf{y}|$. Another choice would be the square metric ρ defined by $\rho(\mathbf{x}, \mathbf{y}) = \max\{|x_1 - y_1|, \ldots, |x_n - y_n|\}$. By Problem 6 from Problem Set 11, the product topology on $\mathbb{R}^n$ is induced by either of these metrics. Furthermore, a set A is bounded with respect to d if and only if it is bounded with respect to ρ (Prove this!). This means that for proofs involving bounded sets, we can use either metric. Note that in the case $n = 1$, the Euclidean metric and the square metric are the same. We will usually choose to use the square metric because it is a little easier to work with; this is the metric we will use in the proof below.

Proof of Theorem 12.14: First suppose that $(A, \mathcal{T}_A)$ is compact. Let $\mathcal{X} = \{B_k(\mathbf{0}; \rho) \mid k \in \mathbb{Z}^+\}$ be the collection of open balls centered at $\mathbf{0}$ and with radius k for each $k \in \mathbb{Z}^+$. Then $\mathbb{R}^n = \cup \mathcal{X}$. Since $(A, \mathcal{T}_A)$ is compact, by Lemma 12.10, there is a finite subcollection $\mathcal{Y}$ of $\mathcal{X}$ that covers A. Suppose that $\mathcal{Y} = \{B_{k_1}(\mathbf{0}; \rho), B_{k_2}(\mathbf{0}; \rho), \ldots, B_{k_m}(\mathbf{0}; \rho)\}$. Let $M = \max\{k_1, k_2, \ldots, k_m\}$. Then $A \subseteq B_M(\mathbf{0}; \rho)$. So, if $\mathbf{x}, \mathbf{y} \in A$, then $\rho(\mathbf{x}, \mathbf{y}) \le \rho(\mathbf{x}, \mathbf{0}) + \rho(\mathbf{0}, \mathbf{y}) \le M + M = 2M$. Therefore, A is bounded by $2M$ in the metric ρ.

By Theorem 12.12, A is closed in $\mathbb{R}^n$.

Conversely, assume that A is closed and bounded in $\mathbb{R}^n$ under the square metric ρ. Since $\emptyset$ is trivially compact, we may also assume that $A \ne \emptyset$. Since A is bounded in $\mathbb{R}^n$ under ρ, there is $M \in \mathbb{N}$ such that for all $\mathbf{x}, \mathbf{y} \in A$, $\rho(\mathbf{x}, \mathbf{y}) \le M$. Since $A \ne \emptyset$, there is $\mathbf{a} \in A$. Let $\rho(\mathbf{a}, \mathbf{0}) = c$. Then for each $\mathbf{x} \in A$, we have $\rho(\mathbf{x}, \mathbf{0}) \le \rho(\mathbf{x}, \mathbf{a}) + \rho(\mathbf{a}, \mathbf{0}) \le M + c$. So, $A \subseteq [-(M+c), M+c]^n$, which is compact by the Heine-Borel Theorem (Theorem 12.6) and Tychonoff's Theorem (theorem 12.5). Since A is closed in $\mathbb{R}^n$, by part (ii) of Problem 7 from Problem Set 9, $A = A \cap ([-(M+c), M+c]^n)$ is closed in $[-(M+c), M+c]^n$. So, by Theorem 12.13, A is compact. $\quad\square$

Compactness in Metric Spaces

We say that a metric space (S, d) is **sequentially compact** if every sequence in S has a convergent subsequence.

Example 12.15:

1. If S is a finite set, say $S = \{a_0, a_1, \ldots, a_n\}$, then for any metric d on S, the metric space (S, d) is sequentially compact. To see this, let (s_n) be a sequence in S. Since there are only finitely many choices for s_n, there must be $k \in \mathbb{N}$ with $k \leq n$ such that $s_n = a_k$ for infinitely many k. It follows that (a_k) is a subsequence of S that clearly converges to a_k.

    Note that the sequence (a_k) is a constant sequence. Here a_k is one specific element of S. This shouldn't be confused with s_n, which potentially can have a different value for each $n \in \mathbb{N}$.

2. Let S be an infinite set and let $d: S \times S \to \mathbb{R}$ be the discrete metric on S defined by
    $$d(x, y) = \begin{cases} 0 & \text{if } x = y. \\ 1 & \text{if } x \neq y. \end{cases}$$

    Then (S, d) is **not** sequentially compact. Indeed, if (s_n) is a sequence in S, where the coordinates are all distinct, then $d(s_n, s_m) = 1$ whenever $n \neq m$, and therefore, (s_n) has no convergent subsequence.

3. $(\mathbb{R}, d)$, where d is the Euclidean metric on $\mathbb{R}$, is **not** sequentially compact. Indeed, the sequence (n) in $\mathbb{R}$ has no convergent subsequence because if $n \neq m$, then $d(n, m) \geq 1$.

4. Let $(A, \mathcal{T}_A)$ be the topological space, where $A \subseteq \mathbb{R}$ is defined by $A = \left\{ \frac{1}{n} \mid n \in \mathbb{Z}^+ \right\} \cup \{0\}$, and as usual $\mathcal{T}_A$ is the subspace topology. Then $(A, \mathcal{T}_A)$ is sequentially compact. To see this, let (s_n) be a sequence in A. If (s_n) takes on a single value for infinitely many values of n, then (s_n) has a constant subsequence, which converges to that constant value. Otherwise, (s_n) has a subsequence of the form $\left(\frac{1}{n_k} \right)$, which converges to 0.

    Note that if we remove 0 from the set A to get the set $B = \left\{ \frac{1}{n} \mid n \in \mathbb{Z}^+ \right\}$, then we wind up with a subspace of $\mathbb{R}$ that is **not** sequentially compact. The sequence $\left(\frac{1}{n} \right)$ converges to 0 in $\mathbb{R}$, and therefore, any subsequence of $\left(\frac{1}{n} \right)$ converges to 0 in $\mathbb{R}$. However, since $0 \notin B$, $\left(\frac{1}{n} \right)$ has no convergent subsequence in B.

5. The open interval $(0, 1)$ considered as a subspace of $\mathbb{R}$ is **not** sequentially compact. The same counterexample given in the previous paragraph for $B = \left\{ \frac{1}{n} \mid n \in \mathbb{Z}^+ \right\}$ can be used here too.

If you compare examples 12.1 and 12.15, you should notice a direct correlation between compact spaces and sequentially compact spaces. Indeed, our main goal of this section is to prove that for metrizable spaces, sequential compactness is equivalent to compactness.

Note: In part 6 of Example 12.1, we showed that any set S with the cofinite topology is compact. You may be wondering why we excluded this space from our discussion in Example 12.15. The reason is simply because this space is not metrizable. Recall from part 2 of Example 10.8 that this space is **not** a T_2-space. However, all metrizable spaces are T_2-spaces (in fact, metrizable spaces are T_4-spaces by problem 7 in Problem Set 11).

Before we begin the proof of the equivalence of compactness with sequential compactness for metrizable spaces, we will require a few preliminary definitions and theorems.

Let $(S, \mathcal{T})$ be a topological space and let $\mathcal{C}$ be an open covering of S. We say that $\delta \in \mathbb{R}^+$ is a **Lebesgue number** for $\mathcal{C}$ if every subset A of S with diam $A < \delta$ is contained in at least one member of $\mathcal{C}$.

Example 12.16:

1. Let's look at the open interval $(0, 1)$ considered as a subspace of $\mathbb{R}$ and the open covering $\mathcal{C} = \left\{ \left(\frac{1}{n}, 1 \right) \,\middle|\, n \in \mathbb{Z}^+ \right\}$. This open covering has no Lebesgue number. To see this, let $\delta \in \mathbb{R}^+$. The set $\left(0, \frac{\delta}{2} \right)$ has diameter $\frac{\delta}{2} < \delta$ and $\left(0, \frac{\delta}{2} \right) \nsubseteq \left(\frac{1}{n}, 1 \right)$ for any $n \in \mathbb{Z}^+$. To verify this last claim, let $n \in \mathbb{Z}^+$ and use the Archimedean Property of $\mathbb{R}$ to find a natural number m such that $\frac{2}{\delta} < m$, or equivalently, $\frac{1}{m} < \frac{\delta}{2}$. Let $K = \max\{m, n+1\}$. Then $\frac{1}{K} \leq \frac{1}{m} < \frac{\delta}{2}$, and so, $\frac{1}{K} \in \left(0, \frac{\delta}{2} \right)$. Also, $\frac{1}{K} \leq \frac{1}{n+1} < \frac{1}{n}$, and so, $\frac{1}{K} \notin \left(\frac{1}{n}, 1 \right)$. This example shows that an open covering in a metrizable space that is not compact (or sequentially compact) does not necessarily have a Lebesgue number.

2. Let's now look at the closed interval $[0, 1]$ considered as a subspace of $\mathbb{R}$ and the open covering $\mathcal{C} = \left\{ \left(\frac{1}{n}, 1 \right] \,\middle|\, n \in \mathbb{Z}^+ \right\} \cup \left\{ \left[0, \frac{1}{100} \right) \right\}$. Let $\delta = \frac{1}{100}$. Then δ is a Lebesgue number for $\mathcal{C}$. To see this, let $A \subseteq [0, 1]$ with diam $A < \delta$. If $a = \inf A > 0$, then $A \subseteq [a, 1]$ and by the Archimedean Property of $\mathbb{R}$, we can find $n \in \mathbb{N}$ with $n > \frac{1}{a}$, or equivalently, $\frac{1}{n} < a$. Then $A \subseteq \left(\frac{1}{n}, 1 \right]$, which is in $\mathcal{C}$. Otherwise, $\inf A = 0$, and in this case, $A \subseteq \left[0, \frac{1}{100} \right)$, which is also in $\mathcal{C}$. Although this specific example doesn't prove anything, it turns out that open covers in compact metrizable spaces (as well as sequentially compact topological spaces) always have Lebesgue numbers.

Given an arbitrary topological space, a given open covering for that space need not necessarily have a Lebesgue number (as we just saw in part 1 of Example 12.16). However, for a sequentially compact metrizable space, every open covering does indeed have a Lebesgue number.

Theorem 12.17 (Lebesgue's Covering Lemma): Let $(S, \mathcal{T})$ be a metrizable, sequentially compact topological space and let $\mathcal{C}$ be an open covering of S. Then $\mathcal{C}$ has a Lebesgue number.

You will be asked to prove Theorem 12.17 in Problem 15 below.

Let (S, d) be a metric space and let $r \in \mathbb{R}^+$. A finite subset A of S is called an **r-net** if $S = \cup \{ B_r(x) \mid x \in A \}$. If the metric space has an r-net for every $r \in \mathbb{R}^+$, then it is said to be **totally bounded**.

Example 12.18:

1. Let S be any infinite set and let $d: S \times S \to \mathbb{R}$ be the discrete metric on S defined by
$$d(x,y) = \begin{cases} 0 & \text{if } x = y. \\ 1 & \text{if } x \neq y. \end{cases}$$

   Then (S,d) is **bounded** because for all $x, y \in S$, $d(x,y) \leq 1$. However, (S,d) is **not** totally bounded. To see this let $r = 1$. Then for any $x \in S$, $B_r(x) = \{x\}$. It follows that S cannot be the union of finitely many balls of radius 1, and so, (S,d) has no 1-net.

2. Any totally bounded metric space is bounded. You will be asked to prove this in Problem 6 below. By contrapositive, it follows that any metric space that is unbounded is **not** totally bounded. For example, $\mathbb{R}^n$ with the Euclidean or square metric is not totally bounded.

3. For subspaces of $\mathbb{R}^n$ (with the Euclidean or square metric), bounded and totally bounded are equivalent. You will be asked to prove this in Problem 11 below.

Theorem 12.19: Let (S,d) be a sequentially compact metric space. Then (S,d) is totally bounded.

You will be asked to prove Theorem 12.19 in Problem 8 below.

We are now ready to prove the equivalence of compactness with sequential compactness.

Theorem 12.20: Let $(S,\mathcal{T})$ be a metrizable space. The following are equivalent.

1. $(S,\mathcal{T})$ is compact.

2. Every infinite subset of S has an accumulation point.

3. $(S,\mathcal{T})$ is sequentially compact.

Proof: For 1→2, assume that $(S,\mathcal{T})$ is compact and let A be an infinite subset of S. Assume toward contradiction that A has no accumulation point. Let $x \in A$. Since x is not an accumulation point of A, there is an open ball $B_{r_x}(x)$ with center x such that $B_{r_x}(x) \cap A = \{x\}$. The collection $\mathcal{C} = \{B_{r_x}(x) \mid x \in A\}$ is an open cover of A by open sets in S. By the compactness of $(S,\mathcal{T})$, $\mathcal{C}$ has a finite subcover $\mathcal{D}$. Now, let $y \in A$. Since $\mathcal{D}$ covers A, there is $B_{r_x}(x) \in \mathcal{D}$ with $y \in B_{r_x}(x)$. Since $B_{r_x}(x) \cap A = \{x\}$, $y = x$. It follows that $A \subseteq \{x \mid B_{r_x}(x) \in \mathcal{D}\}$, which is a finite set. So, A is finite, contradicting our assumption that A is infinite.

For 2→3, assume that every infinite subset of S has an accumulation point and let (s_n) be an arbitrary sequence in S. If there is a fixed $s \in S$ such that $s_n = s$ for infinitely many $n \in \mathbb{Z}^+$, then (s) is a constant subsequence of (s_n) with limit s. Otherwise, no coordinate of (s_n) is infinitely repeated. The set $A = \{s_n \mid n \in \mathbb{Z}^+\}$ is then infinite, and so, by our assumption, A has an accumulation point x. So, for each $k \in \mathbb{Z}^+$, there is $s_{n_k} \in B_{\frac{1}{n}}(x)$ with $n_1 < n_2 < \cdots < n_k < \cdots$. It is easily checked that $s_{n_k} \to x$.

Finally, for 3→1, let $(S,\mathcal{T})$ be sequentially compact and let $\mathcal{C}$ be an open covering of S. By Theorem 12.17, $\mathcal{C}$ has a Lebesgue number δ. Let $r = \frac{\delta}{3}$. By Theorem 12.19, S has an r-net $N = \{s_0, s_1, \ldots, s_k\}$. For each $n = 0, 1, \ldots, k$, we have diam $B_r(s_n) \leq 2r = \frac{2\delta}{3} < \delta$. Since δ is a Lebesgue number, for each n, we can find $D_n \in \mathcal{C}$ such that $B_r(s_n) \subseteq D_n$. Since every point of S belongs to $B_r(s_n)$ for some $n \in \mathbb{Z}^+$ with $n \leq k$, $\{D_0, D_1, \ldots, D_n\}$ is a finite subcover of $\mathcal{C}$. Therefore, $(S,\mathcal{T})$ is compact. $\qquad \square$

Locally Compact Spaces

A topological space $(S, \mathcal{T})$ is **locally compact** if for every $x \in S$, there is a compact subspace $(K, \mathcal{T}_K)$ such that K contains a neighborhood of x.

Example 12.21:

1. If $(S, \mathcal{T})$ is compact, then it is also locally compact. If $x \in S$, then $(S, \mathcal{T}_S) = (S, \mathcal{T})$ is a compact subspace of $(S, \mathcal{T})$ and S is a neighborhood of x with $S \subseteq S$.

2. $(\mathbb{R}, \mathcal{T})$, where $\mathcal{T}$ is the standard topology on $\mathbb{R}$ is locally compact. If $x \in \mathbb{R}$, then we have
$$x \in (x - 1, x + 1) \subseteq [x - 1, x + 1].$$

3. $(\mathbb{R}^n, \mathcal{T})$, where $\mathcal{T}$ is the product topology on $\mathbb{R}^n$ is locally compact. If $\mathbf{x} \in \mathbb{R}^n$ with $\mathbf{x} = (x_1, x_2, \ldots, x_n)$, then we have
$$\mathbf{x} \in (x_1 - 1, x_1 + 1) \times (x_2 - 1, x_2 + 1) \times \cdots \times (x_n - 1, x_n + 1)$$
$$\subseteq [x_1 - 1, x_1 + 1] \times [x_2 - 1, x_2 + 1] \times \cdots \times [x_n - 1, x_n + 1].$$

4. $(\mathbb{Q}, \mathcal{T}_\mathbb{Q})$ is **not** locally compact. In fact, given any $x \in \mathbb{Q}$, any neighborhood U of x contains an irrational number y. If B is any set containing U, then y is an accumulation point of B that is not in B. Thus, B is not closed in $\mathbb{R}$ and therefore, by the Heine-Borel Theorem, B is not compact.

5. $(^\mathbb{N}\mathbb{R}, \mathcal{T})$, where $\mathcal{T}$ is the product topology on $^\mathbb{N}\mathbb{R}$ is **not** locally compact. Let $\mathbf{x} \in {}^\mathbb{N}\mathbb{R}$ with $\mathbf{x} = (x_k)$ and let U be a basic open set containing $\mathbf{x}$. Any set B containing U has infinitely many coordinate sets equal to $\mathbb{R}$, and therefore, B would not be compact.

In Lesson 8, we formed the Extended Complex Plane $\overline{\mathbb{C}}$ by adding a single "point at infinity" (∞) to $\mathbb{C}$. We then extended the standard topology on $\mathbb{C}$ to include sets of the form $(\mathbb{C} \setminus K) \cup \{\infty\}$ for each compact set $K \subseteq \mathbb{C}$ (by the Generalized Heine-Borel Theorem, the compact subsets of $\mathbb{C}$ are the same as the closed and bounded subsets of $\mathbb{C}$). This procedure gave us a compact topological space that can be visualized as the Riemann sphere. More precisely, the Extended Complex Plane with the topology just defined is homeomorphic to the Riemann sphere in the subspace topology of $\mathbb{R}^3$.

This procedure can be generalized to any locally compact T_2-space. Specifically, if $(S, \mathcal{T})$ is a locally compact T_2-space, we let $\overline{S} = S \cup \{\infty\}$, where ∞ is a new element called the **point at infinity**. We then let $\overline{\mathcal{T}} = \mathcal{T} \cup \{S \setminus K \mid (K, \mathcal{T}_K) \text{ is a compact subspace of } (S, \mathcal{T})\}$. $(\overline{S}, \overline{\mathcal{T}})$ is called the **one-point compactification** of S.

Theorem 12.22: Let $(S, \mathcal{T})$ be a locally compact T_2-space. Then the one-point compactification $(\overline{S}, \overline{\mathcal{T}})$ of $(S, \mathcal{T})$ is a compact T_2-space.

The proof of Theorem 12.22 is not too difficult, but it is a bit tedious. Therefore, I leave it as an exercise for the reader (see Problem 13 below).

Baire Category

Let $(S, \mathcal{T})$ be a topological space and let A be a subset of S. The **interior** of A, written A°, is the union of all open sets U such that $U \subseteq A$. If $A^\circ = \emptyset$, then we say that A has **empty interior**. A subset $A \subseteq S$ is **nowhere dense** in S if the closure of A has empty interior. Symbolically, $A \subseteq S$ is nowhere dense in S if $\left(\overline{A} \right)^\circ = \emptyset$.

Example 12.23:

1. Consider $(\mathbb{R}, \mathcal{T})$, where $\mathcal{T}$ is the standard topology on $\mathbb{R}$. The set of rational numbers, $\mathbb{Q}$, has empty interior. However, $\mathbb{Q}$ is **not** nowhere dense in $\mathbb{R}$. Indeed, $\overline{\mathbb{Q}} = \mathbb{R}$ ($\mathbb{Q}$ is dense in $\mathbb{R}$), and so, $\left(\overline{\mathbb{Q}} \right)^\circ = \mathbb{R} \neq \emptyset$.

   On the other hand, the set of integers, $\mathbb{Z}$, is nowhere dense in $\mathbb{R}$. Indeed, $\overline{\mathbb{Z}} = \mathbb{Z}$, and therefore, $\left(\overline{\mathbb{Z}} \right)^\circ = \emptyset$.

2. Let $(S, \mathcal{T})$ be a topological space, where $\mathcal{T}$ is the discrete topology on S. If $A \subseteq S$ is nonempty, then A does **not** have empty interior. Indeed, if $x \in A$, then $x \in A^\circ$ because $\{x\}$ is an open set with $\{x\} \subseteq A$. Therefore, A cannot be nowhere dense.

3. Let $(S, \mathcal{T})$ be a topological space, let $A \subseteq S$ be nowhere dense in S, and let $B \subseteq A$. Then B is also nowhere dense in S. To see this, first note that since $B \subseteq A$, $\overline{B} \subseteq \overline{A}$. Then note that if U is an open set with $U \subseteq \overline{B}$, then since $\overline{B} \subseteq \overline{A}$, U is also an open set with $U \subseteq \overline{A}$. But then $\left(\overline{A} \right)^\circ$ contains U, contradicting that A is nowhere dense.

4. Let $(S, \mathcal{T})$ be a topological space, let $A \subseteq S$ be nowhere dense in S. Then $\overline{A}$ is also nowhere dense in S. Indeed, since $\overline{\overline{A}} = \overline{A}$, we have $\left(\overline{\overline{A}} \right)^\circ = \left(\overline{A} \right)^\circ = \emptyset$.

Example 12.24: In part (ii) of Problem 16 below, you will be asked to prove that the union of finitely many nowhere dense sets is nowhere dense. We cannot replace "finitely many" by "countably many" here. For example, $\mathbb{Q}$ is countable, and therefore, $\mathbb{Q} = \bigcup \{ \{x\} \mid x \in \mathbb{Q} \}$ is a countable union of nowhere dense sets in the standard topology of $\mathbb{R}$ (each singleton set $\{x\}$ is clearly nowhere dense). However, by part 1 of Example 12.23 above, $\mathbb{Q}$ is **not** nowhere dense in $\mathbb{R}$ (and in fact, $\mathbb{Q}$ is dense in $\mathbb{R}$).

Theorem 12.25: Let $(S, \mathcal{T})$ be a topological space and let A be a subset of S. Then A is nowhere dense in S if and only if $(S \setminus A)^\circ$ is dense in S.

You will be asked to prove Theorem 12.25 in part (iii) of Problem 16 below.

If $(S, \mathcal{T})$ is a topological space, then $A \subseteq S$ is called **meagre** (or a set of **first category**) if A can be expressed as a countable union of nowhere dense sets. A set that is not meagre is called **nonmeager** (or a set of **second category**). If $A \subseteq S$ is meagre, then $S \setminus A$ is called **comeagre**.

Note that by Theorem 12.25, a comeagre set is a set that can be expressed as a countable intersection of sets, each of which has dense interior. Indeed, if $A = \bigcup \{ B_n \mid n \in \mathbb{N} \}$ with $\left(\overline{B_n} \right)^\circ = \emptyset$ for each $n \in \mathbb{N}$, then by De Morgan's law, we have $S \setminus A = \bigcap \{ S \setminus B_n \mid n \in \mathbb{N} \}$, and by Theorem 12.25, $(S \setminus B_n)^\circ$ is dense in S for each $n \in \mathbb{N}$.

Theorem 12.26 (Baire Category Theorem): Let $(S, \mathcal{T})$ be a locally compact T_2-space or a completely metrizable space. Then we have the following:

1. Every comeagre subset of S is dense in S.

2. S is nonmeagre.

I leave the proof of the Baire Category Theorem as a challenging exercise for the interested reader (see Problem 19 below).

Example 12.27: In Example 12.24, we saw that the set of rational numbers, $\mathbb{Q}$, is a meagre subset of $\mathbb{R}$. What about the set of irrational numbers, $\mathbb{R} \setminus \mathbb{Q}$? Is $\mathbb{R} \setminus \mathbb{Q}$ meagre in $\mathbb{R}$? The answer to this question does not immediately follow from the definition of meagre. However, the Baire Category Theorem tells us that $\mathbb{R} \setminus \mathbb{Q}$ **cannot** be meagre. If it were, then $\mathbb{R} = \mathbb{Q} \cup (\mathbb{R} \setminus \mathbb{Q})$ would also be meagre, contradicting part 2 of the Baire Category Theorem.

Example 12.28: Consider $\mathbb{Q}$ as a subspace of $\mathbb{R}$ with the standard topology. Since $\mathbb{Q}$ is meagre (by Example 12.24), the Baire Category Theorem tells us that $(\mathbb{Q}, \mathcal{T}_{\mathbb{Q}})$ is not locally compact or completely metrizable. (Note that part 4 of Example 12.21 above provides a direct method for showing that $(\mathbb{Q}, \mathcal{T}_{\mathbb{Q}})$ is not locally compact.)

Problem Set 12

Full solutions to these problems are available for free download here:

www.SATPrepGet800.com/TFBZLF

LEVEL 1

1. Let I be an open or half-open interval of real numbers. Prove that $(I, \mathcal{T}_I)$ is **not** compact.

2. Let $(S, \mathcal{T})$ be a topological space. Prove that a finite union of compact subspaces of S is compact.

3. Prove that the Cantor set is compact and nowhere dense in $\mathbb{R}$.

LEVEL 2

4. Let $(A, \mathcal{T}_A)$ be a compact subspace of a T_1-space $(S, \mathcal{T})$. Is A necessarily closed in S? (Compare this problem with Theorem 12.12.)

5. Prove that the following are equivalent for a topological space $(S, \mathcal{T})$:

   (i) $(S, \mathcal{T})$ is compact.

   (ii) Every covering of S by basic open sets contains a finite subcollection that still covers S.

   (iii) For every collection $\mathcal{C}$ of basic closed sets in S with the finite intersection property, we have $\bigcap \mathcal{C} \neq \emptyset$.

LEVEL 3

6. Prove that a totally bounded metric space is bounded.

7. Give a direct proof that if $(S, \mathcal{T})$ is a sequentially compact metrizable space, then every infinite subset of S has an accumulation point.

8. Prove that every sequentially compact metric space is totally bounded.

9. Prove that a metric space is compact if and only if it is complete and totally bounded.

LEVEL 4

10. Let $(S, \mathcal{T})$ be a compact T_2-space and let K, L be subsets of S that are each compact in the subspace topology. Prove that there are disjoint open sets U and V such that $K \subseteq U$ and $L \subseteq V$. Use this result to prove that a compact T_2-space is a T_4-space.

11. Let (A, d) be a subspace of $(\mathbb{R}^n, d)$, where d is the Euclidean metric. Prove that A is bounded if and only if A is totally bounded.

12. Prove that every compact metrizable space is separable. Is every compact topological space separable?

13. Prove that the one-point compactification of a locally compact T_2-space is a compact T_2-space.

LEVEL 5

14. Prove that a locally compact T_2-space is a T_3-space.

15. Prove Lebesgue's Covering Lemma (Theorem 12.17).

16. Let $(S, \mathcal{T})$ be a topological space. Prove each of the following:

   (i) A is nowhere dense in S if and only if for any open set $U \subseteq S$, there is a nonempty open set V with $V \subseteq U$ and $V \cap A = \emptyset$.

   (ii) The union of finitely many sets that are nowhere dense in S is nowhere dense in S.

   (iii) A is nowhere dense in S if and only if $(S \setminus A)^\circ$ is dense in S.

CHALLENGE PROBLEMS

17. Prove that the following are equivalent for a topological space $(S, \mathcal{T})$:

   (i) $(S, \mathcal{T})$ is compact.

   (ii) Every covering of S by subbasic open sets contains a finite subcollection that still covers S.

   (iii) For every collection $\mathcal{C}$ of subbasic closed sets in S with the finite intersection property, we have $\cap \mathcal{C} \neq \emptyset$.

   Note that the proof of this result will require the Axiom of Choice. There are many equivalent versions of the Axiom of Choice. The most useful version for proving this result is called Zorn's Lemma. The statement of Zorn's Lemma is provided below, right after a few preliminary definitions.

   If $(P, \leq)$ is a partially ordered set, then $(C, \leq)$ is a **chain** in P if $C \subseteq P$ and $(C, \leq)$ is linearly ordered. An **upper bound** of the chain $(C, \leq)$ in P is an element $s \in P$ such that for all $y \in C$, $y \leq s$. An element $m \in P$ is a **maximal element** of $(P, \leq)$ if there is no $y \in P$ such that $m < y$.

   Zorn's Lemma (ZL): Let $(P, \leq)$ be a partially ordered set such that each chain in P has an upper bound in P. Then P contains at least one maximal element.

18. Let $\mathbb{S}^1 \subseteq \mathbb{R}^2$ be the unit circle. Prove that the one-point compactification of $\mathbb{R}$ is homeomorphic to $\mathbb{S}^1$.

19. Prove the Baire Category Theorem (Theorem 12.26).

LESSON 13
CONTINUITY AND HOMEOMORPHISMS

Continuous Functions

Recall from Lesson 4 that if $f: X \to Y$ and $A \subseteq X$, then the **image of A under f** is the set $f[A] = \{f(x) \mid x \in A\}$. Similarly, if $B \subseteq Y$, then the **inverse image of B under f** is the set $f^{-1}[B] = \{x \in X \mid f(x) \in B\}$.

Let $(X, \mathcal{T})$ and $(Y, \mathcal{U})$ be topological spaces. A function $f: X \to Y$ is **continuous** if for each $V \in \mathcal{U}$, we have $f^{-1}[V] \in \mathcal{T}$.

Notes: (1) In words, a function from one topological space to another is continuous if the inverse image of each open set is open.

(2) Continuity of a function may depend just as much on the two given topologies as it does on the function f.

(3) As an example of Note 2, if X is given the discrete topology, then any function $f: X \to Y$ is continuous. After all, **every** subset of X is open in X, and therefore every subset of X of the form $f^{-1}[V]$, where V is an open set in Y, is open in X.

(4) As another example, if $X = \{a, b\}$ is given the trivial topology, and $Y = \{a, b\}$ is given the discrete topology, then the identity function $i_X: X \to Y$ is **not** continuous. To see this, just note that $\{a\}$ is open in Y (because **every** subset of Y is open), but $i_X^{-1}(\{a\}) = \{a\}$ is **not** open in X (because $\{a\} \neq \emptyset$ and $\{a\} \neq X$).

(5) Constant functions are **always** continuous. Indeed, let $b \in Y$ and suppose that $f: X \to Y$ is defined by $f(x) = b$ for all $x \in X$. Let $B \subseteq Y$. If $b \in B$, then $f^{-1}[B] = X$ and if $b \notin B$, then $f^{-1}[B] = \emptyset$. Since X and $\emptyset$ are open in any topology on X, f is continuous.

(6) If $\mathcal{B}$ is a basis for $\mathcal{U}$, then to determine if f is continuous, we need only check that for each $V \in \mathcal{B}$, we have $f^{-1}[V] \in \mathcal{T}$. To see this, assume that for each $V \in \mathcal{B}$, we have $f^{-1}[V] \in \mathcal{T}$, and let $O \in \mathcal{U}$. Since $\mathcal{B}$ is a basis for $\mathcal{U}$, $O = \bigcup X$, for some subset X of $\mathcal{B}$. So, $f^{-1}[O] = f^{-1}[\bigcup X] = \bigcup\{f^{-1}[V] \mid V \in X\}$ (by part (iii) of Problem 15 from Problem Set 4). Since $\mathcal{T}$ is a topology, it is closed under taking arbitrary unions, and therefore, $\bigcup\{f^{-1}[V] \mid V \in X\} \in \mathcal{T}$.

Similarly, if $\mathcal{S}$ is a subbasis for $\mathcal{U}$, then to determine if f is continuous, we need only check that for each $V \in \mathcal{S}$, we have $f^{-1}[V] \in \mathcal{T}$. To see this, let's assume that for each $V \in \mathcal{S}$, we have $f^{-1}[V] \in \mathcal{T}$ and let $\mathcal{B}$ be the collection of all finite intersections of sets in $\mathcal{S}$. Then $\mathcal{B}$ is a basis for $\mathcal{U}$. Let $A \in \mathcal{B}$. Then $A = \bigcap X$ for some finite subset X of $\mathcal{S}$. So, $f^{-1}[A] = f^{-1}[\bigcap X] = \bigcap\{f^{-1}[V] \mid V \in X\}$ (by part (iv) of Problem 15 from Problem Set 4). Since $\mathcal{T}$ is a topology, it is closed under taking finite intersections, and so, $\bigcap\{f^{-1}[V] \mid V \in X\} \in \mathcal{T}$.

Example 13.1:

1. Let $(A, \mathcal{T})$ and $(B, \mathcal{U})$ be the topological spaces with sets $A = \{a, b\}$ and $B = \{1, 2, 3\}$ and topologies $\mathcal{T} = \{\emptyset, \{a\}, \{a, b\}\}$ and $\mathcal{U} = \{\emptyset, \{1, 2\}, \{1, 2, 3\}\}$. The function $f: A \to B$ defined by $f(a) = 1$ and $f(b) = 3$ is continuous because $f^{-1}[\{1, 2\}] = \{a\}$, which is open in $(A, \mathcal{T})$. On the other hand, the function $g: A \to B$ defined by $g(a) = 3$ and $g(b) = 1$ is **not** continuous because $g^{-1}[\{1, 2\}] = \{b\}$, which is **not** open in $(A, \mathcal{T})$. We can visualize these two functions as follows:

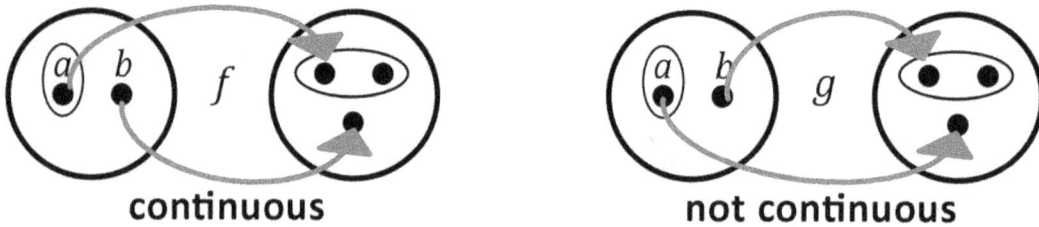

continuous · not continuous

2. Consider $(\mathbb{R}, \mathcal{T})$ and $(\mathbb{R}, \mathcal{U})$, where $\mathcal{T}$ is the standard topology on $\mathbb{R}$ and $\mathcal{U}$ is the topology generated by the basis $\{(a, \infty) \mid a \in \mathbb{R}\}$. To avoid confusion, let's use the notation $\mathbb{R}_{\mathcal{T}}$ and $\mathbb{R}_{\mathcal{U}}$ to indicate that we are considering $\mathbb{R}$ with the topologies $\mathcal{T}$ and $\mathcal{U}$, respectively. The identity function $i_1: \mathbb{R}_{\mathcal{T}} \to \mathbb{R}_{\mathcal{U}}$ is continuous because $i_1^{-1}[(a, \infty)] = (a, \infty)$ is open in $(\mathbb{R}, \mathcal{T})$ for every $a \in \mathbb{R}$. However, the identity function $i_2: \mathbb{R}_{\mathcal{U}} \to \mathbb{R}_{\mathcal{T}}$ is **not** continuous because $(0, 1)$ is open in $(\mathbb{R}, \mathcal{T})$, but $i_2^{-1}[(0, 1)] = (0, 1)$ is **not** open in $(\mathbb{R}, \mathcal{U})$.

3. Consider $(\mathbb{R}, \mathcal{T})$ and $(S, \mathcal{U})$, where $\mathcal{T}$ is the standard topology on $\mathbb{R}$, $S = \{a, b, c\}$, and $\mathcal{U}$ is the topology $\{\emptyset, \{a\}, \{a, b\}, \{a, b, c\}\}$. The function $f: \mathbb{R} \to S$ defined by $f(x) = \begin{cases} b & \text{if } x < 0 \\ c & \text{if } x \geq 0 \end{cases}$ is continuous because $f^{-1}[\{a\}] = \emptyset$ and $f^{-1}[\{a, b\}] = (-\infty, 0)$ are both open in $(\mathbb{R}, \mathcal{T})$.

    If we replace the topology $\mathcal{U}$ by the topology $\mathcal{V} = \{\emptyset, \{c\}, \{a, b, c\}\}$, then the same function f is **not** continuous because $f^{-1}[\{c\}] = [0, \infty)$, which is **not** open in $(\mathbb{R}, \mathcal{T})$.

Let $(X, \mathcal{T})$ and $(Y, \mathcal{U})$ be topological spaces. A function $f: X \to Y$ is **continuous** at $x \in X$ if for each $V \in \mathcal{U}$ with $f(x) \in V$, there is $U \in \mathcal{T}$ with $x \in U$ such that $f[U] \subseteq V$.

Example 13.2:

1. Consider the functions f and g from part 1 of Example 13.1. They are pictured below.

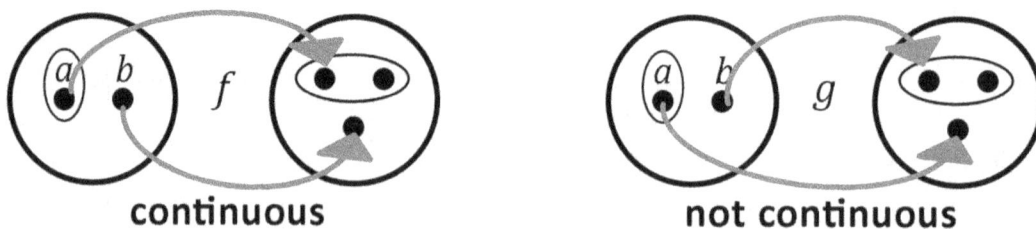

continuous · not continuous

Let's check that f is continuous at a. There are two open sets containing $f(a) = 1$. The first one is $\{1, 2\}$. The set $\{a\}$ is open and $f[\{a\}] = \{1\} \subseteq \{1, 2\}$. The second open set containing 1 is $\{1, 2, 3\}$. We can use the open set $\{a\}$ again because $f[\{a\}] = \{1\} \subseteq \{1, 2, 3\}$. Alternatively, we can use the open set $\{a, b\}$ because $f[\{a, b\}] = \{1, 3\} \subseteq \{1, 2, 3\}$.

Let's also check that f is continuous at b. The only open set containing $f(b) = 3$ is $\{1, 2, 3\}$. We have $b \in \{a, b\}$ and $f[\{a, b\}] = \{1, 3\} \subseteq \{1, 2, 3\}$.

The function g is continuous at a because the only open set containing $g(a) = 3$ is $\{1, 2, 3\}$ and we have $a \in \{a\}$ and $g[\{a\}] = \{3\} \subseteq \{1, 2, 3\}$.

The function g is **not** continuous at b. The open set $\{1, 2\}$ contains $g(b) = 1$. However, the only open set containing b is $\{a, b\}$ and $g[\{a, b\}] = \{1, 3\} \nsubseteq \{1, 2\}$.

2. Define $f: \mathbb{R} \to \mathbb{R}$ by $f(x) = \begin{cases} x & \text{if } x < 0 \\ x + 1 & \text{if } x \geq 0 \end{cases}$. Then f is **not** continuous at 0. To see this, note that $f(0) = 1 \in (0, 2)$ and if $0 \in (a, b)$, then $f[(a, b)] = (a, 0) \cup [1, b + 1) \nsubseteq (0, 2)$ because $\frac{a}{2} \in (a, 0)$, so that $\frac{a}{2} < 0$, and therefore, $\frac{a}{2} \notin (0, 2)$.

If $a > 0$, then f is continuous at a. To see this, let (c, d) be an open interval containing $f(a) = a + 1$. Then $c < a + 1 < d$, and so, $c - 1 < a < d - 1$. Let $k = \max\{0, c - 1\}$. Then we have $k < a < d - 1$. So, $a \in (k, d - 1)$. Since $k > 0$, $f[(k, d - 1)] = (k + 1, d)$. We now show that $(k + 1, d) \subseteq (c, d)$. Let $y \in (k + 1, d)$. Then $k + 1 < y < d$. Since $k \geq c - 1$, $k + 1 \geq c$. Thus, $c < y < d$, and therefore, $y \in (c, d)$. It follows that $f[(k, d - 1)] \subseteq (c, d)$.

Also, if $a < 0$, then f is continuous at a. To see this, let (c, d) be an open interval containing $f(a) = a$. Then $c < a < d$. Let $k = \min\{0, d\}$. Then we have $c < a < k$. So, $a \in (c, k)$. Finally, note that $f[(c, k)] = (c, k) \subseteq (c, d)$.

We will see in Theorem 13.4 below that if $f: \mathbb{R} \to \mathbb{R}$, where $\mathbb{R}$ is given the standard topology, then the topological definition of continuity here agrees with all the equivalent definitions of continuity from Lesson 8.

Theorem 13.3: Let $(A, \mathcal{T})$ and $(B, \mathcal{U})$ be topological spaces and let $f: A \to B$. Then f is continuous if and only if f is continuous at each $x \in A$.

Proof: Let $(A, \mathcal{T})$ and $(B, \mathcal{U})$ be topological spaces and let $f: A \to B$. First, suppose that f is continuous. Let $x \in A$ and let $V \in \mathcal{U}$ with $f(x) \in V$. Since f is continuous, $f^{-1}[V] \in \mathcal{T}$. If we let $U = f^{-1}[V]$, then $U \in \mathcal{T}$, $x \in U$, and by Problem 5 from Problem Set 4 we have $f[U] = f[f^{-1}[V]] \subseteq V$.

Conversely, suppose that f is continuous at each $x \in A$. Let $V \in \mathcal{U}$. If $f^{-1}[V] = \emptyset$, then $f^{-1}[V] \in \mathcal{T}$ because every topology contains the empty set. If $f^{-1}[V] \neq \emptyset$, let $x \in f^{-1}[V]$. Then $f(x) \in V$. So, there is $U_x \in \mathcal{T}$ with $x \in U_x$ such that $f[U_x] \subseteq V$. Let $U = \bigcup\{U_x \mid x \in f^{-1}[V]\}$. Since U is a union of open sets, $U \in \mathcal{T}$. We will show that $U = f^{-1}[V]$. Let $z \in U$. Then there is $x \in A$ with $z \in U_x$. So, we have $f(z) \in f[U_x]$. Since $f[U_x] \subseteq V$, $f(z) \in V$. Thus, $z \in f^{-1}[V]$. Since $z \in U$ was arbitrary, we have shown that $U \subseteq f^{-1}[V]$. Now, let $z \in f^{-1}[V]$. Then $f(z) \in V$. So, $z \in U_z$. Since $U_z \subseteq U$, we have $z \in U$. Since $z \in f^{-1}[V]$ was arbitrary, we have shown that $f^{-1}[V] \subseteq U$. Since $U \subseteq f^{-1}[V]$ and $f^{-1}[V] \subseteq U$, we have $U = f^{-1}[V]$. $\square$

Continuity in Metrizable Spaces

We now give an $\epsilon - \delta$ definition of continuity for metrizable topological spaces.

Theorem 13.4: Let $(A, \mathcal{T})$ and $(B, \mathcal{U})$ be metrizable topological spaces, where $\mathcal{T}$ and $\mathcal{U}$ are induced by the metrics d and ρ, respectively. $f: A \to B$ is continuous at $x \in A$ if and only if for all $\epsilon > 0$ there is $\delta > 0$ such that $d(x, y) < \delta$ implies $\rho\big(f(x), f(y)\big) < \epsilon$.

Proof: Let $(A, \mathcal{T})$ and $(B, \mathcal{U})$ be topological spaces with corresponding metrics d and ρ and let $x \in A$.

First, suppose that $f: A \to B$ is continuous at $x \in A$ and let $\epsilon > 0$. $f(x) \in B_\epsilon\big(f(x)\big)$ and $B_\epsilon\big(f(x)\big)$ is open in $\mathcal{U}$. Since f is continuous at x, there is $U \in \mathcal{T}$ with $x \in U$ such that $f[U] \subseteq B_\epsilon\big(f(x)\big)$. Since the open balls form a basis for $\mathcal{U}$, we can find $\delta > 0$ such that $B_\delta(x) \subseteq U$ (Why?). It follows that $f[B_\delta(x)] \subseteq f[U]$, and so, $f[B_\delta(x)] \subseteq B_\epsilon\big(f(x)\big)$. Now, if $d(x, y) < \delta$, then $y \in B_\delta(x)$. So, $f(y) \in f[B_\delta(x)]$. Since $f[B_\delta(x)] \subseteq B_\epsilon\big(f(x)\big)$, we have $f(y) \in B_\epsilon\big(f(x)\big)$. So, $\rho\big(f(x), f(y)\big) < \epsilon$.

Conversely, suppose that for all $\epsilon > 0$ there is $\delta > 0$ such that $d(x, y) < \delta$ implies $\rho\big(f(x), f(y)\big) < \epsilon$. Let $V \in \mathcal{U}$ with $f(x) \in V$. Since the open balls form a basis for V, there is $\epsilon > 0$ such that $f(x) \in B_\epsilon\big(f(x)\big)$ and $B_\epsilon\big(f(x)\big) \subseteq V$ (Why?). Choose $\delta > 0$ such that $d(x, y) < \delta$ implies $\rho\big(f(x), f(y)\big) < \epsilon$. Let $U = B_\delta(x)$. Then $U \in \mathcal{T}$ and $x \in U$. We show that $f[U] \subseteq V$. Let $y \in f[U]$. Then there is $z \in U$ with $y = f(z)$. Since $z \in U = B_\delta(x)$, $d(x, z) < \delta$. Therefore, $\rho\big(f(x), f(z)\big) < \epsilon$. So, $f(z) \in B_\epsilon\big(f(x)\big)$. Since $B_\epsilon\big(f(x)\big) \subseteq V$, $f(z) \in V$. Since $y = f(z)$, we have $y \in V$, as desired. $\quad\square$

Note: If we consider a function $f: \mathbb{R} \to \mathbb{R}$ with the metric $d(x, y) = |x - y|$, Theorem 13.4 shows that all our definitions of continuity given in Lesson 8 are equivalent to the topological definitions given here. The same goes for functions from $\mathbb{C}$ to $\mathbb{C}$, from $\mathbb{R}^n$ to $\mathbb{R}^m$, and from $\mathbb{C}^n$ to $\mathbb{C}^m$. More generally still, the domains of these functions need not be the whole space. For example, if we consider functions $f: A \to \mathbb{R}$, where $A \subseteq \mathbb{R}$, then all the definitions we've given here and in Lesson 8 are equivalent.

Theorem 13.5: Let $(A, \mathcal{T})$ and $(B, \mathcal{U})$ be metrizable topological spaces where $\mathcal{T}$ and $\mathcal{U}$ are induced by the metrics d and ρ, respectively. Let $f: A \to B$ and let $x \in A$. The following are equivalent:

1. f is continuous at x

2. For every open ball $B_\epsilon(f(x); \rho)$ with center $f(x)$ and radius $\epsilon > 0$, there is an open ball $B_\delta(x; d)$ with center x and radius $\delta > 0$ such that $f[B_\delta(x; d)] \subseteq B_\epsilon(f(x); \rho)$.

3. Whenever $x_n \to x$, it follows that $f(x_n) \to f(x)$.

Proof: Let $(A, \mathcal{T})$ and $(B, \mathcal{U})$ be topological spaces with corresponding metrics d and ρ and let $x \in A$.

(1→2) Suppose that f is continuous at x and let $B_\epsilon(f(x); \rho)$ be an open ball with center $f(x)$ and radius $\epsilon > 0$. Since f is continuous, by Theorem 13.4, there is $\delta > 0$ such that $d(x, y) < \delta$ implies $\rho\big(f(x), f(y)\big) < \epsilon$. It then follows that

$$f(y) \in f[B_\delta(x; d)] \to y \in B_\delta(x; d) \to d(x, y) < \delta \to \rho\big(f(x), f(y)\big) < \epsilon \to f(y) \in B_\epsilon(f(x); \rho).$$

It follows that $f[B_\delta(x; d)] \subseteq B_\epsilon(f(x); \rho)$, as desired.

(2→3) Now, suppose that 2 holds and that $x_n \to x$. Let $B_\epsilon(f(x); \rho)$ be an open ball with center $f(x)$ and radius ϵ. By 2, there is an open ball $B_\delta(x; d)$ with center x and radius $\delta > 0$ such that $f[B_\delta(x; d)] \subseteq B_\epsilon(f(x); \rho)$. Since $x_n \to x$, by Theorem 11.4, there is $K \in \mathbb{N}$ such that $n > K$ implies $x_n \in B_\delta(x; d)$. So, $n > K$ implies $f(x_n) \in f[B_\delta(x; d)]$. Since $f[B_\delta(x; d)] \subseteq B_\epsilon(f(x); \rho)$, it follows that $n > K$ implies $f(x_n) \in B_\epsilon(f(x); \rho)$. Again, by Theorem 11.4, $f(x_n) \to f(x)$.

(3→1) Suppose that f is **not** continuous at x. Then there is $\epsilon > 0$ such that for each $n \in \mathbb{N}$, there is $x_n \in X$ with $d(x_n, x) < \frac{1}{n}$ and $\rho\big(f(x_n), f(x)\big) \geq \epsilon$. Then $x_n \to x$, but $f(x_n) \nrightarrow f(x)$. □

Note: 1 implies 3 even if the topological spaces are not metrizable. 3 implies 1 is true as long as $(A, \mathcal{T})$ is first-countable (see Problem 4 below).

The definition of uniform continuity from Lesson 8 (see Note 2 following Example 8.12) can be generalized to arbitrary metrizable spaces as follows.

Let $(A, \mathcal{T})$ and $(B, \mathcal{U})$ be metrizable topological spaces where $\mathcal{T}$ and $\mathcal{U}$ are induced by the metrics d and ρ, respectively. We say that a function $f: A \to B$ is **uniformly continuous** on A if

$$\forall \epsilon > 0 \, \exists \delta > 0 \, \forall a, b \in A \, \big(d(a, b) < \delta \to \rho\big(f(a), f(b)\big) < \epsilon\big).$$

Uniformly continuous functions will be explored in the Problem Set below.

Homeomorphisms

Let $(A, \mathcal{T})$ and $(B, \mathcal{U})$ be topological spaces. A function $f: A \to B$ is a **homeomorphism** if f is a bijection such that $O \in \mathcal{T}$ if and only if $f[O] \in \mathcal{U}$.

Notes: (1) If $f: A \to B$ is a bijection, then every subset $V \subseteq B$ can be written as $f[O]$ for exactly one subset $O \subseteq A$. If f is also continuous, then given $O \subseteq A$ with $f[O] \in \mathcal{U}$, we have $O = f^{-1}[f[O]] \in \mathcal{T}$. Conversely, suppose that f is a bijection such that for every subset O of A, $f[O] \in \mathcal{U}$ implies $O \in \mathcal{T}$. Then, given $V \in \mathcal{U}$, since there is $O \subseteq A$ with $V = f[O]$, by our assumption, we have $f^{-1}[V] = f^{-1}[f[O]] = O \in \mathcal{T}$, showing that f is continuous. It follows that f is a continuous bijection if and only if f is a bijection such that $\forall O \subseteq A(f[O] \in \mathcal{U} \to O \in \mathcal{T})$.

(2) Similarly, $f: A \to B$ is a bijective function with continuous inverse $f^{-1}: B \to A$ if and only if f is a bijection such that $\forall O \subseteq A(O \in \mathcal{T} \to f[O] \in \mathcal{U})$.

(3) Notes 1 and 2 tell us that $f: A \to B$ is a homeomorphism if and only if f is a continuous bijective function with a continuous inverse.

(4) Since a homeomorphism is bijective, it provides a one to one correspondence between the elements of A and the elements of B. However, a homeomorphism does much more than this. It also provides a one to one correspondence between the sets in $\mathcal{T}$ and the sets in $\mathcal{U}$.

(5) A homeomorphism between two topological spaces is analogous to an isomorphism between two algebraic structures (see the Note following Theorem 6.7). From the topologists point of view, if there is a homeomorphism from one space to another, the two topological spaces are indistinguishable.

We say that two topological spaces $(A, \mathcal{T})$ and $(B, \mathcal{U})$ are **homeomorphic** or **topologically equivalent** if there is a homeomorphism $f: A \to B$.

Example 13.6:

1. Let $S = \{a, b\}$, $\mathcal{T} = \{\emptyset, \{a\}, \{a, b\}\}$, and $\mathcal{U} = \{\emptyset, \{b\}, \{a, b\}\}$. The map $f: S \to S$ defined by $f(a) = b$ and $f(b) = a$ is a homeomorphism from $(S, \mathcal{T})$ to $(S, \mathcal{U})$. Notice that the inverse image of the open set $\{b\} \in \mathcal{U}$ is the open set $\{a\} \in \mathcal{T}$. This shows that f is continuous. Conversely, the image of the open set $\{a\} \in \mathcal{T}$ is the open set $\{b\} \in \mathcal{U}$. This shows that f^{-1} is continuous. Since f is also a bijection, we have shown that f is a homeomorphism. On the other hand, the identity function $g: S \to S$ defined by $g(a) = a$ and $g(b) = b$ is **not** a homeomorphism because it is not continuous. For example, the inverse image of the open set $\{b\} \in \mathcal{U}$ is the set $\{b\}$ which is **not** in the topology $\mathcal{T}$. We can visualize these two functions as follows:

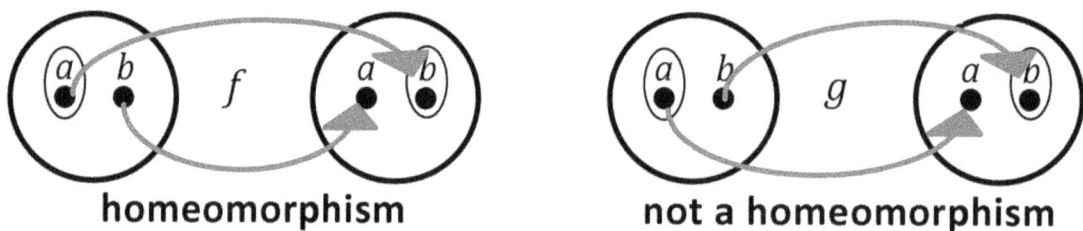

Notice that f and g are both bijections from S to S, but only the function f also gives a one to one correspondence between the open sets of the topology $(S, \mathcal{T})$ and the open sets of the topology $(S, \mathcal{U})$.

The homeomorphism f shows that $(S, \mathcal{T})$ and $(S, \mathcal{U})$ are topologically equivalent. So, up to topological equivalence, there are only three topologies on a set with two elements: the trivial topology, the discrete topology, and the topology with exactly three open sets.

2. Let $S = \{a, b, c\}$, $\mathcal{T} = \{\emptyset, \{b\}, \{a, b\}, \{a, b, c\}\}$, and $\mathcal{U} = \{\emptyset, \{a, b\}, \{a, b, c\}\}$. Then the identity function $f: S \to S$ is a continuous bijection from $(S, \mathcal{T})$ to $(S, \mathcal{U})$. Indeed, the inverse image of the open set $\{a, b\} \in \mathcal{U}$ is the open set $\{a, b\} \in \mathcal{T}$. However, f is **not** a homeomorphism because f^{-1} is not continuous. The set $\{b\}$ is open in $\mathcal{T}$, but its image $f[\{b\}] = \{b\}$ is **not** open in $\mathcal{U}$.

3. We saw in part 3 of Example 9.1 that there are 29 topologies on a set with three elements. However, up to topological equivalence, there are only 9. Below is a visual representation of the 9 distinct topologies on the set $S = \{a, b, c\}$, up to topological equivalence.

The dedicated reader should verify that each of the other 20 topologies are topologically equivalent to one of these and that no two topologies displayed here are topologically equivalent.

4. Consider $\mathbb{R}$ together with the standard topology. Define $f: \mathbb{R} \to \mathbb{R}$ by $f(x) = 2x + 3$. Let's check that f is a homeomorphism. If $x \neq y$, then $2x \neq 2y$, and so, $2x + 3 \neq 2y + 3$. Therefore, $\forall x, y \in \mathbb{R}\big(x \neq y \to f(x) \neq f(y)\big)$. That is, f is injective. Next, if $y \in \mathbb{R}$, let $x = \frac{y-3}{2}$. Then $f(x) = f\left(\frac{y-3}{2}\right) = 2\left(\frac{y-3}{2}\right) + 3 = (y - 3) + 3 = y$. So, $\forall y \in \mathbb{R}\, \exists x \in \mathbb{R}(f(x) = y)$. That is, f is surjective. Now, let (a, b) be a bounded open interval. $f^{-1}[(a,b)] = \left(\frac{a-3}{2}, \frac{b-3}{2}\right)$, which is open. So, f is continuous. Also, $f[(a,b)] = (2a + 3, 2b + 3)$, which is open. So, f^{-1} is continuous. Since f is a continuous bijection with a continuous inverse, f is a homeomorphism.

5. Consider $(\mathbb{R}, \mathcal{T})$ and $(\mathbb{R}, \mathcal{U})$, where $\mathcal{T}$ is the standard topology on $\mathbb{R}$ and $\mathcal{U}$ is the topology generated by the basis $\{(a, \infty) \mid a \in \mathbb{R}\}$. We saw in part 2 of Example 13.1 that the identity function $i: \mathbb{R}_{\mathcal{T}} \to \mathbb{R}_{\mathcal{U}}$ is continuous because $i^{-1}[(a, \infty)] = (a, \infty)$ is open in $(\mathbb{R}, \mathcal{T})$ for every $a \in \mathbb{R}$. However, this function is **not** a homeomorphism because i^{-1} is not continuous. For example, $(0, 1)$ is open in $(\mathbb{R}, \mathcal{T})$, but $i[(0,1)] = (0,1)$ is **not** open in $(\mathbb{R}, \mathcal{U})$.

A **topological property** or **topological invariant** is a property that is preserved under homeomorphisms. More specifically, we say that property P is a topological property if whenever the topological space $(A, \mathcal{T})$ has property P and $(B, \mathcal{U})$ is topologically equivalent to $(A, \mathcal{T})$, then $(B, \mathcal{U})$ also has property P.

In Problem 6 below, you will be asked to show that compactness is a topological property. As another example, let's show that the property of being a T_2-space is a topological property.

Theorem 13.7: Let $(A, \mathcal{T})$ be a T_2-space and let $(B, \mathcal{U})$ be topologically equivalent to $(A, \mathcal{T})$. Then $(B, \mathcal{U})$ is a T_2-space.

Proof: Let $(A, \mathcal{T})$ be a T_2-space and let $f: A \to B$ be a homeomorphism. Let $x, y \in B$ with $x \neq y$. Since f is bijective, there are $z, w \in A$ with $z \neq w$ such that $f(z) = x$ and $f(w) = y$. Since $(A, \mathcal{T})$ is a T_2-space, there are open sets $U, V \in \mathcal{T}$ with $z \in U$, $w \in V$, and $U \cap V = \emptyset$. Since f is a homeomorphism, $f[U], f[V] \in \mathcal{U}$. We also have $x = f(z) \in f[U]$ and $y = f(w) \in f[V]$. We show that $f[U] \cap f[V] = \emptyset$. If not, there is $c \in f[U] \cap f[V]$. So, there are $a \in U$ and $b \in V$ with $f(a) = c$ and $f(b) = c$. So, $f(a) = f(b)$. Since f is injective, $a = b$. But then $a \in U \cap V$, contradicting that $U \cap V = \emptyset$. It follows that $f[U] \cap f[V] = \emptyset$. Therefore, $(X, \mathcal{U})$ is a T_2-space. $\square$

The dedicated reader might want to show that the other separation axioms (T_0 through T_4) are topological properties, that the countability axioms are topological properties, and that metrizability is a topological property.

Tychonoff Spaces

A topological space $(S, \mathcal{T})$ is a **Tychonoff space** if $(S, \mathcal{T})$ is a T_1-space and for every $x \in S$ and closed set A with $A \subseteq S \setminus \{x\}$, there is a continuous function $f: S \to [0, 1]$ such that $f(x) = 0$ and $f[A] = \{1\}$.

Note: If we drop the condition that $(S, \mathcal{T})$ is a T_1-space, then we say that $(S, \mathcal{T})$ is a **Completely Regular** space or $T_{3\frac{1}{2}}$-**space**.

Example 13.8:

1. Every Tychonoff space $(S, \mathcal{T})$ is a T_3-space. To see this, let $(S, \mathcal{T})$ be a Tychonoff space, let $x \in S$ and let A be a closed set with $A \subseteq S \setminus \{x\}$. Since $(S, \mathcal{T})$ is a Tychonoff space, there is a continuous function $f: S \to [0, 1]$ such that $f(x) = 0$ and $f[A] = \{1\}$. Now, $\left[0, \frac{1}{2}\right)$ and $\left(\frac{1}{2}, 1\right]$ are disjoint open sets in $[0, 1]$. Since f is continuous, $f^{-1}\left(\left[0, \frac{1}{2}\right)\right)$ and $f^{-1}\left(\left(\frac{1}{2}, 1\right]\right)$ are open in S. Since $f(x) = 0$, $x \in f^{-1}\left(\left[0, \frac{1}{2}\right)\right)$. Since $f[A] = \{1\}$, $A \subseteq f^{-1}\left(\left(\frac{1}{2}, 1\right]\right)$. Finally, we have $f^{-1}\left(\left[0, \frac{1}{2}\right)\right) \cap f^{-1}\left(\left(\frac{1}{2}, 1\right]\right) = \emptyset$ because if $x \in f^{-1}\left(\left[0, \frac{1}{2}\right)\right) \cap f^{-1}\left(\left(\frac{1}{2}, 1\right]\right)$, then we would have $f(x) \in \left[0, \frac{1}{2}\right) \cap \left(\frac{1}{2}, 1\right]$, which of course is impossible.

2. Every T_4-space is a Tychonoff space. This will follow immediately from Urysohn's Lemma (see Theorem 13.9 and Corollary 13.10 below).

3. In the definition of a Tychonoff space, we could just have easily required $f(x) = 1$ and $f[A] = \{0\}$. Indeed, if we let $g(x) = 1 - f(x)$, then g is continuous if and only if f is To see this, observe that for $0 \le a < b \le 1$, we have $g^{-1}[(a, b)] = f^{-1}[(1 - b, 1 - a)]$, $g^{-1}[[0, b)] = f^{-1}[(1 - b, 1]]$, and $g^{-1}[(a, 1]] = f^{-1}[[0, 1 - a)]$ (we actually need only check the latter two because $\{[0, b) \mid 0 < b \le 1\} \cup \{(a, 1] \mid 0 \le a < 1\}$ is a subbasis for $[0, 1]$).

4. If $(S, \mathcal{T})$ is a Tychonoff space and $(B, \mathcal{T}_B)$ is a subspace of $(S, \mathcal{T})$, then $(B, \mathcal{T}_B)$ is also a Tychonoff space. To see this, let $x \in B$ and let A be a closed set in B with $A \subseteq B \setminus \{x\}$. Let $\overline{A}$ be the closure of A in S. By Problem 13 from Problem Set 9, $A = \overline{A} \cap B$. So, $x \notin \overline{A}$. Since $(S, \mathcal{T})$ is a Tychonoff space, there is a continuous function $f: S \to [0, 1]$ such that $f(x) = 0$ and $f[\overline{A}] = \{1\}$. Let $g: B \to [0, 1]$ be defined by $g(b) = f(b)$ for all $b \in B$. Then $g(x) = f(x) = 0$ and $g[A] = f[A] = \{1\}$. We need to show that g is continuous. To this end, let U be an open set in $[0, 1]$. Then $g^{-1}[U] = f^{-1}[U] \cap B$. Since f is continuous, $f^{-1}[U]$ is open in S, and therefore, $f^{-1}[U] \cap B$ is open in the subspace topology relative to B.

5. Let K be an index set and let $(S_k, \mathcal{T}_k)$ be a Tychonoff space for each $k \in K$. Then $\prod S_k$ is a Tychonoff space. To see this, let $\mathbf{x} \in \prod S_k$ and let $A \subseteq \prod S_k \setminus \{\mathbf{x}\}$ be a closed set in $\prod S_k$. Let $\prod U_k$ be a basis element in $\prod S_k$ containing $\mathbf{x}$ with $\prod U_k \cap A = \emptyset$. Let $L = \{k \in K \mid U_k \ne S_k\}$. By the definition of the product topology, L is finite. For each $k \in L$, let $f_k: S_k \to [0, 1]$ be a continuous function such that $f_k(x_k) = 0$ and $f_k[S_k \setminus U_k] = \{1\}$. Define $g_k: \prod S_k \to [0, 1]$ for each $k \in L$ by $g_k(\mathbf{y}) = f_k(y_k)$. Then we have for each $k \in K$, $g_k(\mathbf{x}) = f_k(x_k) = 0$ and $g_k[\prod S_j \setminus \prod U_j] = f_k[S_k \setminus U_k] = \{1\}$. Define $f: \prod S_k \to [0, 1]$ by $f(\mathbf{y}) = \prod g_k(\mathbf{y})$, where the product is taken over g_k for $k \in L$ (so that $f(\mathbf{y})$ is a finite product). Then $f(\mathbf{x}) = \prod g_k(\mathbf{x}) = 0$, $f[\prod S_k \setminus \prod U_k] = \prod g_k[\prod S_j \setminus \prod U_j] = \{1\}$, and f is continuous (see Notes 1 through 7 below for the outline of a proof that f is continuous.).

Notes: (1) Suppose that $(A, \mathcal{T})$, $(B, \mathcal{U})$, and $(C, \mathcal{V})$ are topological spaces, and that $h: A \to B$, $k: B \to C$ are continuous. Then the composite function $k \circ h: A \to C$ is also continuous. To see this, let W be open in C. By the continuity of k, $k^{-1}[W]$ is open in B. By the continuity of h, $h^{-1}[k^{-1}[W]]$ is open in A. Now just observe that $h^{-1}[k^{-1}[W]] = (k \circ h)^{-1}[W]$ because $x \in h^{-1}[k^{-1}[W]]$ if and only if $h(x) \in k^{-1}[W]$ if and only if $(k \circ h)(x) = k(h(x)) \in W$ if and only if $x \in (k \circ h)^{-1}[W]$.

(2) Let $(S, \mathcal{T})$ be a topological space and let $h, k: S \to [0, 1]$ be continuous functions. Then the function $j: S \to [0, 1] \times [0,1]$ defined by $j(x) = \big(h(x), k(x)\big)$ is continuous. To see this, let $U \times V$ be open in $[0, 1] \times [0,1]$. Then $j^{-1}[U \times V] = h^{-1}[U] \cap k^{-1}[V]$. Indeed, $x \in j^{-1}[U \times V]$ if and only if $\big(h(x), k(x)\big) = j(x) \in U \times V$ if and only if $h(x) \in U$ and $k(x) \in V$ if and only if $x \in h^{-1}[U]$ and $x \in k^{-1}[V]$ if and only if $x \in h^{-1}[U] \cap k^{-1}[V]$. Since h and k are continuous, $h^{-1}[U]$ and $k^{-1}[V]$ are open in S. Since finite intersections of open sets are open, $j^{-1}[U \times V] = h^{-1}[U] \cap k^{-1}[V]$ is open in S.

(3) More generally, if $h_k: S \to [0, 1]$ is continuous for $k = 1, 2, \dots, n$, then $j: S \to [0, 1]^n$ defined by $j(x) = \big(h_1(x), h_2(x), \dots, h_n(x)\big)$ is continuous. The argument is similar to what was done in Note 2. I leave the details to the reader.

(4) The function $b: [0, 1] \times [0,1] \to [0, 1]$ defined by $b(x, y) = xy$ is continuous. This can be proved with a standard $\epsilon - \delta$ argument using the techniques learned in Lesson 8. See Problem 24 in Problem Set 8.

(5) More generally, the function $b: [0, 1]^n \to [0, 1]$ defined by $b(x_1, x_2, \dots, x_n) = x_1 x_2 \cdots x_n$ is continuous. See Problem 24 in Problem Set 8.

(6) Let $(S, \mathcal{T})$ be a topological space and let $f, g: S \to [0, 1]$ be continuous functions. Let's prove that the function fg defined by $(fg)(x) = f(x)g(x)$ is continuous. We first define $h: S \to [0, 1] \times [0,1]$ by $h(x) = \big(f(x), g(x)\big)$ and $b: [0, 1] \times [0,1] \to [0, 1]$ by $b(x, y) = xy$. By Note 2 above, h is continuous. By Note 4 above, b is continuous. So, by Note 1 above, $b \circ h: S \to [0, 1]$ is continuous. If $x \in S$, then $(b \circ h)(x) = b\big(h(x)\big) = b\big(f(x), g(x)\big) = f(x)g(x)$. Therefore, $b \circ h = fg$, proving that fg is continuous.

(7) More generally, let $(S, \mathcal{T})$ be a topological space and let $g_k: S \to [0, 1]$ be continuous for each $k \in L$, where L is a finite set. Let's prove that the function f defined by $f(x) = \prod g_k(x)$ is continuous. We first define $h: S \to {}^L[0, 1]$ by $h(x) = \big(g_k(x)\big)$ and $b: {}^L[0, 1] \to [0, 1]$ by $b(s_k) = \prod s_k$. By Note 3 above, h is continuous. By Note 5 above, b is continuous. So, by Note 1 above, $b \circ h: S \to [0, 1]$ is continuous. If $x \in S$, then $(b \circ h)(x) = b\big(h(x)\big) = b\big(g_k(x)\big) = \prod g_k(x) = f(x)$. Therefore, $b \circ h = f$, proving that f is continuous.

(8) The types of arguments presented in Notes 1 through 7 above are typical of proofs involving combinations of continuous functions. Advanced text books in mathematics will usually skip over these types of proofs and say things like "By a standard continuity argument, it follows that..." It's understandable why they would do this. The arguments are usually similar to what we have done here. They are generally straightforward, using the same techniques that come up repeatedly. However, they require multiple steps, and so, writing out the details can be tedious.

In part 2 of Example 13.8 above, we made the claim that every T_4-space is a Tychonoff space. This result is a direct consequence of Urysohn's Lemma, which we now prove.

Theorem 13.9 (Urysohn's Lemma): Let $(S, \mathcal{T})$ be a T_4-space and let A and B be disjoint closed subsets of S. Then there is a continuous function $f: S \to [0, 1]$ such that $f[A] = \{0\}$ and $f[B] = \{1\}$.

Proof: Let $(S, \mathcal{T})$ be a T_4-space and let A and B be disjoint closed subsets of S. $S \setminus B$ is an open set containing A. Therefore, by Problem 4 from Problem Set 10, there is an open set $U_{\frac{1}{2}}$ such that

$$A \subseteq U_{\frac{1}{2}} \subseteq \overline{U}_{\frac{1}{2}} \subseteq S \setminus B.$$

Since $U_{\frac{1}{2}}$ is an open set containing A and $S \setminus B$ is an open set containing $\overline{U}_{\frac{1}{2}}$, there are open sets $U_{\frac{1}{4}}$ and $U_{\frac{3}{4}}$ such that

$$A \subseteq U_{\frac{1}{4}} \subseteq \overline{U}_{\frac{1}{4}} \subseteq U_{\frac{1}{2}} \subseteq \overline{U}_{\frac{1}{2}} \subseteq U_{\frac{3}{4}} \subseteq \overline{U}_{\frac{3}{4}} \subseteq S \setminus B.$$

We continue in this fashion so that whenever $x < y$, $x = \frac{k_1}{2^n}$, and $y = \frac{k_2}{2^n}$, where $n, k_1, k_2 \in \mathbb{Z}^+$ and $1 \leq k_1 < k_2 \leq 2^n - 1$, we get open sets U_x and U_y such that

$$x < y \Rightarrow A \subseteq U_x \subseteq \overline{U}_x \subseteq U_y \subseteq \overline{U}_y \subseteq S \setminus B.$$

Define $f : S \to [0, 1]$ by $f(s) = \begin{cases} 0 & \text{if } s \text{ is in every } U_x. \\ \sup\{x \mid s \notin U_x\} & \text{otherwise.} \end{cases}$

Since $A \subseteq U_x$ for all $x \in S$, $f[A] = \{0\}$. Since $U_x \subseteq S \setminus B$ for all $x \in S$, $f[B] = 1$.

We finish the proof by showing that f is continuous. $\mathcal{F} = \{[0, b) \mid 0 < b \leq 1\} \cup \{(a, 1] \mid 0 \leq a < 1\}$ is a subbasis for $[0, 1]$. So, we need only check that for each $a \in [0, 1)$ and $b \in (0, 1]$, $f^{-1}[[0, b)]$ and $f^{-1}[(a, 1]]$ are open in S. Now, $s \in f^{-1}[[0, b)]$ if and only if $f(s) \in [0, b)$ if and only if $0 \leq f(s) < b$ if and only if $s \in U_x$ for some $x < b$ if and only if $s \in \bigcup\{U_x \mid x < b\}$. So, $f^{-1}[[0, b)] = \bigcup\{U_x \mid x < b\}$, a union of open sets in S. Therefore, $f^{-1}[[0, b)]$ is open in S. Similarly, $s \in f^{-1}[(a, 1]]$ if and only if $f(s) \in (a, 1]$ if and only if $a < f(s) \leq 1$ if and only if $s \notin \overline{U}_x$ for some $x > a$ if and only if $s \in S \setminus \overline{U}_x$ for some $x > a$ if and only if $s \in \bigcup\{S \setminus \overline{U}_x \mid y > a\}$. So, $f^{-1}[(a, 1]] = \bigcup\{S \setminus \overline{U}_x \mid x > a\}$, a union of open sets in S. Therefore, $f^{-1}[(a, 1]]$ is open in S. $\square$

Corollary 13.10: Every T_4-space is a Tychonoff space.

Proof: Let $(S, \mathcal{T})$ be a T_4-space. Then by definition, $(S, \mathcal{T})$ is a T_1-space. Let $x \in S$ and let A be a closed set with $A \subseteq S \setminus \{x\}$. Since $(S, \mathcal{T})$ is a T_1-space, $\{x\}$ is closed in S. By Urysohn's Lemma, there is a continuous function $f : S \to [0, 1]$ such that $f[\{x\}] = \{0\}$ and $f[A] = \{1\}$. But $f[\{x\}] = \{0\}$ is equivalent to $f(x) = 0$. Therefore, $(S, \mathcal{T})$ is a Tychonoff space. $\square$

We now conclude this lesson with the Tietze Extension Theorem, which allows us to extend continuous functions defined on a closed subspace of a T_4-space to the whole space.

Theorem 13.12 (Tietze Extension Theorem): Let $(S, \mathcal{T})$ be a T_4-space, let $(C, \mathcal{T}_C)$ be a subspace of $(S, \mathcal{T})$ with C closed in S, let $a, b \in \mathbb{R}$ with $a < b$, and let $f : C \to [a, b]$ be continuous. Then f can be extended to a continuous function $g : S \to [a, b]$.

I leave the proof of the Tietze Extension Theorem as a Challenge Problem for the reader (see Problem 14 below).

Problem Set 13

Full solutions to these problems are available for free download here:

www.SATPrepGet800.com/TFBZLF

LEVEL 1

1. Let $(A, \mathcal{T})$ and $(B, \mathcal{U})$ be topological spaces and let $f: A \to B$. Prove that f is continuous if and only if for each C that is closed in B, the set $f^{-1}[C]$ is closed in A.

2. Prove that the one-point compactification of a locally compact T_2-space is a Tychonoff space.

LEVEL 2

3. Prove that $\mathbb{C}^n$ and $\mathbb{R}^{2n}$ with their standard topologies are topologically equivalent.

4. Let $(A, \mathcal{T})$ and $(B, \mathcal{U})$ be topological spaces, let $x \in A$, and let $f: A \to B$. Prove that if f is continuous at x, then then for every sequence x_n converging to x, the sequence $f(x_n)$ converges to $f(x)$. Then prove that if $(A, \mathcal{T})$ is a first-countable space, then the converse holds.

5. Let $(A, \mathcal{T})$ and $(B, \mathcal{U})$ be metrizable topological spaces. Let $f, g: A \to B$ be continuous and let $C \subseteq A$. Suppose that $f(x) = g(x)$ for all $x \in A$. Prove that $f(x) = g(x)$ for all $x \in \overline{A}$.

LEVEL 3

6. Let $(K, \mathcal{T})$ and $(L, \mathcal{U})$ be topological spaces with $(K, \mathcal{T})$ compact and let $f: K \to L$ be a homeomorphism. Prove that $(L, \mathcal{U})$ is compact.

7. Prove that the image of a Cauchy sequence under a uniformly continuous function is a Cauchy sequence. If we replace "uniformly continuous" by "continuous," is the result still true?

8. Let $(S, \mathcal{T})$ be a T_4-space, let A and B be disjoint closed subsets of S, and let a and b be real numbers with $a < b$. Prove that there is a continuous function $f: S \to [a, b]$ such that $f[A] = \{a\}$ and $f[B] = \{b\}$.

LEVEL 4

9. Let $(A, \mathcal{T})$ and $(B, \mathcal{U})$ be topological spaces and let $f: A \to B$. Prove that f is continuous if and only if for each $X \subseteq A$, $f[\overline{X}] \subseteq \overline{f[X]}$.

10. Prove that an injective continuous function from a compact space onto a T_2-space is a homeomorphism.

LEVEL 5

11. Let $(K, \mathcal{T})$ and $(B, \mathcal{U})$ be metrizable topological spaces with $(K, \mathcal{T})$ compact and let $f: K \to B$ be continuous. Prove that f is uniformly continuous.

12. Let $(A, \mathcal{T})$ be a metrizable topological space and let $(B, \mathcal{U})$ be a completely metrizable topological space. Let $C \subseteq A$ be dense in A and suppose that $f: C \to B$ is uniformly continuous. Prove that f can be extended uniquely to a uniformly continuous function $g: A \to B$. If we replace "uniformly continuous" by "continuous," is the result still true?

CHALLENGE PROBLEMS

13. Prove that a second-countable T_3-space is metrizable. (This is known as Urysohn's Metrization Theorem.)

14. Prove the Tietze Extension Theorem (Theorem 13.12).

LESSON 14
CONNECTEDNESS

Connected Spaces

Let $(S, \mathcal{T})$ be a topological space. A **disconnection** of S is a pair (U, V), where U and V are nonempty open sets such that $U \cap V = \emptyset$ and $U \cup V = S$.

Note: Some authors use the expression "**separation**" instead of "**disconnection**." Since we already have the "separation axioms" and "separable spaces," it should be easy to understand why I prefer to use the word "disconnection."

A topological space $(S, \mathcal{T})$ is **disconnected** if there exists a disconnection of S. Otherwise the topological space is **connected**.

Example 14.1:

1. Let S be any set and let $\mathcal{T}$ be the trivial topology on S. Then $(S, \mathcal{T})$ is connected. In this space, the only nonempty open set is S itself, and so, there does not exist a disconnection of S.

2. Let S be any set with $|S| \geq 2$ and let $\mathcal{T}$ be the discrete topology on S. Then $(S, \mathcal{T})$ is disconnected. Any partition of S consisting of two sets is a separation of S. As a specific example, consider $(\mathbb{Z}, \mathcal{T})$, where $\mathcal{T}$ is the discrete topology on $\mathbb{Z}$. Let $\mathbb{E}$ be the set of even integers and let $\mathbb{O}$ be the set of odd integers. Then $\mathbb{E}$ and $\mathbb{O}$ are nonempty open sets (all subsets of $\mathbb{Z}$ are open in the discrete toplogy), $\mathbb{E} \cap \mathbb{O} = \emptyset$, and $\mathbb{E} \cup \mathbb{O} = \mathbb{Z}$. So, $(\mathbb{E}, \mathbb{O})$ is a disconnection of $\mathbb{Z}$.

3. Let's take a look at $\mathbb{Q}$ as a subspace of $\mathbb{R}$. The space $\left(\mathbb{Q}, \mathcal{T}_{\mathbb{Q}}\right)$ is disconnected. To see this, simply observe that $\left(-\infty, \sqrt{2}\right) \cap \mathbb{Q}$ and $\left(\sqrt{2}, \infty\right) \cap \mathbb{Q}$ are disjoint nonempty subsets of $\mathbb{Q}$ whose union is $\mathbb{Q}$. It follows that $\left(\left(-\infty, \sqrt{2}\right) \cap \mathbb{Q}, \left(\sqrt{2}, \infty\right) \cap \mathbb{Q}\right)$ is a disconnection of $\mathbb{Q}$.

4. Consider $[0, 1] \cup [9, 10]$ as a subspace of $\mathbb{R}$. This subspace is disconnected. $[0, 1]$ is open in the subspace topology because $(-1, 2) \cap ([0, 1] \cup [9, 10]) = [0, 1]$. Similarly, $[9, 10]$ is open in the subspace topology. So, $([0, 1], [9, 10])$ is a disconnection of $[0, 1] \cup [9, 10]$.

5. For each $n \in \mathbb{N}$, the spaces $\mathbb{R}^n$ and $\mathbb{C}^n$ are connected. This will follow immediately from Theorem 14.3, Problem 13 below, and Problem 3 from Problem Set 13.

Recall from Lesson 10 that a set is **clopen** if it is both open and closed.

Theorem 14.2: A topological space $(S, \mathcal{T})$ is connected if and only if the only clopen sets in S are S and $\emptyset$.

Proof: Let $(S, \mathcal{T})$ be a topological space. First assume that $(S, \mathcal{T})$ is not connected and let (U, V) be a disconnection of S. Then U is nonempty, $U \neq S$, and U is clopen in S. It is open in S by the definition of a disconnection and it is closed in S because $U = S \setminus V$ and V is open in S.

Conversely, suppose that B is clopen in S, $B \neq S$, and $B \neq \emptyset$. Let $U = B$ and $V = S \setminus B$. Then U and V are nonempty, disjoint open sets such that $U \cup V = S$. So, (U, V) is a disconnection of S. Therefore, $(S, \mathcal{T})$ is not connected. $\qquad \square$

Connected Subspaces of the Real Line

Our goal in this section is to show that the connected subspaces of the real line are precisely the intervals (see Lesson 3 for the 9 types of intervals in $\mathbb{R}$). Let's begin by showing that every interval (including $\mathbb{R}$ itself) is a connected subspace of $\mathbb{R}$.

Theorem 14.3: Let $I \subseteq \mathbb{R}$ be an interval. Then $(I, \mathcal{T}_I)$ is a connected subspace of $(\mathbb{R}, \mathcal{T})$, where $\mathcal{T}$ is the standard topology on $\mathbb{R}$.

Proof: Let $I \subseteq \mathbb{R}$ be an interval and suppose toward contradiction that (U, V) is a disconnection of I. Since U and V are nonempty, we can choose $a \in U$ and $b \in V$. Since $U \cap V = \emptyset$, $a \neq b$. Without loss of generality, assume that $a < b$. Since I is an interval, $[a, b] \subseteq I$. Since $U \cup V = I$, each element of I is in either U or V. Let $c = \sup(U \cap [a, b])$. Since $a \leq c \leq b$, $c \in I$. Since V is open in $\mathbb{R}$, $U = \mathbb{R} \setminus V$ is closed in $\mathbb{R}$. By Lemma 12.9, $c \in U$. Since $b \in V$ and $U \cap V = \emptyset$, we must have $c < b$. For every $r \in \mathbb{R}^+$ with $c + r \leq b$, by the definition of c, we must have $c + r \in V$. Therefore, c is an accumulation point of V. Since U is open in $\mathbb{R}$, $V = \mathbb{R} \setminus U$ is closed in $\mathbb{R}$. By Theorem 7.24, $c \in V$. So, we have $c \in U \cap V$, contradicting $U \cap V = \emptyset$. It follows that there does not exist a disconnection of I. Therefore, I is connected. $\qquad \square$

We now show that there are no other connected subspaces of $\mathbb{R}$.

Theorem 14.4: Let $A \subseteq \mathbb{R}$ with $(A, \mathcal{T}_A)$ a connected subspace of $(\mathbb{R}, \mathcal{T})$, where $\mathcal{T}$ is the standard topology on $\mathbb{R}$. Then A is an interval.

Proof: Let $A \subseteq \mathbb{R}$ and assume that A is **not** an interval. Then there are $a, b, c \in \mathbb{R}$ with $a < c < b$, $a, b \in A$, and $c \notin A$. Let $U = A \cap (-\infty, c)$ and $V = A \cap (c, \infty)$. We will show that (U, V) is a disconnection of A. Since $a \in A$ and $a < c$, $a \in U$. Since $b \in A$ and $c < b$, $b \in V$. It follows that U and V are both nonempty. U and V are both open by the definition of the subspace topology. If $x \in U \cap V$, then $x < c$ and $c < x$, from which it follows that $x < x$. Since $<$ is a strict linear ordering on $\mathbb{R}$, this is impossible. So, $U \cap V = \emptyset$. Finally, let $x \in A$. If $x < c$, then $x \in U$. If $x > c$, then $x \in V$. So, $x \in U$ or $x \in V$. Therefore, $x \in U \cup V$, and so, $U \cup V = A$. Since (U, V) is a disconnection of A, $(A, \mathcal{T}_A)$ is a **not** a connected subspace of $(\mathbb{R}, \mathcal{T})$. $\qquad \square$

Theorem 14.5: Let $(A, \mathcal{T})$ and $(B, \mathcal{U})$ be topological spaces with $(A, \mathcal{T})$ connected and let $f: A \to B$ be a continuous function. Then $f[A]$ is connected.

Proof: Let $(A, \mathcal{T})$ and $(B, \mathcal{U})$ be topological spaces with $(A, \mathcal{T})$ connected, let $f: A \to B$ be a continuous function, and suppose that $f[A]$ is not connected. Let (U, V) be a disconnection of $f[A]$. Then U and V are open sets in $f[A]$ such that $U \cap V = \emptyset$ and $U \cup V = f[A]$. Since f is continuous, $f^{-1}[U]$ and $f^{-1}[V]$ are open in A. If we could find $x \in f^{-1}[U] \cap f^{-1}[V]$, then $f(x) \in U \cap V$, contradicting $U \cap V = \emptyset$. So, $f^{-1}[U] \cap f^{-1}[V] = \emptyset$. If $x \in A$ is arbitrary, then $f(x) \in f[A] = U \cup V$. So, $f(x) \in U$ or $f(x) \in V$. Therefore, $x \in f^{-1}[U]$ or $x \in f^{-1}[V]$. So, $A = f^{-1}[U] \cup f^{-1}[V]$. It follows that $(f^{-1}[U], f^{-1}[V])$ is a disconnection of A. Therefore, A is not connected. $\qquad \square$

It follows from Theorem 14.5 that connectedness is a topological property. In other words, suppose that $(A, \mathcal{T})$ and $(B, \mathcal{U})$ are topologically equivalent (there is a homeomorphism $f: A \to B$). If $(A, \mathcal{T})$ is connected, then so is $(B, \mathcal{U})$. This follows immediately from Theorem 14.5 together with the fact that a homeomorphism is continuous and surjective.

Corollary 14.6 (Intermediate Value Theorem): Let $(S, \mathcal{T})$ be a connected topological space and let $f: S \to \mathbb{R}$ be a continuous function. Then $f[S]$ is an interval.

Proof: Let $(S, \mathcal{T})$ be a connected topological space and let $f: S \to \mathbb{R}$ be continuous. By Theorem 14.5, $f[S]$ is a connected subspace of $\mathbb{R}$. By Theorem 14.4, $f[S]$ is an interval. $\qquad\square$

Connected Components

Theorem 14.7: Let $(S, \mathcal{T})$ be a topological space and let $\mathcal{A} = \{(A_k, \mathcal{T}_{A_k}) \mid k \in K\}$ be a collection of connected subspaces of $(S, \mathcal{T})$ such that $\cap\{A_k \mid k \in K\} \neq \emptyset$. Let $A = \cup\{A_k \mid k \in K\}$ and let $\mathcal{T}^*$ be the topology generated by $\cup\{\mathcal{T}_{A_k} \mid k \in K\}$. Then $(A, \mathcal{T}^*)$ is a connected subspace of $(S, \mathcal{T})$.

Proof: Let $(S, \mathcal{T})$ be a topological space, let $\mathcal{A} = \{(A_k, \mathcal{T}_{A_k}) \mid k \in K\}$ be a collection of connected subspaces of $(S, \mathcal{T})$ such that $\cap\{A_k \mid k \in K\} \neq \emptyset$, and assume toward contradiction that $(A, \mathcal{T}^*)$ is disconnected. Let (U, V) be a disconnection of A. So, U and V are nonempty open sets in A such that $U \cap V = \emptyset$ and $U \cup V = A$. For each $k \in K$, $(A_k, \mathcal{T}_k)$ is connected and $A_k \subseteq A = U \cup V$. Since U and V are open in A, we have $U \cap A_k$ and $V \cap A_k$ are open in A_k. Also, we have the following:

$$(U \cap A_k) \cap (V \cap A_k) = (U \cap V) \cap A_k = \emptyset \cap A_k = \emptyset.$$
$$(U \cap A_k) \cup (V \cap A_k) = (U \cup V) \cap A_k = A \cap A_k = A_k.$$

Fix $k \in K$. If both $U \cap A_k$ and $V \cap A_k$ were nonempty, then $(U \cap A_k, V \cap A_k)$ would be a disconnection of A_k, contradicting that $(A_k, \mathcal{T}_{A_k})$ is connected. So, we must have $U \cap A_k = \emptyset$ or $V \cap A_k = \emptyset$. Without loss of generality, assume that $V \cap A_k = \emptyset$. Then $A_k \subseteq U$ (if $x \in A_k$, then $x \notin V$, and so, $x \in U$).

Since $\cap\{A_k \mid k \in K\} \neq \emptyset$, if there is $k \in \mathbb{N}$ with $A_k \subseteq U$, then for all $k \in \mathbb{N}$, we must have $A_k \subseteq U$. Therefore, $A = \cup\{A_k \mid k \in K\} \subseteq U$, contradicting that (U, V) is a disconnection of A.

It follows that $(A, \mathcal{T}^*)$ is a connected subspace of $(S, \mathcal{T})$. $\qquad\square$

Let $(S, \mathcal{T})$ be a topological space and define a relation $\sim$ on S by $x \sim y$ if and only if $x, y \in A$ for some connected subspace $(A, \mathcal{T}_A)$ of $(S, \mathcal{T})$. In Problem 2 below, you will be asked to prove that $\sim$ is an equivalence relation on S (you will need to use Theorem 14.7). The equivalence classes of $\sim$ are called the **connected components** of S (or more simply, the **components** of S).

Theorem 14.8: Let $(S, \mathcal{T})$ be a topological space.

1. Every $x \in S$ is contained in exactly one component of S.

2. If $(A, \mathcal{T}_A)$ is a connected subspace of $(S, \mathcal{T})$, then A is contained in a component of S.

3. If $(A, \mathcal{T}_A)$ is a connected subspace of $(S, \mathcal{T})$ and A is both open and closed in S, then A is a component of S.

4. Every component of S is closed in S.

You will be asked to prove Theorem 14.8 in Problem 11 below.

A topological space $(S, \mathcal{T})$ is **totally disconnected** if for all $x, y \in S$ with $x \neq y$, there is a disconnection (U, V) of S with $x \in U$ and $y \in V$.

Example 14.9:

1. Let S be any set and let $\mathcal{T}$ be the discrete topology on S, then $(S, \mathcal{T})$ is totally disconnected. Indeed, if $x, y \in S$ with $x \neq y$, then $(\{x\}, S \setminus \{x\})$ is a disconnection of S with $x \in \{x\}$ and $y \in S \setminus \{x\}$.

2. Consider $\mathbb{Q}$ as a subspace of $\mathbb{R}$ with the standard topology. Then $(\mathbb{Q}, \mathcal{T}_{\mathbb{Q}})$ is totally disconnected. To see this, let $x, y \in \mathbb{Q}$ with $x < y$. By Problem 19 in Problem Set 6, there is an irrational number r with $x < r < y$. $((-\infty, r) \cap \mathbb{Q}, (r, \infty) \cap \mathbb{Q})$ is a disconnection of $\mathbb{Q}$, $x \in (-\infty, r) \cap \mathbb{Q}$, and $y \in (r, \infty) \cap \mathbb{Q}$.

3. $(\mathbb{R} \setminus \mathbb{Q}, \mathcal{T}_{\mathbb{R} \setminus \mathbb{Q}})$ (the space of irrational numbers) is also totally disconnected. The argument is similar to that given for $(\mathbb{Q}, \mathcal{T}_{\mathbb{Q}})$ in part 2 above.

4. Consider the Cantor set C as a subspace of $\mathbb{R}$ with the standard topology. Then $(C, \mathcal{T}_C)$ is totally disconnected. I leave the details of the proof to the reader (see Problem 9 below).

5. It is easy to see that a totally disconnected space with more than one point is disconnected. However, a space with just one point is both connected and totally disconnected.

Locally Connected Spaces

A topological space $(S, \mathcal{T})$ is **locally connected** if for every $x \in S$ and every open set U containing x, there is an open set V containing x such that $(V, \mathcal{T}_V)$ is connected and $V \subseteq U$.

Example 14.10:

1. $\mathbb{R}$ is both connected and locally connected. The same is true for any interval of real numbers (see Theorem 14.3).

2. $[0, 1] \cup [9, 10]$ as a subspace of $\mathbb{R}$ is not connected (see part 4 of Example 14.1). However, it is easy to see that it is locally connected.

3. $\mathbb{Q}$ as a subspace of $\mathbb{R}$ is not connected (see part 3 of Example 14.1). It is also **not** locally connected. Given $x \in \mathbb{Q}$, if U is an open set containing x, then there is no interval I with $x \in I$ and $I \subseteq U$. By Theorem 14.4, $\mathbb{Q}$ is not locally connected.

Note: A connected space does **not** need to be locally connected. The reader is encouraged to try to produce an example of such a space (see Problem 14 below).

Problem Set 14

Full solutions to these problems are available for free download here:

www.SATPrepGet800.com/TFBZLF

LEVEL 1

1. Prove that a topological space $(S, \mathcal{T})$ is disconnected if and only if there is a surjective continuous function $f: S \to \{0, 1\}$, where $\{0, 1\}$ is given the discrete topology.

2. Let $(S, \mathcal{T})$ be a topological space and define the relation $\sim$ on S by $x \sim y$ if and only if $x, y \in A$ for some connected subspace $(A, \mathcal{T}_A)$ of $(S, \mathcal{T})$. Prove that $\sim$ is an equivalence relation on S.

LEVEL 2

3. Let $(S_1, \mathcal{T}_1)$ and $(S_2, \mathcal{T}_2)$ be connected topological spaces. Prove that $(S_1 \times S_2, \mathcal{T})$ is connected, where $\mathcal{T}$ is the product topology on $S_1 \times S_2$.

4. Let $(S, \mathcal{T})$ be a topological space. Prove that $(S, \mathcal{T})$ is locally connected if and only if each point of S has a basis $\mathcal{B}$ consisting of connected subspaces of S ($\mathcal{B}$ is a basis at x if (i) $V \in \mathcal{B} \to x \in V$; and (ii) for any open set U containing x, there is a $V \in \mathcal{B}$ such that $V \subseteq U$).

5. Prove that local connectedness is a topological property.

LEVEL 3

6. Prove that the components of a totally disconnected space are its points.

7. Prove that an arbitrary product of totally disconnected spaces is totally disconnected.

8. Let $(S, \mathcal{T})$ be a T_2-space and let $\mathcal{B}$ be a basis for $\mathcal{T}$ such that each set in $\mathcal{B}$ is both open and closed. Prove that $(S, \mathcal{T})$ is totally disconnected.

LEVEL 4

9. Consider the Cantor set C as a subspace of $\mathbb{R}$ with the standard topology. Prove that $(C, \mathcal{T}_C)$ is totally disconnected.

10. Let $(S, \mathcal{T})$ be a topological space, let $(A, \mathcal{T}_A)$ be a connected subspace of $(S, \mathcal{T})$, and let $(B, \mathcal{T}_B)$ be a subspace of $(S, \mathcal{T})$ such that $A \subseteq B \subseteq \overline{A}$. Prove that $(B, \mathcal{T}_B)$ is connected.

11. Let $(S, \mathcal{T})$ be a topological space. Prove each of the following:

    (i)   Every $x \in S$ is contained in exactly one component of S.

    (ii)   If $(A, \mathcal{T}_A)$ is a connected subspace of $(S, \mathcal{T})$, then A is contained in a component of S.

    (iii) If $(A, \mathcal{T}_A)$ is a connected subspace of $(S, \mathcal{T})$ and A is both open and closed in S, then A is a component of S.

    (iv) Every component of S is closed in S.

LEVEL 5

12. Let $(S, \mathcal{T})$ be a compact T_2-space. Prove that $\mathcal{T}$ has a basis $\mathcal{B}$ such that each set in $\mathcal{B}$ is both open and closed if and only if $(S, \mathcal{T})$ is totally disconnected.

13. Prove that an arbitrary product of connected spaces is connected.

CHALLENGE PROBLEM

14. Describe a topological space that is connected, but **not** locally connected.

LESSON 15
FUNCTION SPACES

Vector Spaces

To give some motivation for the definition of a vector space, let's begin with an example.

Example 15.1: Consider the set $\mathbb{C}$ of complex numbers together with the usual definition of addition. Let's also consider another operation, which we will call **scalar multiplication**. For each $k \in \mathbb{R}$ and $z = a + bi \in \mathbb{C}$, we define kz to be $ka + kbi$.

The operation of scalar multiplication is a little different from other types of operations we have looked at previously because instead of multiplying two elements from $\mathbb{C}$ together, we are multiplying an element of $\mathbb{R}$ with an element of $\mathbb{C}$. In this case, we will call the elements of $\mathbb{R}$ **scalars**.

Let's observe that we have the following properties:

1. **$(\mathbb{C}, +)$ is a commutative group.** In other words, for addition in $\mathbb{C}$, we have closure, associativity, commutativity, an identity element (called 0), and the inverse property (the inverse of $a + bi$ is $-a - bi$). This follows immediately from the fact that $(\mathbb{C}, +, \cdot)$ is a field (see Lesson 6 for the definition of a field). When we choose to think of $\mathbb{C}$ as a vector space, we will "forget about" the multiplication in $\mathbb{C}$, and just consider $\mathbb{C}$ together with addition. In doing so, we lose much of the field structure of the complex numbers, but we retain the group structure of $(\mathbb{C}, +)$.

2. **$\mathbb{C}$ is closed under scalar multiplication.** That is, **for all $k \in \mathbb{R}$ and $z \in \mathbb{C}$, we have $kz \in \mathbb{C}$.** To see this, let $z = a + bi \in \mathbb{C}$ and let $k \in \mathbb{R}$. Then, by definition, $kz = ka + kbi$. Since $a, b \in \mathbb{R}$, and $\mathbb{R}$ is closed under multiplication, $ka \in \mathbb{R}$ and $kb \in \mathbb{R}$. It follows that $ka + kbi \in \mathbb{C}$.

3. **$1z = z$.** To see this, consider $1 \in \mathbb{R}$ and let $z = a + bi \in \mathbb{C}$. Then, since 1 is the multiplicative identity for $\mathbb{R}$, we have $1z = 1a + 1bi = a + bi = z$.

4. **For all $j, k \in \mathbb{R}$ and $z \in \mathbb{C}$, $(jk)z = j(kz)$ (Associativity of scalar multiplication).** To see this, let $j, k \in \mathbb{R}$ and $z = a + bi \in \mathbb{C}$. Then since multiplication is associative in $\mathbb{R}$, we have

$$(jk)z = (jk)(a + bi) = (jk)a + (jk)bi = j(ka) + j(kb)i = j(ka + kbi) = j(kz).$$

5. **For all $k \in \mathbb{R}$ and $z, w \in \mathbb{C}$, $k(z + w) = kz + kw$ (Distributivity of 1 scalar over 2 vectors).** To see this, let $k \in \mathbb{R}$ and $z = a + bi, w = c + di \in \mathbb{C}$. Then since multiplication distributes over addition in $\mathbb{R}$, we have

$$k(z + w) = k\big((a + bi) + (c + di)\big) = k\big((a + c) + (b + d)i\big) = k(a + c) + k(b + d)i$$
$$= (ka + kc) + (kb + kd)i = (ka + kbi) + (kc + kdi) = k(a + bi) + k(c + di) = kz + kw.$$

6. **For all $j, k \in \mathbb{R}$ and $z \in \mathbb{C}$, $(j + k)z = jz + kz$ (Distributivity of 2 scalars over 1 vector).** To see this, let $j, k \in \mathbb{R}$ and $z = a + bi \in \mathbb{C}$. Then since multiplication distributes over addition in $\mathbb{R}$, we have

$$(j + k)z = (j + k)(a + bi) = (j + k)a + (j + k)bi = (ja + ka) + (jb + kb)i$$
$$= (ja + jbi) + (ka + kbi) = j(a + bi) + k(a + bi) = jz + kz.$$

Notes: (1) Since the properties listed in 1 through 6 above are satisfied, we say that $\mathbb{C}$ is a **vector space** over $\mathbb{R}$. We will give the formal definition of a vector space below.

(2) Note that a vector space consists of (i) a set of **vectors** (in this case $\mathbb{C}$), (ii) a field (in this case $\mathbb{R}$), and (iii) two operations called **addition** and **scalar multiplication**.

(3) The operation of addition is a binary operation on the set of vectors, and the set of vectors together with this binary operation forms a commutative group. In the previous example (Example 15.1), we have that $(\mathbb{C}, +)$ is a commutative group.

(4) Scalar multiplication is **not** a binary operation on the set of vectors. It takes pairs of the form (k, v), where k is in the field and v is a vector to a vector kv. Formally speaking, scalar multiplication is a function $f \colon \mathbb{F} \times V \to V$, where $\mathbb{F}$ is the field of scalars and V is the set of vectors.

(5) We started with the example of $\mathbb{C}$ as a vector space over $\mathbb{R}$ because it has a geometric interpretation where we can draw simple pictures to visualize what the vector space looks like. Recall from Lesson 5 that we can think of the complex number $a + bi$ as a directed line segment (which from now on we will call a **vector**) in the complex plane that begins at the origin and terminates at the point (a, b).

For example, pictured to the right, we can see the vectors $i = 0 + 1i$, $1 + 2i$, and $2 = 2 + 0i$ in the complex plane.

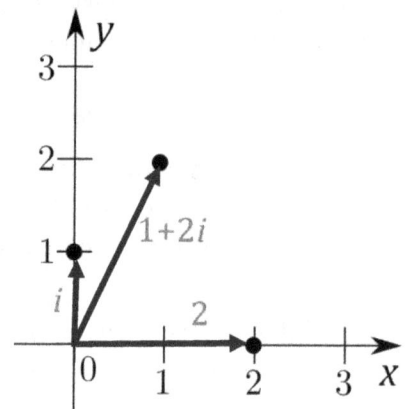

We can visualize the sum of two vectors as the vector starting at the origin that is the diagonal of the parallelogram formed from the original vectors. We see this in the first figure on the left below. In this figure, we have removed the complex plane and focused on the vectors $1 + 2i$ and 2, together with their sum $(1 + 2i) + (2 + 0i) = (1 + 2) + (2 + 0)i = 3 + 2i$.

A second way to visualize the sum of two vectors is to translate one of the vectors so that its initial point coincides with the terminal point of the other vector. The sum of the two vectors is then the vector whose initial point coincides with the initial point of the "unmoved" vector and whose terminal point coincides with the terminal point of the "moved" vector. We see two ways to do this in the center and rightmost figures below.

Technically speaking, the center figure shows the sum $(1 + 2i) + 2$ and the rightmost figure shows the sum $2 + (1 + 2i)$. If we superimpose one figure on top of the other, we can see strong evidence that commutativity holds for addition.

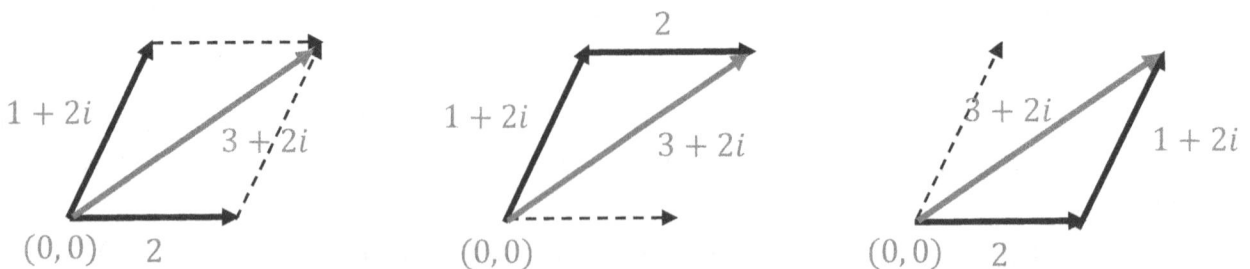

We can visualize a scalar multiple of a vector as follows: (i) if k is a positive real number and $z \in \mathbb{C}$, then the vector kz points in the same direction as z and has a length that is k times the length of z; (ii) if k is a negative real number and $z \in \mathbb{C}$, then the vector kz points in the direction opposite of z and has a length that is $|k|$ times the length of z; (iii) if $k = 0$ and $z \in \mathbb{C}$, then kz is a point.

In the figures below, we have a vector $z \in \mathbb{C}$, together with several scalar multiples of z.

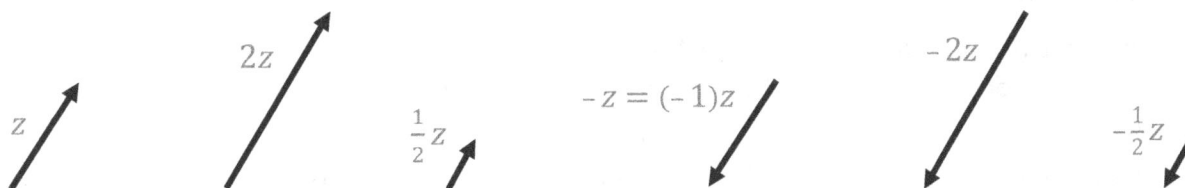

We are now ready for the general definition of a vector space.

A **vector space** (or **linear space**) over a field $\mathbb{F}$ is a set V together with a binary operation $+$ on V (called **addition**) and an operation called **scalar multiplication** satisfying:

(1) $(V, +)$ is a commutative group.

(2) **(Closure under scalar multiplication)** For all $k \in \mathbb{F}$ and $v \in V$, $kv \in V$.

(3) **(Scalar multiplication identity)** If 1 is the multiplicative identity of $\mathbb{F}$ and $v \in V$, then $1v = v$.

(4) **(Associativity of scalar multiplication)** For all $j, k \in \mathbb{F}$ and $v \in V$, $(jk)v = j(kv)$.

(5) **(Distributivity of 1 scalar over 2 vectors)** For all $k \in \mathbb{F}$ and $v, w \in V$, $k(v + w) = kv + kw$.

(6) **(Distributivity of 2 scalars over 1 vector)** For all $j, k \in \mathbb{F}$ and $v \in V$, $(j + k)v = jv + kv$.

Notes: (1) Recall from Lesson 6 that $(V, +)$ a commutative group means the following:

- **(Closure)** For all $v, w \in V$, $v + w \in V$.

- **(Associativity)** For all $v, w, u \in V$, $(v + w) + u = v + (w + u)$.

- **(Commutativity)** For all $v, w \in V$, $v + w = w + v$.

- **(Identity)** There exists an element $0 \in V$ such that for all $v \in V$, $0 + v = v + 0 = v$.

- **(Inverse)** For each $v \in V$, there is $-v \in V$ such that $v + (-v) = (-v) + v = 0$.

(2) The fields that we are familiar with are $\mathbb{Q}$ (the field of rational numbers), $\mathbb{R}$ (the field of real numbers), and $\mathbb{C}$ (the field of complex numbers). For our purposes here, we can always assume that $\mathbb{F}$ is one of these three fields.

Let's look at some basic examples of vector spaces.

Example 15.2:

1. Let $\mathbb{R}^2$ be the set of all ordered pairs of real numbers. That is, $\mathbb{R}^2 = \{(a, b) \mid a, b \in \mathbb{R}\}$ We define **addition** by $(a, b) + (c, d) = (a + c, b + d)$. We define **scalar multiplication** by $k(a, b) = (ka, kb)$ for each $k \in \mathbb{R}$. With these definitions, $\mathbb{R}^2$ is a vector space over $\mathbb{R}$.

245

Notice that $\mathbb{R}^2$ looks just like $\mathbb{C}$. In fact, (a, b) is sometimes used as another notation for $a + bi$. Therefore, the verification that $\mathbb{R}^2$ is a vector space over $\mathbb{R}$ is nearly identical to what we did in Example 15.1 above.

We can visualize elements of $\mathbb{R}^2$ as points or vectors in the Cartesian plane (see part 4 of Example 3.3) in exactly the same way that we visualize complex numbers as points or vectors in the complex plane.

2. $\mathbb{R}^3 = \{(a, b, c) \mid a, b, c \in \mathbb{R}\}$ is a vector space over $\mathbb{R}$, where we define addition and scalar multiplication by $(a, b, c) + (d, e, f) = (a + d, b + e, c + f)$ and $k(a, b, c) = (ka, kb, kc)$, respectively.

   We can visualize elements of $\mathbb{R}^3$ as points in space in a way similar to visualizing elements of $\mathbb{R}^2$ and $\mathbb{C}$ as points in a plane.

3. More generally, we can let $\mathbb{R}^n = \{(a_1, a_2, \ldots, a_n) \mid a_i \in \mathbb{R} \text{ for each } i = 1, 2, \ldots, n\}$. Then $\mathbb{R}^n$ is a vector space over $\mathbb{R}$, where we define addition and scalar multiplication by

$$(a_1, a_2, \ldots, a_n) + (b_1, b_2, \ldots, b_n) = (a_1 + b_1, a_2 + b_2, \ldots, a_n + b_n).$$
$$k(a_1, a_2, \ldots, a_n) = (ka_1, ka_2, \ldots, ka_n).$$

4. More generally still, if $\mathbb{F}$ is any field (for our purposes, we can think of $\mathbb{F}$ as $\mathbb{Q}$, $\mathbb{R}$, or $\mathbb{C}$), we let $\mathbb{F}^n = \{(a_1, a_2, \ldots, a_n) \mid a_i \in \mathbb{F} \text{ for each } i = 1, 2, \ldots, n\}$. Then $\mathbb{F}^n$ is a vector space over $\mathbb{F}$, where we define addition and scalar multiplication by

$$(a_1, a_2, \ldots, a_n) + (b_1, b_2, \ldots, b_n) = (a_1 + b_1, a_2 + b_2, \ldots, a_n + b_n).$$
$$k(a_1, a_2, \ldots, a_n) = (ka_1, ka_2, \ldots, ka_n).$$

   You will be asked to verify that $\mathbb{F}^n$ is a vector space over the field $\mathbb{F}$ in Problem 3 below. Unless stated otherwise, we will always consider the vector space $\mathbb{F}^n$ to be over the field $\mathbb{F}$.

5. $^{\mathbb{R}}\mathbb{R} = \{f \mid f \colon \mathbb{R} \to \mathbb{R}\}$ is a vector space over $\mathbb{R}$, where we define addition and scalar multiplication as follows:

   If $f, g \in {}^{\mathbb{R}}\mathbb{R}$, then $f + g \in {}^{\mathbb{R}}\mathbb{R}$ is defined by $(f + g)(x) = f(x) + g(x)$.

   If $f \in {}^{\mathbb{R}}\mathbb{R}$ and $c \in \mathbb{R}$, then $cf \in {}^{\mathbb{R}}\mathbb{R}$ is defined by $(cf)(x) = c \cdot f(x)$.

   It is straightforward to check that $^{\mathbb{R}}\mathbb{R}$ is a vector space over $\mathbb{R}$. For example, to see that addition is commutative in $^{\mathbb{R}}\mathbb{R}$, simply observe that given $x \in \mathbb{R}$, we have

$$(f + g)(x) = f(x) + g(x) = g(x) + f(x) = (g + f)(x).$$

   Here we used the fact that $\mathbb{R}$ is a field, and therefore, addition is commutative in $\mathbb{R}$. Since x was an arbitrary element of $^{\mathbb{R}}\mathbb{R}$, it follows that $f + g = g + f$.

   Checking each of the other vector space properties is just as easy, and so, I leave the verification to the reader. The zero vector is the function $\mathbf{0}$, where $\mathbf{0}(x) = 0$ for all $x \in \mathbb{R}$. The additive inverse of the function $f \in {}^{\mathbb{R}}\mathbb{R}$ is the function $-f$ defined by $(-f)(x) = -f(x)$ for all $x \in \mathbb{R}$.

   $^{\mathbb{R}}\mathbb{R}$ is an example of a **function space**. A function space is simply a vector space whose elements are functions.

6. More generally, if S is any set and $\mathbb{F}$ is any field (once again, for our purposes, we can think of $\mathbb{F}$ as $\mathbb{Q}$, $\mathbb{R}$, or $\mathbb{C}$), then $\mathcal{F}(S, \mathbb{F}) = {}^S\mathbb{F} = \{f \mid f : S \to \mathbb{F}\}$ is a vector space over $\mathbb{F}$. Addition and scalar multiplication are defined the same way we defined them in part 5 above. Specifically, we have the following:

If $f, g \in \mathcal{F}(S, \mathbb{F})$, then $f + g : S \to \mathbb{F}$ is defined by $(f + g)(x) = f(x) + g(x)$.

If $f \in \mathcal{F}(S, \mathbb{F})$ and $c \in \mathbb{F}$, then $cf : S \to \mathbb{F}$ is defined by $(cf)(x) = c \cdot f(x)$.

I leave it to the reader to verify that $\mathcal{F}(S, \mathbb{F})$ is a vector space over $\mathbb{F}$. As in part 5 above, $\mathcal{F}(S, \mathbb{F})$ is an example of a function space.

Note: Recall that in Lesson 4, we defined ${}^B A$ to be the set of functions from B to A. For example, ${}^\mathbb{R}\mathbb{R}$ is the set of functions from $\mathbb{R}$ to $\mathbb{R}$, and more generally, for an arbitrary set S and a given field $\mathbb{F}$, ${}^S\mathbb{F}$ is the set of functions from S to $\mathbb{F}$. In part 6 of Example 15.2 above, I defined $\mathcal{F}(S, \mathbb{F})$ to be ${}^S\mathbb{F}$. Either of these notations can be used for the vector space of functions from S to $\mathbb{F}$. The reason I introduced this new notation is because it agrees with the notation I use a bit later for some important subsets of this vector space.

Subspaces

Let V be a vector space over a field $\mathbb{F}$. A subset U of V is called a **subspace** of V, written $U \leq V$, if it is also a vector space with respect to the same operations of addition and scalar multiplication as they were defined in V.

Notes: (1) A **universal statement** is a statement that describes a property that is true for all elements without mentioning the existence of any new elements. A universal statement begins with the quantifier $\forall$ ("For all") and never includes the quantifier $\exists$ ("There exists" or "There is").

Properties defined by universal statements are **closed downwards**. This means that if a property defined by a universal statement is true in V and U is a subset of V, then the property is true in U as well.

For example, the statement for commutativity is $\forall v, w(v + w = w + v)$. This is read "For all v and w, $v + w = w + v$." The quantifier $\forall$ is referring to whichever set we are considering. If we are thinking about the set V, then we mean "For all v and w in V, $v + w = w + v$." If we are thinking about the set U, then we mean "For all v and w in U, $v + w = w + v$."

If we assume that $+$ is commutative in V and $U \subseteq V$, we can easily show that $+$ is also commutative in U. To see this, let $v, w \in U$. Since $U \subseteq V$, we have $v, w \in V$. Since $+$ is commutative in V, we have $v + w = w + v$. Since v and w were arbitrary elements in U, we see that $+$ is commutative in U.

(2) Associativity, commutativity, and distributivity are all defined by universal statements, and therefore, when checking if U is a subspace of V, we **do not** need to check any of these properties—they will always be satisfied in the subset U.

(3) The identity property for addition is **not** defined by a universal statement. It begins with the existential quantifier ∃ "There is." Therefore, we **do** need to check that the identity 0 is in a subset U of V when determining if U is a subspace of V. However, once we have checked that 0 is there, we **do not** need to check that it satisfies the property of being an identity. As long as $0 \in U$ (the same 0 from V), then it will behave as an identity because the defining property of 0 contains only the quantifier ∀.

(4) The inverse property for addition will always be true in a subset U of a vector space V that is closed under scalar multiplication. To see this, we use the fact that $-1v = -v$ for all v in a vector space (see Problem 4 (iv) below).

(5) Since the multiplicative identity 1 comes from the field $\mathbb{F}$ and not the vector space V, and we are using the same field for the subset U, we **do not** need to check the scalar multiplication identity when verifying that U is a subspace of V.

(6) The main issue when checking if a subset U of V is a subspace of V is closure. For example, we need to make sure that whenever we add 2 vectors in U, we get a vector that is also in U. If we were to take an arbitrary subset of V, then there is no reason this should happen. For example, let's consider the vector space $\mathbb{C}$ over the field $\mathbb{R}$. Let $A = \{2 + bi \mid b \in \mathbb{R}\}$. A is a subset of $\mathbb{C}$, but A is not a subspace of $\mathbb{C}$. To see this, we just need a single counterexample. $2 + i \in A$, but $(2 + i) + (2 + i) = 4 + 2i \notin A$ (because the real part is 4 and not 2).

(7) Notes 1 through 6 above tell us that to determine if a subset U of a vector space V is a subspace of V, we need only check that $0 \in U$, and U is closed under addition and scalar multiplication.

(8) The statements for closure, as we have written them do look a lot like universal statements. For example, the statement for closure under addition is "For all $v, w \in V$, $v + w \in V$." The issue here is that the set V is not allowed to be explicitly mentioned in the formula. It needs to be understood.

For example, we saw in Note 1 that the statement for commutativity can be written as "$\forall v, w(v + w = w + v)$." The quantifier ∀ (for all) can be applied to any set for which there is a notion of addition defined. We also saw that if the statement is true in V, and U is a subset of V, then the statement will be true in U.

With the statement of closure, to eliminate the set V from the formula, we would need to say something like, "For all x and y, $x + y$ exists." However, there is no way to say "exists" using just logical notation without talking about the set we wish to exist inside of.

We summarize these notes in the following theorem.

Theorem 15.3: Let V be a vector space over a field $\mathbb{F}$ and let $U \subseteq V$. Then $U \leq V$ if and only if (i) $0 \in U$, (ii) for all $v, w \in U$, $v + w \in U$, and (iii) for all $v \in U$ and $k \in \mathbb{F}$, $kv \in U$.

Proof: Let V be a vector space over a field $\mathbb{F}$, and $U \subseteq V$.

If U is a subspace of V, then by definition of U being a vector space, (i), (ii), and (iii) hold.

Now suppose that (i), (ii), and (iii) hold.

By (ii), $+$ is a binary operation on U.

Associativity and commutativity of $+$ are defined by universal statements, and therefore, since they hold in V and $U \subseteq V$, they hold in U.

We are given that $0 \in U$. If $v \in U$, then since $U \subseteq V$, $v \in V$. Since 0 is the additive identity for V, $0 + v = v + 0 = v$. Since $v \in U$ was arbitrary, the additive identity property holds in U.

Let $v \in U$. Since $U \subseteq V$, $v \in V$. Therefore, there is $-v \in V$ such that $v + (-v) = (-v) + v = 0$. By (iii), $-1v \in U$ and by Problem 4 below (part (iv)), $-1v = -v$. Since $v \in U$ was arbitrary, the additive inverse property holds in U.

So, $(U, +)$ is a commutative group.

By (iii), U is closed under scalar multiplication.

Associativity of scalar multiplication and both types of distributivity are defined by universal statements, and therefore, since they hold in V and $U \subseteq V$, they hold in U.

Finally, if $v \in U$, then since $U \subseteq V$, $v \in V$. So, $1v = v$, and the scalar multiplication identity property holds in U.

Therefore, $U \leq V$. $\qquad\qquad\qquad\qquad\qquad\qquad\qquad\qquad\qquad\qquad\qquad\qquad\qquad\qquad$ $\square$

Example 15.4:

1. Let $V = \mathbb{R}^2 = \{(a, b) \mid a, b \in \mathbb{R}\}$ be the vector space over $\mathbb{R}$ with the usual definitions of addition and scalar multiplication and let $U = \{(a, 0) \mid a \in \mathbb{R}\}$. If $(a, 0) \in U$, then $a, 0 \in \mathbb{R}$, and so, $(a, 0) \in V$. Thus, $U \subseteq V$. The 0 vector of V is $(0, 0)$ which is in U. If $(a, 0), (b, 0) \in U$ and $k \in \mathbb{R}$, then $(a, 0) + (b, 0) = (a + b, 0) \in U$ and $k(a, 0) = (ka, 0) \in U$. It follows from Theorem 15.3 that $U \leq V$.

   This subspace U of $\mathbb{R}^2$ looks and behaves just like $\mathbb{R}$, the set of real numbers. More specifically, we say that U is **isomorphic** to $\mathbb{R}$. Most mathematicians identify this subspace U of $\mathbb{R}^2$ with $\mathbb{R}$, and just call it $\mathbb{R}$.

   In general, it is common practice for mathematicians to call various isomorphic copies of certain structures by the same name. As a generalization of this example, if $m < n$, then we can say $\mathbb{R}^m \leq \mathbb{R}^n$ by identifying $(a_1, a_2, \ldots, a_m) \in \mathbb{R}^m$ with the vector $(a_1, a_2, \ldots, a_m, 0, 0, \ldots, 0) \in \mathbb{R}^n$ that has a tail end of $n - m$ zeros. For example, we may say that $(2, \sqrt{2}, 7, -\frac{1}{2}, 0, 0, 0)$ is in $\mathbb{R}^4$, even though it is technically in $\mathbb{R}^7$. With this type of identification, we have $\mathbb{R}^4 \leq \mathbb{R}^7$.

2. Let $V = \mathbb{Q}^3 = \{(a, b, c) \mid a, b, c \in \mathbb{Q}\}$ be the vector space over $\mathbb{Q}$ with the usual definitions of addition and scalar multiplication and let $U = \{(a, b, c) \in \mathbb{Q}^3 \mid c = a + 2b\}$. Let's check that $U \leq V$.

   It's clear that $U \subseteq V$. Since $0 = 0 + 2 \cdot 0$, we see that the zero vector $(0, 0, 0)$ is in U. Let $(a, b, c), (d, e, f) \in U$ and $k \in \mathbb{Q}$. Then we have

$$(a, b, c) + (d, e, f) = (a, b, a + 2b) + (d, e, d + 2e) = \big(a + d, b + e, (a + d) + 2(b + e)\big).$$

$$k(a, b, c) = k(a, b, a + 2b) = (ka, kb, ka + 2kb).$$

These vectors are both in U, and so, by Theorem 15.3, $U \leq V$.

3. Consider $V = \mathbb{C}$ as a vector space over $\mathbb{R}$ in the usual way and let $U = \{z \in \mathbb{C} \mid \operatorname{Re} z = 1\}$. Then $U \subseteq V$, but $U \not\leq V$ because the zero vector is not in U. After all, $0 = 0 + 0i$, and so, $\operatorname{Re} 0 = 0$.

4. Every vector space is a subspace of itself, and the vector space consisting of just the 0 vector from the vector space V is a subspace of V.

   In other words, for any vector space V, $V \leq V$ and $\{0\} \leq V$.

   The empty set, however, can never be a subspace of a vector space because it doesn't contain a zero vector.

5. Let S be a set and let $V = \mathcal{F}(S, \mathbb{R})$ be the set of functions from S to $\mathbb{R}$. We say that a function $f \in \mathcal{F}(S, \mathbb{R})$ is **bounded** if ran f is bounded in $\mathbb{R}$. Equivalently, f is bounded if there is $M \in \mathbb{R}$ such that for all $x \in S$, $|f(x)| \leq M$. Let $\mathcal{B}(S, \mathbb{R}) = \{f \in \mathcal{F}(S, \mathbb{R}) \mid f \text{ is bounded}\}$. Let's check that $\mathcal{B}(S, \mathbb{R}) \leq \mathcal{F}(S, \mathbb{R})$.

   First note that $|\mathbf{0}(x)| = |0| = 0 \leq 0$ for all $x \in S$, and so, $\mathbf{0} \in \mathcal{B}(S, \mathbb{R})$.

   Now, if $f, g \in \mathcal{B}(S, \mathbb{R})$, then there are $M, N \in \mathbb{R}$ such that for all $x \in S$, $|f(x)| \leq M$ and $|g(x)| \leq N$. It follows that $|(f + g)(x)| = |f(x) + g(x)| \leq |f(x)| + |g(x)| \leq M + N$ for all $x \in S$. Therefore, $f + g$ is bounded by the real number $M + N$.

   Also, if $c \in \mathbb{R}$, then $|(cf)(x)| = |c \cdot f(x)| = |c| \cdot |f(x)| \leq |c| \cdot M$, and so, cf is bounded by the real number $|c| \cdot M$.

   By Theorem 15.3, $\mathcal{B}(S, \mathbb{R})$ is a subspace of $\mathcal{F}(S, \mathbb{R})$.

   Note that in this example, we can replace the field $\mathbb{R}$ by any other field. For example, we have $\mathcal{B}(S, \mathbb{Q}) \leq \mathcal{F}(S, \mathbb{Q})$ and $\mathcal{B}(S, \mathbb{C}) \leq \mathcal{F}(S, \mathbb{C})$.

Banach Spaces

Let V be a vector space over a field $\mathbb{F}$. A **norm** on V is a function that assigns to each $v \in V$ a real number $|v|$ satisfying

1. For all $v \in V$, $|v| = 0$ if and only if $v = 0$.

2. For all $v, w \in V$, $|v + w| \leq |v| + |w|$.

3. For all $v \in V$ and $c \in \mathbb{F}$, $|cv| = |c| \, |v|$.

Note: If $v \in V$, then $|v|$ is the **norm** of v, whereas if $c \in \mathbb{F}$, then $|c|$ is the **absolute value** of c. Observe that in clause 3 above, both of these definitions are being used: $|cv|$ and $|v|$ are norms of vectors and $|c|$ is the absolute value of $c \in \mathbb{F}$.

Some authors use the notation $\|v\|$ instead of $|v|$ for the norm of a vector v to avoid confusion with the absolute value of a number. I prefer the less cumbersome notation $|v|$.

V together with a norm is called a **normed vector space** (or **normed linear space**).

A norm on a vector space essentially provides us with a notion of the distance between an arbitrary vector and the zero vector. It is easy to show that for all $v \in V$, $|v| \geq 0$ (this is similar to Problem 1 from Problem Set 11). It is also easy to prove that for all $v \in V$, $|-v| = |v|$ (use part (iv) of Problem 4 below together with clause 3 in the definition of the norm).

If a norm is defined on V, we can define a metric $d: V \times V \to \mathbb{R}$ by $d(v, w) = |v - w|$ (where $v - w$ is defined to be $v + (-w)$). I leave the verification that d satisfies the three properties of a metric to the reader. We say that the norm **induces** the metric d, or that the metric d **is induced by** the norm.

Let V be a normed vector space and let d be the induced metric. We say that V is a **Banach space** if (V, d) is a complete metric space.

Example 15.5:

1. Let $V = \mathbb{R}^n = \{(a_1, a_2, \dots, a_n) \mid a_i \in \mathbb{R} \text{ for each } i = 1, 2, \dots, n\}$ together with the norm

    $$|\mathbf{x}| = \sqrt{x_1^2 + x_2^2 + \cdots + x_n^2} = \left(\sum_{k=1}^{n} x_k^2 \right)^{\frac{1}{2}}.$$

    V is a vector space by Problem 3 below.

    Properties 1 and 3 of a norm are easy to verify and property 2 is precisely the Generalized Triangle Inequality (Theorem 7.36). Therefore, V is a normed vector space.

    By Theorem 11.11, (V, d) is a complete metric space, where d is the metric induced by the norm.

    It follows that $V = \mathbb{R}^n$ is Banach Space. With the norm defined as in this example, this Banach space is called **Euclidean space**.

    Similarly, the **unitary space** $\mathbb{C}^n$ is also a Banach space.

2. Let $V = \mathcal{B}(S, \mathbb{R}) = \{f: S \to \mathbb{R} \mid f \text{ is bounded}\}$ together with the norm
    $$|f| = \sup|f(x)|.$$
    We saw in part 5 of Example 15.4 that $\mathcal{B}(S, \mathbb{R})$ is a vector space (it is a subspace of the vector space $\mathcal{F}(S, \mathbb{R})$). In Problem 9 below, you will be asked to prove that with this norm, $\mathcal{B}(S, \mathbb{R})$ is a Banach space.

    Note that the corresponding metric is defined by $|f - g| = \sup|f(x) - g(x)|$.

    Similarly, $\mathcal{B}(S, \mathbb{C})$ is also a Banach space.

Spaces of Continuous Functions

Let (S, d) be a metric space. We will now restrict our attention to the bounded functions that are also continuous. To this end, we let $\mathcal{C}(S, \mathbb{R}) = \{f \in \mathcal{B}(S, \mathbb{R}) \mid f \text{ is continuous}\}$. Let's begin by showing that $\mathcal{C}(S, \mathbb{R})$ is a subspace of $\mathcal{B}(S, \mathbb{R})$ (and therefore, $\mathcal{C}(S, \mathbb{R})$ is a vector space).

Theorem 15.6: Let (S, d) be a metric space. Then $\mathcal{C}(S, \mathbb{R}) \leq \mathcal{B}(S, \mathbb{R})$.

Proof: Let's first check that $\mathbf{0}$ is continuous. Recall that $\mathbf{0}$ is the function from S to $\mathbb{R}$ defined by $\mathbf{0}(x) = 0$ for all $x \in S$, S is a metric space with metric d, and $\mathbb{R}$ is given the standard metric. Let $x \in S$, let $\epsilon > 0$, and let $\delta = \epsilon$. Then we have

$$d(x, y) < \delta \text{ implies } |\mathbf{0}(x) - \mathbf{0}(y)| = |0 - 0| = |0| = 0 < \delta = \epsilon.$$

By Theorem 13.4, $\mathbf{0}$ is continuous at x. Since $x \in S$ was arbitrary, $\mathbf{0}$ is continuous.

Next, let $f, g \in \mathcal{C}(S, \mathbb{R})$ and let $x \in S$. We need to show that $f + g$ is continuous at x.

Let $\epsilon > 0$. Since f and g are continuous, there are positive real numbers δ_1 and δ_2 such that $d(x, y) < \delta_1$ implies $|f(x) - f(y)| < \frac{\epsilon}{2}$ and such that $d(x, y) < \delta_2$ implies $|g(x) - g(y)| < \frac{\epsilon}{2}$. Let $\delta = \min\{\delta_1, \delta_2\}$ and assume that $d(x, y) < \delta$. Then $d(x, y) < \delta_1$, and so, $|f(x) - f(y)| < \frac{\epsilon}{2}$. Also, $d(x, y) < \delta_2$, and so, $|g(x) - g(y)| < \frac{\epsilon}{2}$. Therefore, we have

$$|(f + g)(x) - (f + g)(y)| = \left|(f(x) + g(x)) - (f(y) + g(y))\right|$$

$$= \left|(f(x) - f(y)) + (g(x) - g(y))\right| \leq |f(x) - f(y)| + |g(x) - g(y)| < \frac{\epsilon}{2} + \frac{\epsilon}{2} = \epsilon.$$

By Theorem 13.4, $f + g$ is continuous at x. Since $x \in S$ was arbitrary, $f + g$ is continuous.

Finally, let $f \in \mathcal{C}(S, \mathbb{R})$, let $c \in \mathbb{R}$, and let $x \in S$. We need to show that cf is continuous at x.

Let $\epsilon > 0$. If $c = 0$, $cf = \mathbf{0}$, which we already proved is continuous above. So, we may assume that $c \neq 0$. Since f is continuous, there is a positive real number δ such that $d(x, y) < \delta$ implies $|f(x) - f(y)| < \frac{\epsilon}{|c|}$. Assume that $d(x, y) < \delta$. Then we have

$$|(cf)(x) - (cf)(y)| = |c \cdot f(x) - c \cdot f(y)| = \left|c \cdot (f(x) - f(y))\right|$$

$$= |c| \cdot |f(x) - f(y)| < |c| \cdot \frac{\epsilon}{|c|} = \epsilon.$$

By Theorem 13.4, cf is continuous at x. Since $x \in S$ was arbitrary, $f + g$ is continuous.

By Theorem 15.3, $\mathcal{C}(S, \mathbb{R}) \leq \mathcal{B}(S, \mathbb{R})$. $\qquad\square$

Note: We also have $\mathcal{C}(S, \mathbb{C}) \leq \mathcal{B}(S, \mathbb{C})$. The proof is nearly identical to the proof of Theorem 15.6.

In Problem 10 below, you will be asked to prove that $\mathcal{C}(S, \mathbb{R})$ is a closed subspace of $\mathcal{B}(S, \mathbb{R})$. Therefore, by the following theorem, $\mathcal{C}(S, \mathbb{R})$ is itself a Banach space with norm defined by $|f| = \sup|f(x)|$ (and similarly for $\mathcal{C}(S, \mathbb{C})$).

Theorem 15.7: A closed subspace of a Banach space is a Banach space.

Proof: Let V be a Banach space with induced metric d and let W be a subspace of V with W closed in V. We need to show that (W, d) is complete.

To see this, let (v_n) be a Cauchy sequence in W. Since $W \subseteq V$, (v_n) is a Cauchy sequence in V. Since (V, d) is complete, there is $v \in V$ with $v_n \to v$. By Theorem 11.4, for every open ball $B_r(v)$ with center v, there is $K \in \mathbb{N}$ such that $n > K$ implies $v_n \in B_r(v)$. Therefore, v is an accumulation point of W. By part 3 of Theorem 9.4, $v \in \overline{W}$. Since W is closed, by part 4 of Theorem 9.4, $W = \overline{W}$. It follows that $v \in W$. Since (v_n) was an arbitrary Cauchy sequence in W, we have shown that every Cauchy sequence in W converges to a point in W.

Therefore, (W, d) is complete. $\qquad\qquad\square$

Let (f_n) be a sequence in $\mathcal{C}(S, \mathbb{R})$. We will say that (f_n) **converges pointwise** to f if for each $x \in S$, $f_n(x) \to f(x)$. In other words, given $x \in S$ and $\epsilon > 0$, there is $K \in \mathbb{N}$ such that $n > K$ implies $|f_n(x) - f(x)| < \epsilon$.

This definition seems natural enough. However, it is unfortunate that there is no guarantee that the limiting function f will be in $\mathcal{C}(S, \mathbb{R})$, as we will see in the following example.

Example 15.8: Let $S = [0, 1]$ and for each $n \in \mathbb{Z}^+$, define $f_n \colon S \to \mathbb{R}$ by $f_n(x) = \begin{cases} nx & \text{if } 0 \le x < \frac{1}{n}. \\ 1 & \text{if } \frac{1}{n} \le x \le 1. \end{cases}$

If $x \in S$ with $x \ne 0$, then $f_n(x) \to 1$, whereas $f_n(0) = 0$ for all $n \in \mathbb{N}$, and so, $f_n(0) \to 0$. It follows that $f_n \to f$, where f is defined by $f(x) = \begin{cases} 0 & \text{if } x = 0. \\ 1 & \text{if } x \ne 0. \end{cases}$ So, in this case, the sequence of functions (f_n) in $\mathcal{C}([0, 1], \mathbb{R})$ converges pointwise to a function f that is **not** in $\mathcal{C}([0, 1], \mathbb{R})$.

The undesirable behavior demonstrated by the pointwise convergent sequence of functions given in Example 15.8 suggests that we may wish to define a stronger notion of convergence of functions. We do this now.

Let (f_n) be a sequence in $\mathcal{C}(S, \mathbb{R})$. We will say that (f_n) **converges uniformly** to f if given $\epsilon > 0$, we can find $K \in \mathbb{N}$ such that $n > K$ implies that for all $x \in S$, $|f_n(x) - f(x)| < \epsilon$.

Theorem 15.9: Let (S, d) be a metric space and let (f_n) be a sequence in $\mathcal{C}(S, \mathbb{R})$ such that (f_n) converges uniformly to f. Then $f \in \mathcal{C}(S, \mathbb{R})$.

Proof: Let (f_n) be a sequence in $\mathcal{C}(S, \mathbb{R})$ such that (f_n) converges uniformly to f and let $x \in S$. We will show that f is continuous at x.

To this end, let $\epsilon > 0$. Since (f_n) converges uniformly to f, there is $K \in \mathbb{N}$ such that $n > K$ implies that for all $y \in S$, $|f_n(y) - f(y)| < \frac{\epsilon}{3}$. In particular, for all $y \in S$, $|f_{K+1}(y) - f(y)| < \frac{\epsilon}{3}$. Since f_{K+1} is continuous, it is continuous at x. So, there is $\delta > 0$ such that $d(x, y) < \delta$ implies $|f_{K+1}(x) - f_{K+1}(y)| < \frac{\epsilon}{3}$. Now, assume that $d(x, y) < \delta$. Then, we have

$$|f(x) - f(y)| = \left|\left(f(x) - f_{K+1}(x)\right) + \left(f_{K+1}(x) - f_{K+1}(y)\right) + \left(f_{K+1}(y) - f(y)\right)\right|$$

$$\le |f(x) - f_{K+1}(x)| + |f_{K+1}(x) - f_{K+1}(y)| + |f_{K+1}(y) - f(y)| < \frac{\epsilon}{3} + \frac{\epsilon}{3} + \frac{\epsilon}{3} = \epsilon.$$

Therefore, f is continuous at x. Since $x \in S$ was arbitrary, $f \in \mathcal{C}(S, \mathbb{R})$. $\qquad\square$

Recall from Lesson 12 that the Generalized Heine-Borel Theorem (Theorem 12.14) says that a subspace of $\mathbb{R}^n$ is compact if and only if it is closed and bounded. It is natural to ask if this "Heine-Borel property" holds for $\mathcal{C}([0,1], \mathbb{R})$. In other words, if $\mathcal{A}$ is a subset of $\mathcal{C}([0,1], \mathbb{R})$, is it true that $\mathcal{A}$ is compact if and only if $\mathcal{A}$ is closed and bounded in $\mathcal{C}([0,1], \mathbb{R})$. Unfortunately, the next example shows that this is not the case.

Example 15.10: The closed unit ball $B = \{f \in \mathcal{C}([0,1], \mathbb{R}) \mid |f| \leq 1\}$ in $\mathcal{C}([0,1], \mathbb{R})$ is closed and bounded. However, B is not compact.

Indeed, consider the sequence $f_n(x) = \begin{cases} nx & \text{if } 0 \leq x < \frac{1}{n} \\ 1 & \text{if } \frac{1}{n} \leq x \leq 1 \end{cases}$ from Example 15.8. As we saw there, for each $n \in \mathbb{Z}^+$, $|f_n| = \sup|f_n(x)| = 1$ (because $f_n(1) = 1$), and so, for each $n \in \mathbb{Z}^+$, $f_n \in B$.

On the other hand, the pointwise limit of (f_n) is $f(x) = \begin{cases} 0 & \text{if } x = 0. \\ 1 & \text{if } x \neq 0. \end{cases}$ So, in this case, the sequence of functions (f_n) in $\mathcal{C}([0,1], \mathbb{R})$ converges to a function f that is not in B (it's not even in $\mathcal{C}([0,1], \mathbb{R})$) because f fails to be continuous. Furthermore, since $f_n \to f$, every subsequence of (f_n) also converges to f.

So, we have found a sequence in B without a convergent subsequence in B. It follows that B is not sequentially compact, and therefore, by Theorem 12.20, B is not compact.

Example 15.10 shows that a closed and bounded subset of $\mathcal{C}([0,1], \mathbb{R})$ is not necessarily compact (unlike the analogous situation for $\mathbb{R}^n$). Another condition is required, which we now describe.

Let (K, d) be a compact metric space and let $\mathcal{A} \subseteq \mathcal{C}(K, \mathbb{R})$. If $f \in \mathcal{A}$, then by Problem 11 from Problem Set 13, f is uniformly continuous. In other words, given $\epsilon > 0$, we can find $\delta > 0$ such that for all $x, y \in K$, $d(x, y) < \delta$ implies $|f(x) - f(y)| < \epsilon$. Although δ does not depend on the elements $x, y \in K$, it may depend on the function f.

For example, let $\mathcal{A} = \{f_n \mid n \in \mathbb{Z}^+\}$, where for each $n \in \mathbb{Z}^+$, $f_n(x) = \begin{cases} nx & \text{if } 0 \leq x < \frac{1}{n}. \\ 1 & \text{if } \frac{1}{n} \leq x \leq 1. \end{cases}$ Each of these functions is uniformly continuous on $[0, 1]$. However, given $\epsilon > 0$, there is no single δ that will work for every $f_n \in \mathcal{A}$. You will be asked to prove this in Problem 6 below.

We will say that $\mathcal{A}$ is **equicontinuous** if given $\epsilon > 0$, we can find $\delta > 0$ such that for all $x, y \in K$ and every $f \in \mathcal{A}$, $d(x, y) < \delta$ implies $|f(x) - f(y)| < \epsilon$.

We are now ready to give an equivalent formulation of compact subsets of $\mathcal{C}([0,1], \mathbb{R})$, just as we did with the Generalized Heine-Borel Theorem for $\mathbb{R}^n$.

Theorem 15.11 (Arzela-Ascoli Theorem): Let (K, d) be a compact metric space and let $\mathcal{A} \subseteq \mathcal{C}(K, \mathbb{R})$ with $\mathcal{A}$ closed in K. $\mathcal{A}$ is compact if and only if $\mathcal{A}$ is bounded and equicontinuous.

I leave the proof of the Arzela-Ascoli Theorem as an exercise for the reader (see Problem 13 below).

Problem Set 15

Full solutions to these problems are available for free download here:

www.SATPrepGet800.com/TFBZLF

LEVEL 1

1. Determine if each of the following subsets of $\mathbb{R}^2$ is a subspace of $\mathbb{R}^2$:

    (i) $A = \{(x, y) \mid x + y = 0\}$

    (ii) $B = \{(x, y) \mid xy = 0\}$

    (iii) $C = \{(x, y) \mid 2x = 3y\}$

    (iv) $D = \{(x, y) \mid x \in \mathbb{Q}\}$

2. Let X be a nonempty subset of a Banach space V. Prove that X is bounded if and only if there is a real number M such that for all $x \in X$, $|x| \leq M$, where $|x|$ is the norm of x in V.

LEVEL 2

3. Let $\mathbb{F}$ be a field. Prove that $\mathbb{F}^n$ is a vector space over $\mathbb{F}$.

4. Let V be a vector space over $\mathbb{F}$. Prove each of the following:

    (i)   For every $v \in V$, $-(-v) = v$.

    (ii) For every $v \in V$, $0v = 0$.

    (iii) For every $k \in \mathbb{F}$, $k \cdot 0 = 0$.

    (iv) For every $v \in V$, $-1v = -v$.

LEVEL 3

5. Let V be a vector space over a field $\mathbb{F}$ and let X be a set of subspaces of V. Prove that $\cap X$ is a subspace of V.

6. Let $\mathcal{A} = \{f_n : [0,1] \rightarrow \mathbb{R} \mid n \in \mathbb{Z}^+\}$, where for each $n \in \mathbb{Z}^+$, $f_n(x) = \begin{cases} nx & \text{if } 0 \leq x < \frac{1}{n}. \\ 1 & \text{if } \frac{1}{n} \leq x \leq 1. \end{cases}$

    Use the definition of equicontinuity to prove that $\mathcal{A}$ is not equicontinuous.

LEVEL 4

7. Let U and W be subspaces of a vector space V. Determine necessary and sufficient conditions for $U \cup W$ to be a subspace of V.

8. Give an example of vector spaces U and V with $U \subseteq V$ such that U is closed under scalar multiplication, but U is not a subspace of V.

LEVEL 5

9. Prove that $\mathcal{B}(S, \mathbb{R})$ together with the norm defined by $|f| = \sup|f(x)|$ is a Banach space.

10. Prove that $\mathcal{C}(S, \mathbb{R})$ is a closed subspace of $\mathcal{B}(S, \mathbb{R})$.

11. Find a sequence (f_n) in $\mathcal{C}([0, 1], \mathbb{R})$ such that f_n converges pointwise to $f \in \mathcal{C}([0, 1], \mathbb{R})$, but (f_n) does **not** converge uniformly to f.

CHALLENGE PROBLEMS

12. Determine with proof whether $\mathcal{C}([0, 1], \mathbb{R})$ is locally compact?

13. Prove the Arzela-Ascoli Theorem (Theorem 15.11).

LESSON 16
ALGEBRAIC TOPOLOGY

Homotopy

Let $(X, \mathcal{T})$ and $(Y, \mathcal{U})$ be topological spaces and let $f, g \colon X \to Y$ be continuous functions.

A **homotopy** from f to g is a continuous function $F \colon X \times [0,1] \to Y$ such that for all $x \in X$, $F(x, 0) = f(x)$ and $F(x, 1) = g(x)$. We will sometimes say that F **deforms** f into g.

f is **homotopic** to g, written $f \simeq g$ if there exists a homotopy from f to g.

Example 16.1:

1. Let's consider $(\mathbb{R}, \mathcal{T})$, where $\mathcal{T}$ is the standard topology on $\mathbb{R}$. Let $f, g \colon \mathbb{R} \to \mathbb{R}$ be the constant functions defined by $f(x) = 0$ and $g(x) = 1$ for all $x \in \mathbb{R}$. We define a function $F \colon \mathbb{R} \times [0,1] \to \mathbb{R}$ by $F(x, t) = t$. Let's check that F is a homotopy from f to g.

   To see that F is continuous, observe that for $a, b \in \mathbb{R}$ with $a < b$, we have $F^{-1}[(a, b)] = \mathbb{R} \times \big((a, b) \cap [0, 1]\big)$. Since $\mathbb{R}$ is open in $\mathbb{R}$ and $(a, b) \cap [0, 1]$ is open in the subspace topology relative to $[0, 1]$, we have that $\mathbb{R} \times \big((a, b) \cap [0, 1]\big)$ is open in the product topology. Since (a, b) was an arbitrary basic open set in the standard topology of $\mathbb{R}$, F is continuous.

   Finally, observe that for all $x \in \mathbb{R}$, $F(x, 0) = 0 = f(x)$ and $F(x, 1) = 1 = g(x)$.

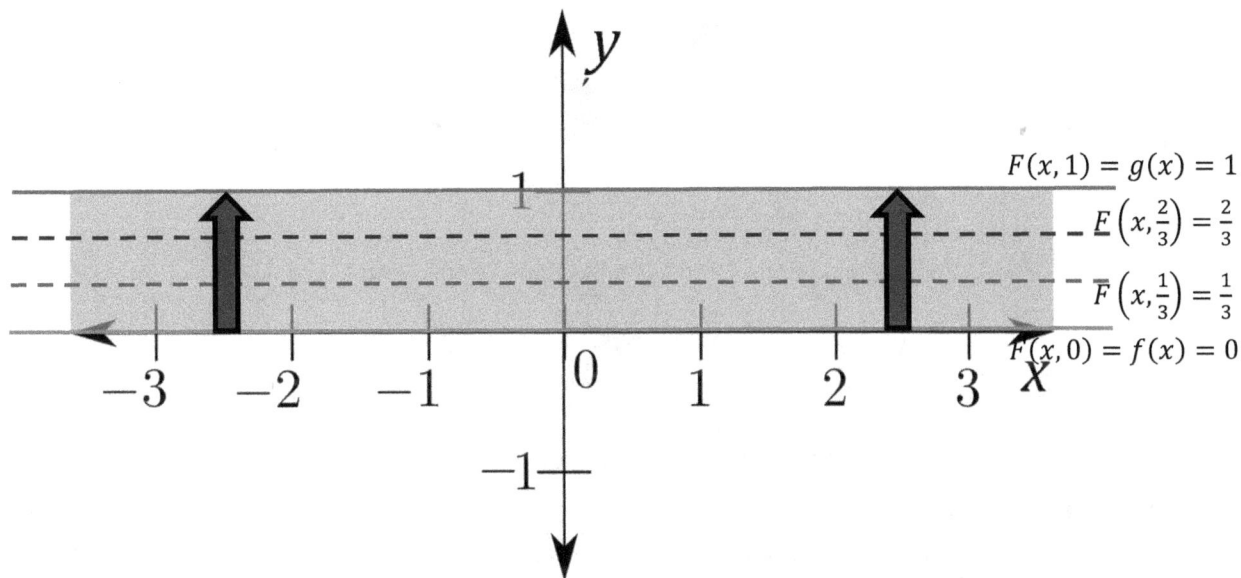

The figure above shows the homotopy F "in action." The horizontal line $f(x) = 0$ is where the homotopy begins. This is $F(x, 0)$. As t increases from 0 to 1, we get the horizontal lines $F(x, t) = t$. In the figure, as examples, we drew $F\left(x, \frac{1}{3}\right) = \frac{1}{3}$, $F\left(x, \frac{2}{3}\right) = \frac{2}{3}$, and then finally $F(x, 1) = g(x) = 1$.

Notice how the line $f(x) = 0$ is continuously deformed into the line $g(x) = 1$ by passing through every line of the form $h_t(x) = t$.

2. Let $f, g: \mathbb{R} \to \mathbb{R}$ be any two continuous functions. We define a function $F: \mathbb{R} \times [0, 1] \to \mathbb{R}$ by $F(x, t) = (1 - t)f(x) + tg(x)$. Let's check that F is a homotopy from f to g.

We begin by showing that F is continuous. We first show that $G, H: \mathbb{R} \times [0, 1] \to \mathbb{R}$ defined by $G(x, t) = tg(x)$ and $H(x, t) = (1 - t)f(x)$ are continuous. Let's start with G. Define $J: \mathbb{R} \times [0, 1] \to \mathbb{R} \times [0, 1]$ by $J(x, t) = (g(x), t)$ and $K: \mathbb{R} \times [0, 1] \to \mathbb{R}$ by $K(x, t) = tx$. If $U \times V$ is open in $\mathbb{R} \times [0, 1]$, then $J^{-1}[U \times V] = g^{-1}[U] \times V$, which is open in the product topology of $\mathbb{R} \times [0, 1]$. So, J is continuous. By Problem 24 in Problem Set 8, K is continuous. Since $G = K \circ J$, G is also continuous. The proof that H is continuous is similar.

We now define $M: \mathbb{R} \times [0, 1] \to \mathbb{R} \times \mathbb{R}$ by $M(x, t) = \big(H(x, t), G(x, t)\big)$ and $N: \mathbb{R} \times \mathbb{R} \to \mathbb{R}$ by $N(x, y) = x + y$. If $U \times V$ is open in $\mathbb{R} \times \mathbb{R}$, then $M^{-1}[U \times V] = H^{-1}[U] \times G^{-1}[V]$, which is open in the product topology of $\mathbb{R} \times \mathbb{R}$. So, M is continuous. Using an argument similar to the proof of Theorem 8.13, it can easily be shown that N is continuous. Since $F = N \circ M$, F is also continuous.

Also, $F(x, 0) = (1 - 0)f(x) + 0g(x) = f(x)$ and $F(x, 1) = (1 - 1)f(x) + 1g(x) = g(x)$ for all $x \in \mathbb{R}$.

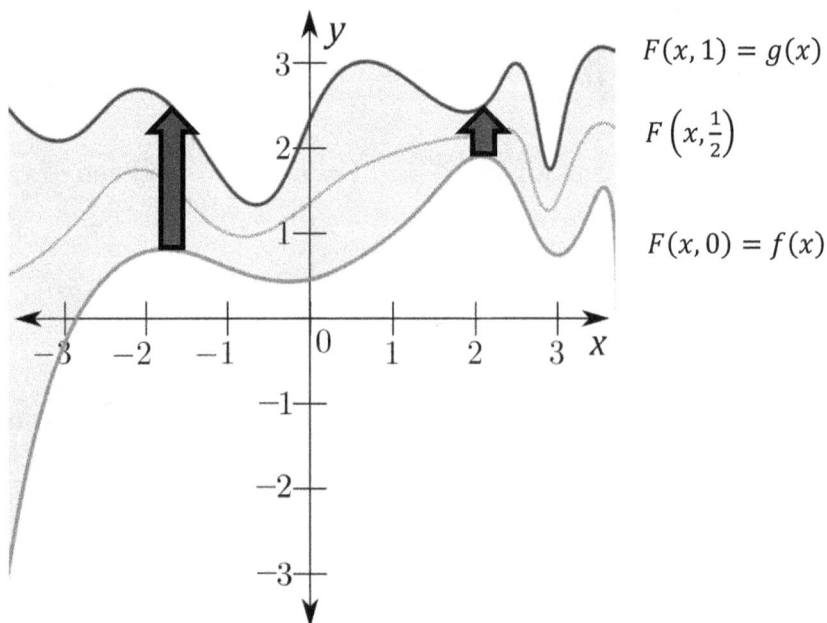

The figure above shows how the homotopy F deforms the function f into the function g.

Note that it is important that for all $t \in [0, 1]$, the expression $(1 - t)f(x) + tg(x)$ is an element of $\mathbb{R}$. To see how things could go wrong, let's let $\mathbb{S}^1$ be the unit circle in the Complex Plane. So, $\mathbb{S}^1 = \{x + yi \mid x^2 + y^2 = 1\}$. Now, let $f, g: (0, 1) \to \mathbb{S}^1$ be defined by $f(x) = e^{\pi xi} = \cos \pi x + i \sin \pi x$ and $g(x) = e^{-\pi xi} = \cos \pi x - i \sin \pi x$. We might be tempted to define $F: (0, 1) \times [0, 1] \to \mathbb{S}^1$ by $F(x, t) = (1 - t)f(x) + tg(x)$. However, we then have

$$F\left(\frac{1}{2}, \frac{1}{2}\right) = \left(1 - \frac{1}{2}\right)f\left(\frac{1}{2}\right) + \frac{1}{2}g\left(\frac{1}{2}\right) = \frac{1}{2}\left(f\left(\frac{1}{2}\right) + g\left(\frac{1}{2}\right)\right) = \frac{1}{2}2\cos\frac{\pi}{2} = 0 + 0i.$$

But $0^2 + 0^2 = 0 \neq 1$, and so, $0 + 0i \notin \mathbb{S}^1$. Therefore, this function F is not a homotopy from f to g.

Note that if we were considering f and g as functions from $(0,1)$ to $\mathbb{C}$, then F **would** be a homotopy from f to g.

The figure below shows the unit circle together with the functions $F(x,0) = f(x)$, $F(x,1) = g(x)$, and $F\left(x, \frac{1}{2}\right)$. Observe that $F\left(x, \frac{1}{2}\right) \notin \mathbb{S}^1$ for all $x \in (0,1)$. In fact, $F(x,t) \notin \mathbb{S}^1$ for all $(x,t) \in (0,1) \times (0,1)$.

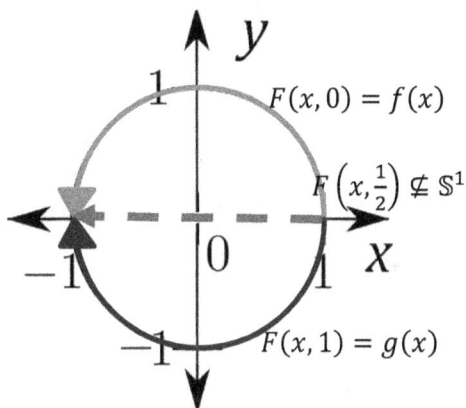

Theorem 16.2 (The Pasting Theorem): Let $(X, \mathcal{T})$ and $(Y, \mathcal{U})$ be topological spaces, let A, B be closed subsets of X such that $X = A \cup B$, let $f: A \to Y$ and $g: B \to Y$ be continuous functions, and suppose that for all $x \in A \cap B$, $f(x) = g(x)$. Let $h: X \to Y$ be defined by $h(x) = \begin{cases} f(x) & \text{if } x \in A. \\ g(x) & \text{if } x \in B. \end{cases}$ Then h is continuous.

You will be asked to prove the Pasting Theorem in Problem 5 below.

Theorem 16.3: $\simeq$ is an equivalence relation.

Proof: If $f: X \to Y$ is a continuous function, define $F: X \times [0,1] \to Y$ by $F(x,t) = f(x)$ for all $x \in X$ and $t \in [0,1]$. To see that F is continuous, let U be open in Y. Then $F^{-1}[U] = f^{-1}[U] \times [0,1]$. Since f is continuous, $f^{-1}[U]$ is open in X. Therefore, $F^{-1}[U]$ is open in the product topology of $X \times [0,1]$. Observe that for all $x \in X$, $F(x,0) = f(x)$ and $F(x,1) = f(x)$. So, $f \simeq f$, and therefore, $\simeq$ is reflexive.

Next, let $f, g: X \to Y$ be continuous functions with $f \simeq g$, and let $F: X \times [0,1] \to Y$ be a homotopy from f to g. Define $G: X \times [0,1] \to Y$ by $G(x,t) = F(x, 1-t)$. Then $G = F \circ H$, where the function $H: X \times [0,1] \to X \times [0,1]$ is defined by $H(x,t) = (x, 1-t)$. Since F and H are both continuous, so is G. Observe that for all $x \in X$, $G(x,0) = F(x,1) = g(x)$ and $G(x,1) = F(x,0) = f(x)$. So, $g \simeq f$, and therefore, $\simeq$ is symmetric.

Finally, let $f, g, h: X \to Y$ be continuous functions with $f \simeq g$ and $g \simeq h$. Let $F: X \times [0,1] \to Y$ and $G: X \times [0,1] \to Y$ be homotopies from f to g and g to h, respectively. Define $H: X \times [0,1] \to Y$ by

$$H(x,t) = \begin{cases} F(x, 2t) & \text{if } t \in \left[0, \frac{1}{2}\right]. \\ G(x, 2t - 1) & \text{if } t \in \left[\frac{1}{2}, 1\right]. \end{cases}$$

First, notice that if $t = \frac{1}{2}$, then $F(x, 2t) = F(x, 1) = g(x)$ and $G(x, 2t - 1) = G(x, 0) = g(x)$. So, H is well-defined. By reasoning similar to what was done in the last paragraph, H is continuous on the closed subsets $X \times \left[0, \frac{1}{2}\right]$ and $X \times \left[\frac{1}{2}, 1\right]$ of $X \times [0, 1]$. By the Pasting Theorem, H is continuous on all of $X \times [0, 1]$. Observe that for all $x \in X$, $H(x, 0) = F(x, 0) = f(x)$ and $H(x, 1) = G(x, 1) = h(x)$. So, $f \simeq h$, and therefore, $\simeq$ is transitive.

Since $\simeq$ is reflexive, symmetric, and transitive, $\simeq$ is an equivalence relation. $\qquad\square$

Path Homotopy

So far, we have been looking at arbitrary topological spaces $(X, \mathcal{T})$ and $(Y, \mathcal{U})$ together with continuous functions $f, g: X \to Y$.

In the special case where $X = [0, 1]$, f and g are called paths. More specifically, if $a, b \in Y$, then a **path** in Y from a (the **initial point**) to b (the **terminal point**) is a continuous function $f: [0,1] \to Y$ such that $f(0) = a$ and $f(1) = b$.

We can visualize a path in Y as an unbroken curve with starting point $f(0) = a$ and ending point $f(1) = b$ (see the figure to the right).

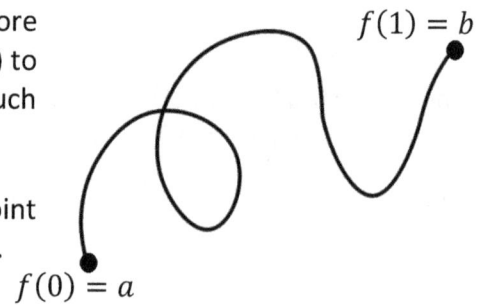

We say that a topological space $(Y, \mathcal{U})$ is **path connected** if for every $a, b \in Y$, there is a path from a to b.

Example 16.4:

1. $\mathbb{R}$ with the standard topology is path connected. Indeed, if $a, b \in \mathbb{R}$, then define $f: [0, 1] \to \mathbb{R}$ by $f(t) = (1 - t)a + tb$. Then f is continuous, $f(0) = a$, and $f(1) = b$. We could visualize this path as a line segment in the xy-plane. However, we will try to stick with the practice of visualizing a path as a subset of the codomain (in this case, the codomain is $\mathbb{R}$). The figure below demonstrates such a visualization.

    Essentially the same argument also shows that $\mathbb{C}$ is path connected. In this case, the path $f: [0, 1] \to \mathbb{C}$ defined by $f(t) = (1 - t)a + tb$ can be visualized as a line segment in the Complex Plane with initial point $f(0) = a$ and terminal point $f(1) = b$. The figure below gives a visualization of this path from $a = 3 + 2i$ to $b = -1 - 2i$.

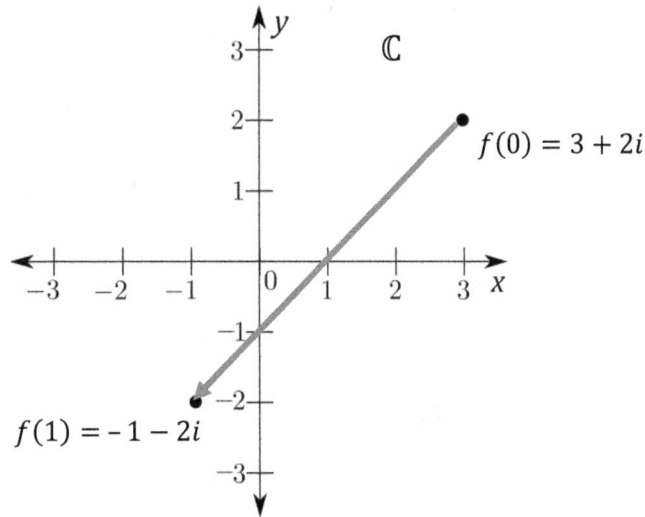

$f(0) = 3 + 2i$

$f(1) = -1 - 2i$

2. Recall that the open ball in $\mathbb{C}$ centered at w with radius r is $B_r(w) = \{z \in \mathbb{C} \mid |w - z| < r\}$. Each of these open balls is path connected. To see this, let $a, b \in B_r(w)$ be arbitrary, and define $f: [0,1] \to \mathbb{C}$ by $f(t) = (1-t)a + tb$. As in part 1 above, f is continuous, $f(0) = a$, and $f(1) = b$. We need to check that for all $t \in [0,1]$, $f(t) \in B_r(w)$. To see this, simply note that

$$|w - f(t)| = |w - ((1-t)a + tb)|$$
$$= |(1-t)(w-a) + t(w-b)|$$
$$\leq (1-t)|w-a| + t|w-b| < (1-t)r + tr$$
$$= r - tr + tr = r.$$

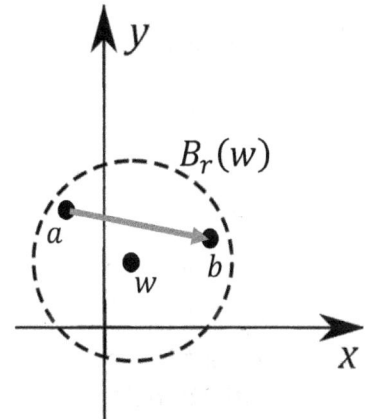

$B_r(w)$

3. We say that a topological space $(X, \mathcal{T})$ is **convex** if whenever $x, y \in X$, $(1-t)x + ty \in X$ for all $t \in [0,1]$. Note that there must be a notion of "scalar multiplication" defined on X for this definition to make sense. For example, this makes sense if X is a subset of a vector space over $\mathbb{R}$. In words, $(X, \mathcal{T})$ is convex if whenever $x, y \in X$, the whole line segment between them is also in X. Geometrically, a convex set has "no holes," "no breaks," and "no bends." For example, the convex subspaces of $\mathbb{R}$ are the intervals. The computation in part 2 above shows that every open ball in $\mathbb{C}$ is convex. The star-shaped figure to the right provides an example of a subspace of $\mathbb{C}$ that is **not** convex.

not convex

Every convex topological space $(X, \mathcal{T})$ is path connected. Indeed, if $a, b \in X$, simply let $f: [0,1] \to X$ be defined by $f(t) = (1-t)a + tb$. Clearly, f is a path from a to b.

4. Let $X = \mathbb{C} \setminus \{0\}$ be the **punctured Complex Plane**. Then X is path connected. To see this, let $a, b \in X$ be arbitrary. If $(1-t)a + tb \neq 0$ for all $t \in [0,1]$, define $f: [0,1] \to X$ by $f(t) = (1-t)a + tb$. If there is $k \in [0,1]$ such that $(1-k)a + kb = 0$, choose $z \in \mathbb{C}$ so that $(1 - 2t)a + 2tz \neq 0$ for all $t \in \left[0, \frac{1}{2}\right]$ and $(2 - 2t)z + (2t - 1)b \neq 0$ for all $t \in \left[\frac{1}{2}, 1\right]$. Then

$$\text{define } f: [0,1] \to X \text{ by } f(t) = \begin{cases} (1 - 2t)a + 2tz & \text{if } x \in \left[0, \frac{1}{2}\right]. \\ (2 - 2t)z + (2t - 1)b & \text{if } x \in \left[\frac{1}{2}, 1\right]. \end{cases}$$

261

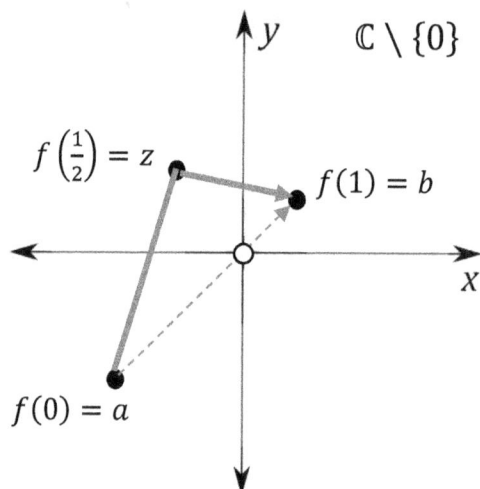

5. Let $(X, \mathcal{T})$ and $(Y, \mathcal{U})$ be topological spaces with $(X, \mathcal{T})$ path connected. If $g: X \to Y$ is a continuous function, then $(g[X], \mathcal{U}_{g[X]})$ is also path connected. To see this, let $a, b \in g[X]$ and choose $c, d \in X$ with $g(c) = a$ and $g(d) = b$. Let $f: [0, 1] \to X$ be a path from c to d and define $h: [0, 1] \to Y$ by $h(t) = (g \circ f)(t)$. Since the composition of two continuous functions is continuous, h is continuous. Also, we have $h(0) = (g \circ f)(0) = g(f(0)) = g(c) = a$ and $h(1) = (g \circ f)(1) = g(f(1)) = g(d) = b$.

6. The unit circle $\mathbb{S}^1 = \{z \in \mathbb{C} \mid |z| = 1\}$ is path connected. To see this, define $g: \mathbb{C} \setminus \{0\} \to \mathbb{S}^1$ by $g(z) = \frac{z}{|z|}$. Then g is continuous and surjective. By part 4 above, $\mathbb{C} \setminus \{0\}$ is path connected and by part 5 above, the continuous image of a path connected space is path connected. So, $\mathbb{S}^1$ is path connected.

Alternatively, we can show that $\mathbb{S}^1$ is path connected directly. Let $a = e^{i\theta}$ and $b = e^{i\phi}$ be two points on $\mathbb{S}^1$ and define $f: [0, 1] \to \mathbb{S}^1$ by $f(t) = e^{i[(1-t)\theta + t\phi]}$. It's easy to see that $f(0) = a$, $f(1) = b$, and I leave it to the reader to verify that f is continuous.

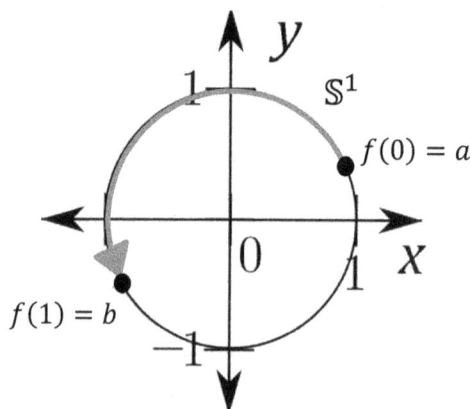

If $f, g: [0, 1] \to Y$ are continuous functions with the same initial and terminal points (say a and b), then a **path homotopy** from f to g is a continuous function $F: [0, 1] \times [0, 1] \to Y$ such that:

- for all $s \in [0, 1]$, $F(s, 0) = f(s)$ and $F(s, 1) = g(s)$;
- for all $t \in [0, 1]$, $F(0, t) = a$ and $F(1, t) = b$.

f is **path homotopic** to g, written $f \simeq_p g$ if there exists a path homotopy from f to g.

Notes: (1) The first bullet point says that F is a homotopy from f to g. In words, F gives a continuous deformation of f to g.

(2) The second bullet point says that for each t between 0 and 1, the function $f_t(s): [0, 1] \to Y$ defined by $f_t(s) = F(s, t)$ is a path from a to b. In other words, the initial and terminal points remain fixed during the deformation.

Example 16.5:

1. Let $f, g: [0, 1] \to \mathbb{R}$ be any two paths from a to b. We define a function $F: [0, 1] \times [0, 1] \to \mathbb{R}$ by $F(s, t) = (1 - t)f(s) + tg(s)$. The proof that F is continuous and for all $s \in [0, 1]$, $F(s, 0) = f(s)$ and $F(s, 1) = g(s)$ is similar to what was done in part 2 of Example 16.1. Also, for all $t \in [0, 1]$, we have $F(0, t) = (1 - t)f(0) + tg(0) = (1 - t)a + ta = a - ta + ta = a$ and $F(1, t) = (1 - t)f(1) + tg(1) = (1 - t)b + tb = b - tb + tb = b$. So, F is a path homotopy from f to g, and therefore, $f \simeq_p g$.

   More generally, if A is a convex subset of $\mathbb{R}$ or $\mathbb{C}$ with the subspace topology, then the same argument given in the last paragraph shows that any two paths $f, g: [0, 1] \to A$ from a to b are path homotopic.

   Note that this argument says nothing about spaces that are not convex. For example, the star-shaped figure drawn in part 3 of Example 16.4 is **not** convex. However, any two paths with the same starting point and ending point are path homotopic. We will see later that in the circle $\mathbb{S}^1$, not all paths with the same starting and ending points are path homotopic.

2. Let $X = \mathbb{C} \setminus \{0\}$ be the punctured Complex Plane, and let $f, g: [0, 1] \to X$ be defined by $f(s) = \cos \pi s + i \sin \pi s$ and $g(s) = \cos \pi s - i \sin \pi s$. The function $F: [0, 1] \times [0, 1] \to X$ defined by $F(s, t) = (1 - t)f(s) + tg(s)$ is **not** a path homotopy from f to g. Indeed, the straight line $F\left(s, \frac{1}{2}\right) = \left(1 - \frac{1}{2}\right)f(s) + \frac{1}{2}g(s) = \frac{1}{2}\left(f(s) + g(s)\right) = \frac{1}{2}(2\cos \pi s) = \cos \pi s + 0i$ passes through the origin when $s = \frac{1}{2}$ (because $\cos \frac{\pi}{2} = 0$).

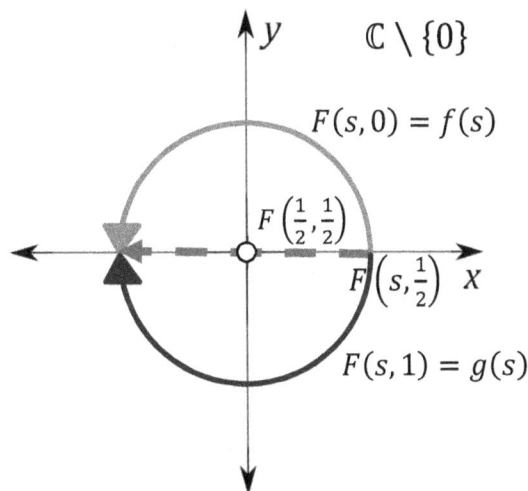

263

It can be shown that there is **no** path homotopy from f to g. Although this is difficult to prove formally, intuitively it is not too hard to see that this should be the case. It is simply impossible to deform f through the hole at the origin in a continuous manner. At some point, we will need to "jump over" the hole to deform to g.

Note that many pairs of paths h and k **are** path homotopic in X. The function F defined in the paragraph above will work to show that h and k are homotopic as long as h and k do not lie on opposite sides of the origin.

Theorem 16.6: $\simeq_p$ is an equivalence relation.

Proof: If $f: [0,1] \to Y$ is a path from a to b, define $F: [0,1] \times [0,1] \to Y$ by $F(s,t) = f(s)$ for all $s, t \in [0,1]$. We already saw in the proof of Theorem 16.3 that F is continuous and for all $s \in [0,1]$, $F(s,0) = f(s)$ and $F(s,1) = f(s)$. Observe also, that for all $t \in [0,1]$, $F(0,t) = f(0) = a$ and $F(1,t) = f(1) = b$. So, $f \simeq_p f$, and therefore, $\simeq_p$ is reflexive.

Next, let $f, g: [0,1] \to Y$ be paths from a to b with $f \simeq_p g$, and let $F: [0,1] \times [0,1] \to Y$ be a path homotopy from f to g. Define $G: [0,1] \times [0,1] \to Y$ by $G(s,t) = F(s, 1-t)$. We already saw in the proof of Theorem 16.3 that G is continuous and for all $s \in [0,1]$, $G(s,0) = g(s)$ and $G(s,1) = f(s)$. Observe also, that for all $t \in [0,1]$, $G(0,t) = F(0, 1-t) = a$ and $G(1,t) = F(1, 1-t) = b$. So, $g \simeq_p f$, and therefore, $\simeq_p$ is symmetric.

Finally, let $f, g, h: X \to Y$ be paths from a to b with $f \simeq g$ and $g \simeq h$. Let $F: [0,1] \times [0,1] \to Y$ and $G: [0,1] \times [0,1] \to Y$ be path homotopies from f to g and g to h, respectively. As in the proof of Theorem 16.3, define $H: [0,1] \times [0,1] \to Y$ by

$$H(s,t) = \begin{cases} F(s, 2t) & \text{if } t \in \left[0, \dfrac{1}{2}\right]. \\ G(s, 2t-1) & \text{if } t \in \left[\dfrac{1}{2}, 1\right]. \end{cases}$$

We already saw that H is continuous on $[0,1] \times [0,1]$ and for all $s \in [0,1]$, $H(s,0) = f(s)$ and $H(s,1) = h(s)$. Observe also, that for all $t \in [0,1]$, $H(0,t) = \begin{cases} F(0, 2t) & \text{if } t \in \left[0, \dfrac{1}{2}\right] \\ G(0, 2t-1) & \text{if } t \in \left[\dfrac{1}{2}, 1\right] \end{cases} = a$ and

$H(1,t) = \begin{cases} F(1, 2t) & \text{if } t \in \left[0, \dfrac{1}{2}\right] \\ G(1, 2t-1) & \text{if } t \in \left[\dfrac{1}{2}, 1\right] \end{cases} = b$. So, $g \simeq_p h$, and therefore, $\simeq_p$ is transitive.

Since $\simeq_p$ is reflexive, symmetric, and transitive, $\simeq_p$ is an equivalence relation. $\square$

Products of Paths

If f is a path in Y from a to b and g is a path in Y from b to c, then the product of f and g, written $f \star g$ is the path in Y from a to c defined by $(f \star g)(s) = \begin{cases} f(2s) & \text{if } s \in \left[0, \frac{1}{2}\right] \\ g(2s-1) & \text{if } s \in \left[\frac{1}{2}, 1\right] \end{cases}$.

Notes: (1) Since $f\left(2 \cdot \frac{1}{2}\right) = f(1) = b$ and $g\left(2 \cdot \frac{1}{2} - 1\right) = g(0) = b$, $f \star g$ is well-defined. By the Pasting Theorem, $f \star g$ is continuous. Also, we have $(f \star g)(0) = f(2 \cdot 0) = f(0) = a$ and $(f \star g)(1) = g(2 \cdot 1 - 1) = g(1) = c$. So, $f \star g$ is a path in Y from a to c.

(2) The intuitive idea behind the product of two paths is quite simple. The image of the product path just consists of the image of the first path followed by the image of the second path. The reason the definition looks a bit messy is because we must combine two functions with domain $[0, 1]$ to produce a new function that also has domain $[0, 1]$. We do this by adjusting the definition of each of the original paths so that their images are "drawn twice as fast." Notice that when we substitute $s = 0$ into $f(2s)$, we get $f(0)$, and when we substitute $s = \frac{1}{2}$ into $f(2s)$, we get $f\left(2 \cdot \frac{1}{2}\right) = f(1)$. So, by replacing s by $2s$ in the function f, we now complete the image of f when $s = \frac{1}{2}$ instead of when $s = 1$. The function g requires an "extra horizontal shift" because we want it to begin at $\frac{1}{2}$ instead of at 0. Notice that when we substitute $s = \frac{1}{2}$ into $g(2s - 1)$, we get

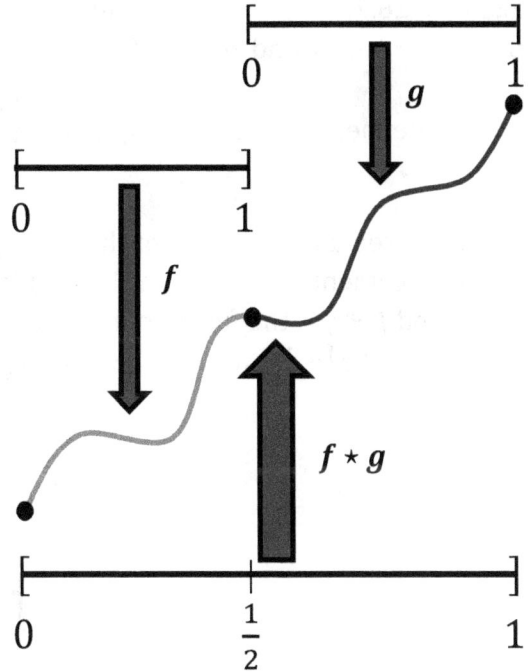

$g\left(2 \cdot \frac{1}{2} - 1\right) = g(1 - 1) = g(0)$, and when we substitute $s = 1$ into $g(2s - 1)$, we get $g(2 \cdot 1 - 1) = g(2 - 1) = g(1)$. So, by replacing s by $2s - 1$ in the function g, we have shifted the domain of g from $[0, 1]$ to $\left[\frac{1}{2}, 1\right]$.

(3) An equivalent definition of $f \star g$ is $(f \star g)(s) = \begin{cases} (f \circ h)(s) & \text{if } s \in \left[0, \frac{1}{2}\right] \\ (g \circ k)(s) & \text{if } s \in \left[\frac{1}{2}, 1\right] \end{cases}$, where $h: \left[0, \frac{1}{2}\right] \to [0, 1]$ is defined by $h(s) = 2s$ and $k: \left[\frac{1}{2}, 1\right] \to [0, 1]$ is defined by $k(s) = 2s - 1$.

Notice that h and k are bijections and their graphs in the xy-plane are lines with positive slope.

(4) In general, for any two intervals $[a, b]$ and $[c, d]$, we define the **natural positive linear bijection** from $[a, b]$ to $[c, d]$, written $L_{[a,b,c,d]}$ to be the line segment in $\mathbb{R} \times \mathbb{R}$ with endpoints (a, c) and (b, d). Specifically, we define $L_{[a,b,c,d]}: [a, b] \to [c, d]$ by $L_{[a,b,c,d]}(x) = \frac{d-c}{b-a}(x - a) + c$.

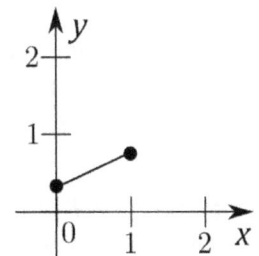

A visualization of $L_{\left[0,1,\frac{1}{4},\frac{3}{4}\right]}$ is shown to the right.

(5) Let f be a path in Y from a to b, let g be a path in Y from b to c, and let h be a path in Y from c to d. For each pair $(x, y) \in [0, 1] \times [0, 1]$, we define a path $k_{(x,y)}: [0, 1] \to Y$ from a to d as follows:

$$k_{(x,y)}(s) = \begin{cases} (f \circ L_{[0,x,0,1]})(s) & \text{if } s \in [0, x]. \\ (g \circ L_{[x,y,0,1]})(s) & \text{if } s \in [x, y]. \\ (h \circ L_{[y,1,0,1]})(s) & \text{if } s \in [y, 1]. \end{cases}$$

Note that $k_{\left(\frac{1}{2}, \frac{3}{4}\right)} = f \star (g \star h)$ and $k_{\left(\frac{1}{4}, \frac{1}{2}\right)} = (f \star g) \star h$.

Example 16.7: Let's consider $\mathbb{C}$ with the standard topology and let $f, g: [0, 1] \to \mathbb{C}$ be the paths defined by $f(s) = s + si$ and $g(s) = (s + 1) + (1 - s)i$. Note that $f(1) = 1 + i$ and $g(0) = 1 + i$, so that $f \star g$ is defined. We have $(f \star g)(s) = \begin{cases} f(2s) & \text{if } s \in \left[0, \frac{1}{2}\right] \\ g(2s - 1) & \text{if } s \in \left[\frac{1}{2}, 1\right] \end{cases} = \begin{cases} 2s + 2si & \text{if } s \in \left[0, \frac{1}{2}\right]. \\ 2s + (2 - 2s)i & \text{if } s \in \left[\frac{1}{2}, 1\right]. \end{cases}$

Below we see a visualization of the paths f, g and $f \star g$. Observe that f is the function from $[0, 1]$ to the line segment with positive slope, g is the function from $[0, 1]$ to the line segment with negative slope, and $f \star g$ is the function from $[0, 1]$ to **both** line segments: the line segment with positive slope is the image of $\left[0, \frac{1}{2}\right]$ and the line segment with negative slope is the image of $\left[\frac{1}{2}, 1\right]$ under the function $f \star g$.

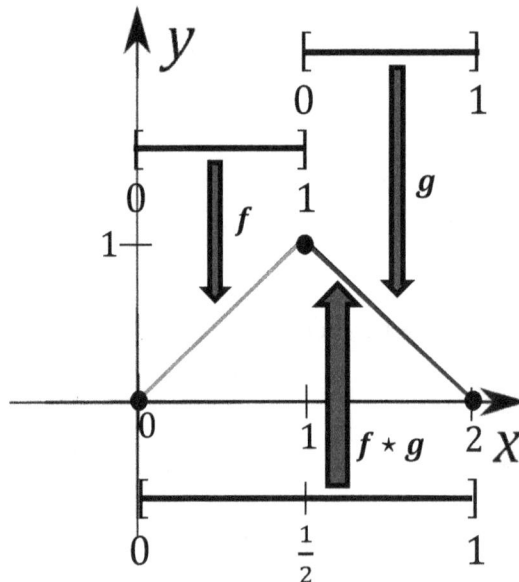

We will need the following two theorems in order to verify some important properties of paths. You will be asked to prove these theorems in Problems 2 and 3 below.

Theorem 16.8: Let $(X, \mathcal{T})$ and $(Y, \mathcal{U})$ be topological spaces, let $h: X \to Y$ be continuous, let $f, g: [0, 1] \to X$ be paths in X, and let $F: [0, 1] \times [0, 1] \to X$ be a path homotopy from f to g. Then $h \circ F: [0, 1] \times [0, 1] \to Y$ is a path homotopy from $h \circ f$ to $h \circ g$.

Theorem 16.9: Let $(X, \mathcal{T})$ and $(Y, \mathcal{U})$ be topological spaces, let $h: X \to Y$ be continuous, let $f, g: [0, 1] \to X$ be paths in X with $f(1) = g(0)$. Then $h \circ (f \star g) = (h \circ f) \star (h \circ g)$.

Theorem 16.10: Let $(Y, \mathcal{T})$ be a topological space. Then $\star$ (as defined above) induces a well-defined operation $\overline{\star}$ on path homotopy classes by $[f] \, \overline{\star} \, [g] = [f \star g]$ with the following properties:

(1) **(Associativity)** If $[f] \, \overline{\star} \, ([g] \, \overline{\star} \, [h])$ is defined, then so is $([f] \, \overline{\star} \, [g]) \, \overline{\star} \, [h]$, and they are equal.

(2) **(Left and Right Identities)** Given $y \in Y$, let $c_y \colon [0,1] \to Y$ be the constant path defined by $c_y(s) = y$. If $f \colon [0,1] \to Y$ is a path in Y from a to b, then $[c_a] \, \overline{\star} \, [f] = [f]$ and $[f] \, \overline{\star} \, [c_b] = [f]$.

(3) **(Inverse)** Given the path $f \colon [0,1] \to Y$ from a to b, define $\overline{f} \colon [0,1] \to Y$ by $\overline{f}(s) = f(1-s)$. Then $\overline{f}$ is a path from b to a such that $[f] \, \overline{\star} \, [\overline{f}] = [c_a]$ and $[\overline{f}] \, \overline{\star} \, [f] = [c_b]$.

Proof: Let's check that $\overline{\star}$ is well-defined. Suppose that $f \simeq_p h$ and $g \simeq_p k$, where f and h are paths from a to b and g and k are paths from b to c. Let $F \colon [0,1] \times [0,1] \to Y$ be a path homotopy from f to h and let $G \colon [0,1] \times [0,1] \to Y$ be a path homotopy from g to k. Let's define $H \colon [0,1] \times [0,1] \to Y$ by

$$H(s,t) = \begin{cases} F(2s, t) & \text{if } s \in \left[0, \frac{1}{2}\right]. \\ G(2s-1, t) & \text{if } s \in \left[\frac{1}{2}, 1\right]. \end{cases}$$

Since $F\left(2 \cdot \frac{1}{2}, t\right) = F(1,t) = b$ and $G\left(2 \cdot \frac{1}{2} - 1\right) = G(0,t) = b$, H is well-defined. By the Pasting Theorem, H is continuous. For all $s \in [0,1]$,

$$H(s,0) = \begin{cases} F(2s, 0) & \text{if } s \in \left[0, \frac{1}{2}\right] \\ G(2s-1, 0) & \text{if } s \in \left[\frac{1}{2}, 1\right] \end{cases} = \begin{cases} f(2s) & \text{if } s \in \left[0, \frac{1}{2}\right] \\ g(2s-1) & \text{if } s \in \left[\frac{1}{2}, 1\right] \end{cases} = (f \star g)(s).$$

$$H(s,1) = \begin{cases} F(2s, 1) & \text{if } s \in \left[0, \frac{1}{2}\right] \\ G(2s-1, 1) & \text{if } s \in \left[\frac{1}{2}, 1\right] \end{cases} = \begin{cases} h(2s) & \text{if } s \in \left[0, \frac{1}{2}\right] \\ k(2s-1) & \text{if } s \in \left[\frac{1}{2}, 1\right] \end{cases} = (h \star k)(s).$$

Also, for all $t \in [0,1]$, $H(0,t) = F(0,t) = a$ and $H(1,t) = G(1,t) = c$. So, $f \star g \simeq_p h \star k$, and therefore, $[f \star g] = [h \star k]$.

Next, we prove (1). Let $f, g, h \colon [0,1] \to Y$ be paths in Y with $f(1) = g(0)$ and $g(1) = h(0)$. Since $k_{\left(\frac{1}{2}, \frac{3}{4}\right)} = f \star (g \star h)$ and $k_{\left(\frac{1}{4}, \frac{1}{2}\right)} = (f \star g) \star h$ (see Note 5 above), it suffices to show that whenever $0 < x < y < 1$ and $0 < z < w < 1$, we have $\left[k_{(x,y)}\right] = \left[k_{(z,w)}\right]$. To see this, let $j \colon [0,1] \to [0,1]$ be defined by

$$j(s) = \begin{cases} L_{[0,x,0,z]}(s) & \text{if } s \in [0, x]. \\ L_{[x,y,z,w]}(s) & \text{if } s \in [x, y]. \\ L_{[y,1,w,1]}(s) & \text{if } s \in [y, 1]. \end{cases}$$

Note that $L_{[0,x,0,z]}(x) = L_{[x,y,z,w]}(x) = z$ and $L_{[x,y,z,w]}(y) = L_{[y,1,w,1]}(y) = w$. So, by the Pasting Theorem, j is continuous. Therefore, j is a path in $[0,1]$ from 0 to 1. Since $[0,1]$ is convex, j is path homotopic to the identity function $i: [0,1] \to [0,1]$. Let $F: [0,1] \times [0,1] \to [0,1]$ be a path homotopy from j to i. By Theorem 16.8, $k_{(z,w)} \circ F$ is a path homotopy in Y from $k_{(z,w)} \circ j$ to $k_{(z,w)} \circ i$. But $k_{(z,w)} \circ j = k_{(x,y)}$ and $k_{(z,w)} \circ i = k_{(z,w)}$. So, $k_{(x,y)} \simeq_p k_{(z,w)}$. Therefore, $[k_{(x,y)}] = [k_{(z,w)}]$, as desired.

Now, we prove (2). Let $f: [0,1] \to Y$ be a path in Y from a to b. Also, consider the constant path $c_0: [0,1] \to [0,1]$ (where for all $t \in [0,1]$, $c_0(t) = 0$) and the identity path $i: [0,1] \to [0,1]$ (where for all $t \in [0,1]$, $i(t) = t$). Since $c_0(1) = 0$ and $i(0) = 0$, $c_0 \star i$ is a well-defined path in $[0,1]$. By an argument similar to part 2 of Example 16.1, any two paths are path homotopic in $[0,1]$. In particular, there is a path homotopy $F: [0,1] \times [0,1] \to [0,1]$ from i to $c_0 \star i$. By Theorem 16.8, $f \circ F: [0,1] \times [0,1] \to Y$ is a path homotopy from $f \circ i$ to $f \circ (c_0 \star i)$. Now, if $t \in [0,1]$, we have $(f \circ i)(t) = f(i(t)) = f(t)$ and $(f \circ c_0)(t) = f(c_0(t)) = f(0) = a = c_a(t)$. So, we have $f \circ i = f$ and $f \circ c_0 = c_a$. By Theorem 16.9, $f \circ (c_0 \star i) = (f \circ c_0) \star (f \circ i) = c_a \star f$. So, we see that $f \circ F$ is a path homotopy from f to $c_a \star f$. Therefore, $[f] = [c_a] \bar{\star} [f]$.

Similarly, using the fact that $i \star c_1$ is path homotopic to i in $[0,1]$, we can show that $[f] = [f] \bar{\star} [c_b]$.

Finally, we prove (3). The path $i \star \bar{i}: [0,1] \to [0,1]$ satisfies $(i \star \bar{i})(0) = i(2 \cdot 0) = i(0) = 0$ and $(i \star \bar{i})(1) = \bar{i}(2 \cdot 1 - 1) = \bar{i}(1) = i(1 - 1) = i(0) = 0$. Since $[0,1]$ is convex, $i \star \bar{i} \simeq_p c_0$. Let $H: [0,1] \times [0,1] \to [0,1]$ be a path homotopy from c_0 to $i \star \bar{i}$. By Theorem 16.8, $f \circ H$ is a path homotopy from $f \circ c_0$ to $f \circ (i \star \bar{i})$. Now, $f \circ c_0 = c_a$ and by Theorem 16.9, we have that $f \circ (i \star \bar{i}) = (f \circ i) \star (f \circ \bar{i}) = f \star \bar{f}$. So, $f \star \bar{f} \simeq_p c_a$. Therefore, $[f] \bar{\star} [\bar{f}] = [c_a]$.

Similarly, using the fact that $\bar{i} \star i$ is path homotopic to c_1 in $[0,1]$, we have $[\bar{f}] \bar{\star} [f] = [c_b]$. $\qquad\square$

The Fundamental Group

The set $\mathcal{P} = \{[f] \mid f$ is a path in $Y\}$ of homotopy classes of paths in Y together with the operation $\bar{\star}$ **almost** forms a group. The only reason that $(\mathcal{P}, \bar{\star})$ fails to be a group is because for arbitrary paths $f, g: [0,1] \to Y$, $f \star g$ (and therefore, $[f] \bar{\star} [g]$) may not be defined. In fact, $f \star g$ is defined if and only if $f(1) = g(0)$. We remove this problem by restricting our attention to paths that begin and end at the same point. We call such paths "loops."

A **loop** in Y based at a is a path in Y that begins and ends at a.

Let $\pi_1(Y, a) = \{[f] \mid f$ is a loop based at $a\}$, where $[f]$ is the path homotopy class of f. The **fundamental group** of Y relative to the **base point** a is $(\pi_1(Y, a), \bar{\star})$.

Theorem 16.11: $(\pi_1(Y, a), \bar{\star})$ is a group.

Proof: Let $[f], [g] \in \pi_1(Y, a)$. Since $f(1) = a$ and $g(0) = a$, $f \star g$ is defined. By Theorem 16.10, $\bar{\star}$ is well-defined on $\pi_1(Y, a)$.

Similarly, if $[f], [g], [h] \in \pi_1(Y, a)$, then $[f] \bar{\star} ([g] \bar{\star} [h])$ is defined, and so, by Theorem 16.10(1), $([f] \bar{\star} [g]) \bar{\star} [h]$ is defined and $[f] \bar{\star} ([g] \bar{\star} [h]) = ([f] \bar{\star} [g]) \bar{\star} [h]$. So, $\bar{\star}$ is associative on $\pi_1(Y, a)$.

By Theorem 16.10 (2), if $[f] \in \pi_1(Y, a)$, then $[c_a] \,\overline{\star}\, [f] = [f]$ and $[f] \,\overline{\star}\, [c_a] = [f]$. So, $[c_a] \in \pi_1(Y, a)$ is an identity for $\overline{\star}$.

Finally, by Theorem 16.10(3), if $[f] \in \pi_1(Y, a)$, then $\left[\overline{f}\right]$ is an inverse of $[f]$.

Therefore, $(\pi_1(Y, a), \,\overline{\star}\,)$ is a group. □

From now on, we will simply refer to $\pi_1(Y, a)$ as the fundamental group of Y relative to the base point a. The operation need not be mentioned explicitly, as it will always be $\overline{\star}$.

Example 16.12: For any $r \in \mathbb{R}$, $\pi_1(\mathbb{R}, r) = \{[c_r]\}$ (the trivial group consisting of just the identity). We showed this in part 1 of Example 16.5, where we saw that any two paths in $\mathbb{R}$ are path homotopic. In fact, we saw in the same example that $\pi_1(A, a)$ is the trivial group for any convex subspace A of $\mathbb{R}$ or $\mathbb{C}$ and any $a \in A$. In particular, intervals in $\mathbb{R}$ and open balls in $\mathbb{C}$ all have trivial fundamental groups. These are examples of **simply connected** spaces. A space $(Y, \mathcal{T})$ is simply connected if it is path connected and $\pi_1(Y, a)$ is the trivial group for **some** base point $a \in Y$. It will follow immediately from Theorem 16.13 below that $(Y, \mathcal{T})$ is simply connected if and only if it is path connected and $\pi_1(Y, a)$ is the trivial group for **every** base point $a \in Y$.

We will now show that for path connected spaces, the fundamental group does not depend upon the base point chosen. First, we will give a precise definition of an isomorphism between two groups.

Let $(G, \star)$ and $(H, \circ)$ be groups. A function $f: G \to H$ is a **homomorphism** if $f(a \star b) = f(a) \circ f(b)$ for all $a, b \in G$. An **isomorphism** is a bijective homomorphism. If there is an isomorphism from a group $(G, \star)$ to a group $(H, \circ)$, then we say that $(G, \star)$ and $(H, \circ)$ are **isomorphic**, and we write $(G, \star) \cong (H, \circ)$. If the operations are understood, we will abbreviate this as $G \cong H$. Mathematicians generally consider isomorphic groups to be the same. Indeed, they behave identically. The only difference between them is the "names" of the elements.

Theorem 16.13: If $(Y, \mathcal{U})$ is a path connected topological space, then for all $a, b \in Y$, $\pi_1(Y, a) \cong \pi_1(Y, b)$.

Proof: Let $a, b \in Y$. Since $(Y, \mathcal{U})$ is path connected, there is a path $g: [0, 1] \to Y$ from a to b. We define a function $\hat{g}: \pi_1(Y, a) \to \pi_1(Y, b)$ by $\hat{g}([f]) = [\overline{g}] \,\overline{\star}\, [f] \,\overline{\star}\, [g]$.

Since $\overline{\star}$ is well-defined, so is $\hat{g}$. Also, if $[f] \in \pi_1(Y, a)$, then f is a loop based at a. It follows that $f: [0, 1] \to Y$ satisfies $f(0) = f(1) = a$. Then $(\overline{g} \star (f \star g))(0) = \overline{g}(0) = g(1 - 0) = g(1) = b$ and $(\overline{g} \star (f \star g))(1) = (f \star g)(1) = g(1) = b$. So, $[\overline{g}] \,\overline{\star}\, [f] \,\overline{\star}\, [g] \in \pi_1(Y, b)$.

Let $[f], [h] \in \pi_1(Y, a)$. Then we have

$$\hat{g}([f]) \,\overline{\star}\, \hat{g}([h]) = ([\overline{g}] \,\overline{\star}\, [f] \,\overline{\star}\, [g]) \,\overline{\star}\, ([\overline{g}] \,\overline{\star}\, [h] \,\overline{\star}\, [g]) = [\overline{g}] \,\overline{\star}\, [f] \,\overline{\star}\, ([g] \,\overline{\star}\, [\overline{g}]) \,\overline{\star}\, [h] \,\overline{\star}\, [g]$$
$$= [\overline{g}] \,\overline{\star}\, ([f] \,\overline{\star}\, [c_a]) \,\overline{\star}\, [h] \,\overline{\star}\, [g] = [\overline{g}] \,\overline{\star}\, ([f] \,\overline{\star}\, [h]) \,\overline{\star}\, [g] = \hat{g}([f] \,\overline{\star}\, [h]).$$

So, $\hat{g}$ is a homomorphism.

To see that $\hat{g}$ is injective, let $[f], [h] \in \pi_1(Y, a)$ with $\hat{g}([f]) = \hat{g}([h])$. Then we have

$$[\overline{g}] \bar{\star} [f] \bar{\star} [g] = [\overline{g}] \bar{\star} [h] \bar{\star} [g]$$

Therefore,

$$[h] = ([g] \bar{\star} [\overline{g}]) \bar{\star} [h] \bar{\star} ([g] \bar{\star} [\overline{g}]) = [g] \bar{\star} ([\overline{g}] \bar{\star} [h] \bar{\star} [g]) \bar{\star} [\overline{g}]$$
$$= [g] \bar{\star} ([\overline{g}] \bar{\star} [f] \bar{\star} [g]) \bar{\star} [\overline{g}] = ([g] \bar{\star} [\overline{g}]) \bar{\star} [f] \bar{\star} ([g] \bar{\star} [\overline{g}]) = [f].$$

To see that $\hat{g}$ is surjective, let $[h] \in \pi_1(Y, b)$, and define $f: [0, 1] \to Y$ by $f(t) = \big(g \star (h \star \overline{g})\big)(t)$. Then $f(0) = g(0) = a$ and $f(1) = (h \star \overline{g})(1) = \overline{g}(1) = g(0) = a$. So, $[f] \in \pi_1(Y, a)$. Finally, we have

$$\hat{g}([f]) = [\overline{g}] \bar{\star} [f] \bar{\star} [g] = [\overline{g}] \bar{\star} ([g] \bar{\star} ([h] \star [\overline{g}]) \bar{\star} [g] = ([\overline{g}] \bar{\star} [g]) \bar{\star} [h] \bar{\star} ([\overline{g}] \bar{\star} [g])$$
$$= [c_b] \bar{\star} ([h] \bar{\star} [c_b]) = [c_b] \bar{\star} [h] = [h].$$

It follows that $\hat{g}$ is an isomorphism, and therefore, $\pi_1(Y, a) \cong \pi_1(Y, b)$. $\qquad\square$

Note: By Theorem 16.13, if $(Y, \mathcal{U})$ is a path connected topological space, then we don't really care which base point we choose. Therefore, we can abbreviate $\pi_1(Y, a)$ as $\pi_1(Y)$.

Let $(Y, \mathcal{U})$ be a topological space and define the relation $\sim$ on Y by $a \sim b$ if and only if there is a path from a to b. In Problem 4 below, you will be asked to prove that $\sim$ is an equivalence relation and that each equivalence class is path connected. These equivalence classes are called the **path components** of Y.

Let $a \in Y$ and let C be the path component of a. If $f: [0, 1] \to Y$ is a loop based at a, then the range of f must be a subset of C. Also, if $f, g: [0, 1] \to Y$ are loops based at a and $F: [0, 1] \times [0, 1] \to Y$ is a path homotopy from f to g, then the range of F must be a subset of C. It follows that $\pi_1(Y, a) \cong \pi_1(C, a)$. For this reason, we will always look at only path connected spaces when computing fundamental groups.

The reader will be asked to prove the following important theorem in Problem 7 below.

Theorem 16.14: Topologically equivalent topological spaces have isomorphic fundamental groups.

The Fundamental Group of the Circle

Computing the fundamental group of a topological space that is not simply connected can be quite difficult. In this last section, I'd like to discuss the fundamental group of the unit circle.

I summarize the main result in the following theorem.

Theorem 16.15: The fundamental group of $\mathbb{S}^1$ is isomorphic to $(\mathbb{Z}, +)$.

For each $n \in \mathbb{N}$, consider the path $d_n: [0, 1] \to \mathbb{S}^1$ defined by $d_n(s) = (\cos 2\pi n s, \sin 2\pi n s)$.

For each $n \in \mathbb{N}$, d_n is a loop in $\mathbb{S}^1$ that "wraps around" the unit circle $|n|$ times.

For example, d_1 is the path that begins at $(1, 0)$ and travels around the unit circle in the counterclockwise direction, stopping as soon as it gets back to the point $(1, 0)$.

On the other hand, d_{-1} travels around the unit circle in the clockwise direction, also stopping as soon as it gets back to the point $(1, 0)$.

In general, if $n > 0$, d_n begins at $(1, 0)$, wraps around the unit circle n times in the counterclockwise direction, ending back at $(1, 0)$. Similarly, if $n < 0$, d_n begins at $(1, 0)$, wraps around the unit circle n times in the clockwise direction, also ending back at $(1, 0)$. For $n = 0$, d_0 is the constant path that stays at $(1, 0)$.

Intuitively, it should seem reasonable that any path in $\mathbb{S}^1$ is homotopic to d_n for some $n \in \mathbb{N}$. It should also seem likely that if $m \neq n$, then d_m and d_n are **not** path homotopic. So, at this point, you might be able to guess a likely candidate for an isomorphism from $(\mathbb{Z}, +)$ to $\pi_1(\mathbb{S}^1)$.

You guessed it: the isomorphism is the function

$$\Phi\colon \mathbb{Z} \to \pi_1(\mathbb{S}^1)$$

defined by $\Phi(n) = [d_n]$.

However, proving that Φ is an isomorphism is not easy. To do this, we will want to "lift" the maps d_n to a space where all paths with the same initial and terminal points are path homotopic. The most natural candidate for this space is $\mathbb{R}$. We begin by formally defining what it means to lift a path.

Let $(X, \mathcal{T})$, $(Y, \mathcal{U})$, and $(Z, \mathcal{V})$ be topological spaces and let $f\colon X \to Y$, $g\colon Z \to Y$ be continuous functions. A **lifting** of f is a continuous function $h_f\colon X \to Z$ such that $g \circ h_f = f$. If h_f is a lifting of f, we say that f **lifts** to h_f. We can visualize a lifting as follows:

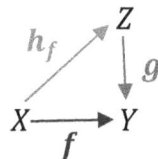

$$
\begin{array}{ccc}
 & & Z \\
 & \nearrow{\scriptstyle h_f} & \downarrow{\scriptstyle g} \\
X & \xrightarrow{\quad f \quad} & Y
\end{array}
$$

Example 16.16: Let $X = [0, 1]$, $Y = \mathbb{S}^1$, and $Z = \mathbb{R}$, all with their standard topologies (X and Y are given the appropriate subspace topologies). Let $g\colon \mathbb{R} \to \mathbb{S}^1$ be defined by $g(x) = (\cos 2\pi x, \sin 2\pi x)$ and let $f\colon [0, 1] \to \mathbb{S}^1$ be the path defined by $f(s) = \left(\cos\frac{\pi}{2}s, \sin\frac{\pi}{2}s\right)$. Then f lifts to the function $h_f\colon [0, 1] \to \mathbb{R}$ defined by $h_f(s) = \frac{s}{4}$. To see this, simply note that

$$(g \circ h_f)(s) = g\left(h_f(s)\right) = g\left(\tfrac{s}{4}\right) = \left(\cos 2\pi\left(\tfrac{s}{4}\right), \sin 2\pi\left(\tfrac{s}{4}\right)\right) = \left(\cos\tfrac{\pi}{2}s, \sin\tfrac{\pi}{2}s\right) = f(s).$$

Using the notation d_n defined earlier, we might say that $f = d_{\frac{1}{4}}$. Note that this particular path is **not** a loop. Its initial point is $(1, 0)$ and its terminal point is $(0, 1)$.

In general, d_n is a loop if and only if n is an integer. For $n = \frac{1}{4}$, the path $d_{\frac{1}{4}}$ travels one quarter of the way around the circle.

Here is a visualization of the lifting:

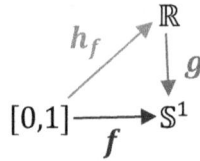

And here is an even more explicit visualization:

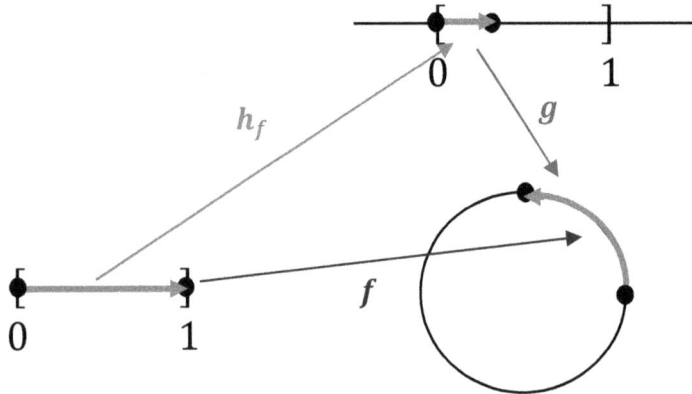

Let's let g be the same function and let's look at some other paths in $\mathbb{S}^1$ beginning at $(1, 0)$ and see if we can lift them.

Let $j: [0, 1] \to \mathbb{S}^1$ be the path defined by $j(s) = (\cos 4\pi s, \sin 4\pi s)$. Then j lifts to the function $h_j: [0, 1] \to \mathbb{R}$ defined by $h_j(s) = 2s$.

Let $k: [0, 1] \to \mathbb{S}^1$ be the path defined by $k(s) = \left(\cos\frac{\pi}{2}s, -\sin\frac{\pi}{2}s\right)$. Then k lifts to the function $h_k: [0, 1] \to \mathbb{R}$ defined by $h_k(s) = -\frac{s}{4}$.

More generally, for any real number r, let $d_r: [0, 1] \to \mathbb{S}^1$ be the path defined by $d_r(s) = (\cos 2\pi rs, \sin 2\pi rs)$. Then d_r lifts to the function $h_{d_r}: [0, 1] \to \mathbb{R}$ defined by $h_{d_r}(s) = rs$.

Notes: (1) $f = d_{\frac{1}{4}}, j = d_2$, and $k = d_{-\frac{1}{4}}$. The dedicated reader should write out the details for these last three examples and draw pictures just as we did for the path f. Observe that the image of j covers the circle twice and the image of k begins at $(1, 0)$ and moves clockwise instead of counterclockwise around the circle.

(2) The function g given in Example 16.16 is called a **covering map** because of the way it continuously wraps around the unit circle repeatedly. We will refer to this as the **standard covering map of $\mathbb{S}^1$**.

It is not a coincidence that in Example 16.16 we were able to lift all the paths d_r (even for noninteger values of r). In fact, we have the following result:

Theorem: 16.17: Let $g: \mathbb{R} \to \mathbb{S}^1$, defined by $g(x) = (\cos 2\pi x, \sin 2\pi x)$, be the standard covering map of $\mathbb{S}^1$ and let $d: [0, 1] \to \mathbb{S}^1$ be any path. Suppose that $d(0) = (x, y) \in \mathbb{S}^1$ and let $z \in g^{-1}[\{x, y\}]$. Then there is a unique lift $k_d: [0, 1] \to \mathbb{R}$ of d such that $k_d(0) = z$.

You will be asked to prove Theorem 16.17 in Problem 9 below.

I will now give a brief outline of the steps for a proof of Theorem 16.15. I leave the details as an exercise for the reader (see Problem 10 below).

Proof outline for Theorem 16.15: Recall that for each $n \in \mathbb{N}$, the path $d_n : [0, 1] \to \mathbb{S}^1$ is defined by $d_n(s) = (\cos 2\pi n s, \sin 2\pi n s)$. We define a function $\Phi : \mathbb{Z} \to \pi_1(\mathbb{S}^1)$ by $\Phi(n) = [d_n]$. To prove that Φ is an isomorphism, you must complete the following steps:

Step 1: Note that if $g : \mathbb{R} \to \mathbb{S}^1$ is the standard covering map of $\mathbb{S}^1$, then for each $n \in \mathbb{N}$, d_n lifts to h_{d_n}. In other words, $d_n = g \circ h_{d_n}$.

Step 2: Show that if $k : [0, 1] \to \mathbb{R}$ is any path in $\mathbb{R}$ from 0 to n, then $\Phi(n) = [g \circ k]$. Use the fact that any two paths in $\mathbb{R}$ with the same initial and terminal points are path homotopic.

Step 3: To prove that Φ is a homomorphism, for each $m \in \mathbb{N}$, define $k_m : \mathbb{R} \to \mathbb{R}$ by $k_m(s) = m + s$, and then use the fact that $h_{d_m} \star (k_m \circ h_{d_n}) : [0, 1] \to \mathbb{R}$ is a path in $\mathbb{R}$ from 0 to $m + n$.

Step 4: To prove that Φ is surjective, use Theorem 16.17 to lift any loop f in $\mathbb{S}^1$ with base point $(1, 0)$ to a path h_f in $\mathbb{R}$ with initial point 0 and show that $h_f(1) = n$ for some $n \in \mathbb{Z}$.

Step 5: To prove that Φ is injective, assume that d_m and d_n are path homotopic. Show that a path homotopy between them lifts to a path homotopy on $\mathbb{R}$, and use this "lifted" homotopy to prove that $m = n$.

Completing these steps will prove the theorem. □

Problem Set 16

Full solutions to these problems are available for free download here:

www.SATPrepGet800.com/TFBZLF

LEVEL 1

1. Prove that every path connected space is connected, but that the converse is false.

LEVEL 2

2. Let $(X, \mathcal{T})$ and $(Y, \mathcal{U})$ be topological spaces, let $h: X \to Y$ be continuous, let $f, g: [0, 1] \to X$ be paths in X, and let $F: [0, 1] \times [0, 1] \to X$ be a path homotopy from f to g. Prove that $h \circ F: [0, 1] \times [0, 1] \to Y$ is a path homotopy from $h \circ f$ to $h \circ g$. (This is Theorem 16.8.)

3. Let $(X, \mathcal{T})$ and $(Y, \mathcal{U})$ be topological spaces, let $h: X \to Y$ be continuous, and let $f, g: [0, 1] \to X$ be paths in X with $f(1) = g(0)$. Prove that $h \circ (f \star g) = (h \circ f) \star (h \circ g)$. (This is Theorem 16.9.)

LEVEL 3

4. Let $(X, \mathcal{T})$ be a topological space and define the relation $\sim$ on X by $a \sim b$ if and only if there is a path from a to b. Prove that $\sim$ is an equivalence relation and that each equivalence class is path connected.

5. Prove the Pasting Theorem (Theorem 16.2).

LEVEL 4

6. Let $(Y, \mathcal{U})$ be a simply connected topological space and let $f, g: [0, 1] \to Y$ be paths in Y having the same initial and terminal points. Prove that $f \simeq_p g$.

LEVEL 5

7. Prove that topologically equivalent topological spaces have isomorphic fundamental groups.

CHALLENGE PROBLEMS

8. Describe a topological space that is connected, but **not** path connected.

9. Let $g: \mathbb{R} \to \mathbb{S}^1$, defined by $g(x) = (\cos 2\pi x, \sin 2\pi x)$, be the standard covering map of $\mathbb{S}^1$ and let $d: [0, 1] \to \mathbb{S}^1$ be any path. Suppose that $d(0) = (x, y) \in \mathbb{S}^1$ and let $z \in g^{-1}[\{x, y\}]$. Prove that there is a unique lift $k_d: [0, 1] \to \mathbb{R}$ of d such that $k_d(0) = z$. (This is Theorem 16.17).

10. Prove that the fundamental group of $\mathbb{S}^1$ is isomorphic to $(\mathbb{Z}, +)$.

INDEX

277

About the Author

Dr. Steve Warner, a New York native, earned his Ph.D. at Rutgers University in Pure Mathematics in May 2001. While a graduate student, Dr. Warner won the TA Teaching Excellence Award.

After Rutgers, Dr. Warner joined the Penn State Mathematics Department as an Assistant Professor and in September 2002, he returned to New York to accept an Assistant Professor position at Hofstra University. By September 2007, Dr. Warner had received tenure and was promoted to Associate Professor. He has taught undergraduate and graduate courses in Precalculus, Calculus, Linear Algebra, Differential Equations, Mathematical Logic, Set Theory, and Abstract Algebra.

From 2003 – 2008, Dr. Warner participated in a five-year NSF grant, "The MSTP Project," to study and improve mathematics and science curriculum in poorly performing junior high schools. He also published several articles in scholarly journals, specifically on Mathematical Logic.

Dr. Warner has nearly two decades of experience in general math tutoring and tutoring for standardized tests such as the SAT, ACT, GRE, GMAT, and AP Calculus exams. He has tutored students both individually and in group settings.

In February 2010, Dr. Warner released his first SAT prep book "The 32 Most Effective SAT Math Strategies," and in 2012 founded Get 800 Test Prep. Since then Dr. Warner has written books for the SAT, ACT, SAT Math Subject Tests, AP Calculus exams, and GRE. In September 2018, Dr. Warner released his first advanced math book "Pure Mathematics for Beginners."

Dr. Steve Warner can be reached at

steve@SATPrepGet800.com

BOOKS BY DR. STEVE WARNER

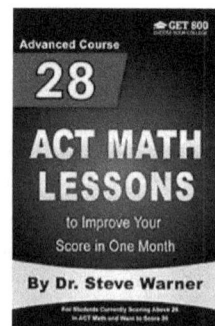

www.ingramcontent.com/pod-product-compliance
Lightning Source LLC
Chambersburg PA
CBHW081240220326
41597CB00023BA/4170